Biotechnology for Sustainability and Social Well Being

Biotechnology for Sustainability and Social Well Being

Editors

Pau Loke Show
Chiaki Ogino
Mohamad Faizal Ibrahim

MDPI • Basel • Beijing • Wuhan • Barcelona • Belgrade • Manchester • Tokyo • Cluj • Tianjin

Editors
Pau Loke Show
University of Nottingham
Malaysia
Malaysia

Chiaki Ogino
Kobe University
Japan

Mohamad Faizal Ibrahim
Universiti Putra Malaysia
Malaysia

Editorial Office
MDPI
St. Alban-Anlage 66
4052 Basel, Switzerland

This is a reprint of articles from the Special Issue published online in the open access journal *Processes* (ISSN 2227-9717) (available at: https://www.mdpi.com/journal/processes/special_issues/biotech_sustainability).

For citation purposes, cite each article independently as indicated on the article page online and as indicated below:

LastName, A.A.; LastName, B.B.; LastName, C.C. Article Title. *Journal Name* **Year**, *Volume Number*, Page Range.

ISBN 978-3-0365-0672-2 (Hbk)
ISBN 978-3-0365-0673-9 (PDF)

Contents

About the Editors

Pau Loke Show is the President of International Bioprocess Association in Malaysia. He is also the director of Sustainable Food Processing Research Centre and Co-director of Future Food Malaysia, Beacon of Excellence, at the University of Nottingham, Malaysia. He is a Full Professor at the Department of Chemical and Environmental Engineering, Faculty of Science and Engineering, University of Nottingham, Malaysia. He successfully obtained his Ph.D. two years after obtaining his bachelor's degree from Universiti Putra Malaysia. He is currently a Professional Engineer (PEng) registered with the Board of Engineers Malaysia, Chartered Engineer (CEng) of the Engineering Council UK (MIChemE) and a Professional Technologist (PTech) registered with the Malaysia Board of Technologists. Prof Ir. Ts. Dr. Show obtained a Post Graduate Certificate of Higher Education (PGCHE) in 2014, and is now a Fellow of the Higher Education Academy (FHEA) UK. Since he started his career in 2012, he has received numerous prestigious academic awards, including the APEC Science Prize for Innovation, Research and Education ("ASPIRE") Malaysia Award 2020, Malaysia Young Scientist 2019 Award, ASEAN–India Research and Training Fellowship 2019, The DaSilva Award 2018, JSPS Fellowship 2018, Top 100 Asian Scientists 2017, Asia's Rising Scientists Award 2017 and was the winner of the Young Researcher in IChemE Award 2016. He has graduated more than 20 Ph.D. and M.Sc. students, and is leading a team of 20 members consisting of postdoctoral, Ph.D. and M.Sc. research students. As of 2020, he has published more than 350 journal papers in less than 8 years of his career. His publications have been cited over 5000 times in the past 5 years. His current h-index is 36, placing him among the top leaders of his chosen field (Microalgae Technology). He is also the Primary Project leader for more than 35 International, National, and Industry Projects, with a total amount of more than MYR5.0 million. He is now serving as an Editor-in-chief at Current Nutrition & Food Science; as an Editor at Scientific Report (IF: 4.011); Biocatalysis and Agricultural Biotechnology (CiteScore: 2.80); as an Associate Editor at Bioengineered (IF: 2.205); Current Biochemical Engineering; Open Microalgae Biotechnology; and as an Editorial board member at Bioresource Technology (IF: 7.539) and the Biochemical Engineering Journal (IF: 3.475). He is also managing lead guest editor for many well respected journals, for example, the Journal of Hazardous Materials, Elsevier (IF: 9.038); Biofuel Research Journal, Elsevier (IF: 7.038); Environmental Pollution, Elsevier (IF: 6.792); Chemosphere, Elsevier (IF: 5.778); Clean Technologies and Environmental Policy, Springer (IF: 2.277); Bioengineered, Taylor & Francis (IF: 2.205); Processes, MDPI (IF: 2.8); Energies, MDPI (IF: 2.707); Biocatalysis and Agricultural Biotechnology, Elsevier; Frontiers in Energy Research, Frontiers; BMC Energy, Springer Nature; Materials Science for Energy Technologies, KeAI; Current Nutrition & Food Science, Bentham Science. Prof Ir. Ts. Dr Show is an active reviewer for more than 100 esteemed international journals published by Elsevier, Wiley, Springer, ACS, RSC and Taylor & Francis publishers. He has recently been awarded as the Top Peer Reviewer 2019 powered by Publons (Top 1% of Reviewers in Global Top Peer Reviewer Awards in Engineering, Global Top Peer Reviewer Awards in Cross-Field, Global Top Peer Reviewer Awards in Chemistry, Global Top Peer Reviewer Awards in Biology and Biochemistry, Global Top Peer Reviewer Awards in Agricultural Sciences). In addition, he has acted as a handling editor for more than 1000 submitted manuscripts in numerous journals.

Chiaki Ogino is a Professor at the Department of Chemical Science and Engineering, Graduate School of Engineering, Kobe University, Japan. He received his B.Sc., M.Sc., and Ph.D. in Chemical Engineering at Kobe University, Japan. He has supervised more than 50 master's students, and over 10 doctoral students at Kobe University. He is the author or co-author of over 250 peer-reviewed journal papers and 40 reviews and book chapters. His research interests are wide and concern: the production of biofuels and chemicals from biomass based on synthetic biotechnology by yeast, fungi and actinomycetes, and the development of novel drug and gene delivery systems based on nanobiotechnology.

Mohamad Faizal Ibrahim is a lecturer at the Department of Bioprocess Technology, Faculty of Biotechnology and Biomolecular Sciences, Universiti Putra Malaysia (UPM), Malaysia. His research interests are in the utilization of biomass for value added products through bioprocessing approaches such as the production and application of enzymes, the production of biofuels and bio-based chemicals, and the extraction of essential oils. He obtained his first degree in Biotechnology in 2009, and later obtained his Ph.D. (Environmental Biotechnology) in 2013. He was appointed as Senior Lecturer by UPM in 2013 and as Associate Professor in 2019. He has published more than 37 articles in cited index journals and 2 book chapters, has edited one book, and has one patent and one trademark. He has also actively contributed as a reviewer to various reputable journals published by Elsevier, Wiley, Springer Nature, ACS and many more. He also works in an industrial capacity on the commercial production of phytase and biocompost, and in heavy metal testing, besides other consultation work on policymaking for the biotechnology industry. In addition, he has actively participated in biotechnology societies. Now, he is the Secretary of the Malaysia Chapter of AFOB as well as organizing several international conferences under the umbrella of AFOB, such as ARS 2014, ACB 2015 and AFOBMCIS 2018 and 2019.

processes

Editorial

Special Issue on "Biotechnology for Sustainability and Social Well Being"

Pau Loke Show

Department of Chemical and Environmental Engineering, Faculty of Science and Engineering, University of Nottingham Malaysia, Broga Road, Semenyih 43500, Malaysia; PauLoke.Show@nottingham.edu.my

Citation: Show, P.L. Special Issue on "Biotechnology for Sustainability and Social Well Being". *Processes* **2021**, *9*, 216. https://doi.org/10.3390/pr9020216

Received: 22 January 2021
Accepted: 22 January 2021
Published: 25 January 2021

Publisher's Note: MDPI stays neutral with regard to jurisdictional claims in published maps and institutional affiliations.

Bioprocessing is a very important part of biotechnology that utilizes living organisms and their components to produce various types of products. The products and services that depend on bioprocessing can be grouped into the following: (1) biopharmaceuticals, which involve the production of therapeutic compounds, vaccines, and diagnostic components; (2) specific bio-based chemicals, such as biofuels, food, and agricultural products, as well as fine chemicals derived from and/or by living organisms and other types of bioproducts; (3) environmental management aids that use bioprocessing to treat, control, or remediate pollutants and toxic components. Bioprocessing is one of the key factors in several emerging industries of biofuels, used in the production of biogas, bioethanol, and biodiesel; industrial enzymes; waste management through biotechnology; new vaccines; and many more. The term bioprocessing is always referred to as a biotechnology method that produces products and provides services that are environmentally friendly, sustainable, and renewable.

The important role of bioprocessing has attracted interest from researchers in terms of finding suitable bioprocesses that can enhance production or process efficiency. The Asian Federation of Biotechnology (AFOB) has set 12 academic divisions that cover all biotechnological areas. Most of these areas require bioprocess technology for the production of their desired products. To promote the recent technologies and findings in biotechnology, AFOB Malaysia Chapter (AFOB-MC) organized the 2nd AFOB Malaysia Chapter International Symposium 2019 (AFOBMCIS 2019), which was held in Putrajaya, Malaysia, from the 20th to 23rd October 2019. The theme of the symposium was "Biotechnology for Sustainability and Social Well Being" and it comprised 12 technical sessions—namely, (1) agricultural and food biotechnology; (2) applied microbiology; (3) biopharmaceutical and medical biotechnology; (4) biocatalysis and protein engineering; (5) bioprocess and bioseparation engineering; (6) bioenergy and biorefinery; (7) environmental biotechnology; (8) marine biotechnology; (9) nanobiotechnology, biosensors, and biochips; (10) systems and synthetic biotechnology; (11) tissue engineering and biomaterials; and (12) bioindustry promotion and bioeducation. All of the authors of the accepted contributions at AFOBMCIS 2019 related to bioprocessing were invited to submit manuscripts to *Processes* under this Special Issue on Bioprocessing.

This Special Issue, "Biotechnology for Sustainability and Social Well Being", invited manuscripts from academicians as well as industry players who are working on biotechnology and green technology-related processes. Authors were invited to submit original research articles covering topics which include, but are not limited to, the following areas: (1) bioprocess; (2) bioproducts; (3) bio-based chemicals; (4) biomaterials; (5) fermentation, etc. The manuscripts were regularly submitted, selected, and reviewed by the regular system and accepted for publication. This Special Issue, "Biotechnology for Sustainability and Social Well Being", aims to incorporate and introduce the advances in bioprocess as well as green technologies to the new generation of academicians and industry players. In this Special Issue on "Biotechnology for Sustainability and Social Well Being", we have accepted and published 17 high-quality and original articles. These research papers cover theoretical, numerical, and experimental approaches on the latest developments in bioprocessing and biotechnology, as well as green technologies that bridge conventional practices

1

and the Industry 4.0 concept. The Special Issue operated a rigorous peer-review process with a single-blind assessment and at least two independent reviewers, hence resulting in our final acceptance of these published high-quality papers.

Apart from that, this Special Issue also attracted one quality review paper and five feature article papers. The review provides insights on the use of renewable sources such as *Jatropha curcas* L. as a biodiesel feedstock in Malaysia [1]. This review paper emphasizes the potential of *Jatropha curcas* as an eco-friendly biodiesel feedstock to promote socio-economic development and meet significantly growing energy demands, even though there are many challenges for its implementation as a national biodiesel and the program might require a long period to be realized. The proposed use of this feedstock is promising, as it shows lower carbon and greenhouse gases emissions. Apart from that, there are five papers selected as feature papers, with topics covering biohydrogen production, rapid monitoring process for bacteria, the optimization of fermentation process for bacteria, nanofiltration and transesterification enhancement by lipase, and liquid biphasic separation for protein recovery. The study on biohydrogen production investigates the effects of alginate and chitosan entrapped in biofilm formations on activated carbon [2]. A positive response to the higher growth of hydrogen-producing bacteria was obtained in the work. Another study on the cultivation of *Lactobacillus pentosus* was conducted with rapid monitoring through the spectrophotometer method [3]. The method developed can rapidly determine the lag phase of *Lactobacillus* in breast milk, which is useful in assessing the bacteria growth curve and growth behavior of the strain. Furthermore, one of the studies applied the isolation, identification, and optimization of a new γ-aminobutyric acid (GABA)-producing *Bacillus cereus* strain from soy sauce [4]. Response surface methodology was used to obtain a high concentration of the GABA strain under optimal fermentation conditions. There is also a study on nanofiltration membrane separation to concentrate lipase for biodiesel production [5]. The nanofiltration membrane technology developed showed a comparable fatty acid methyl ester (FAME) composition to that of commercial lipase, which improves its applicability and scalability. The last featured study shows the application of a Liquid Biphasic Flotation (LBF) system to extract proteins from *Persicaria tenella* leaf [6]. The efficient, environmental friendly, and cost-effective liquid separation method showed a good reliability with a high protein recovery and separation efficiency. All these five feature papers were selected based on a rigorous review process carried out by international independent reviewers assigned by the journal's office.

On the other hand, there are 16 technical research papers which were accepted in this Special Issue. All these selected 16 technical research papers were contributed by participants who attended the 2nd Asian Federation of Biotechnology (AFOB) Malaysia Chapter International Symposium 2019 (AFOBMCIS 2019). The Asian Federation of Biotechnology (AFOB) is a non-profit organization established in 2008. Its incorporation was agreed upon by delegates from Asian countries during the IBS 2008 conference, 12–17 October held in Dalian, China. Preparative meetings for the formation of AFOB had been held four times prior to its official establishment (27th April, Songdo Technopark, Incheon, Korea, KSBB 2007 National Spring Meeting; 5th and 7th November, Taipei International Convention Center, Taipei, Taiwan, APBioChEC 2007 Meeting; April 18th, Jeonju City, Korea, KSBB 2008 (National Spring Meeting)). The delegates from the participating Asian countries discussed the launching of the AFOB during the meetings. The International Federation bears the name Asian Federation of Biotechnology, abbreviated to "AFOB", hereafter referred to as "The Federation". All the accepted papers in this Special Issue are of high quality and were specially selected by international experts before being invited to submit a paper to this issue.

We strongly believe that the novel bioprocesses and green biotechnologies presented in this issue will be useful in assisting the global community in working towards fulfilling the Sustainable Development Goals (SDG) of the United Nations. The guest editors thank the authors for their contribution to this new knowledge and the reviewers for their valuable time and effort given to the review process. Besides this, we would like to thank the editorial office, MS Wency Xiang, MS Jamie Li, and the whole *Processes'* Team for their help and support in completing this Special Issue, especially during the COVID-19 pandemic.

Funding: This work was supported by the Fundamental Research Grant Scheme.

Conflicts of Interest: It is declared that there is no conflict of interest.

References

1. Che Hamzah, N.H.; Khairuddin, N.; Siddique, B.M.; Hassan, M.A. Potential of Jatropha curcas L. as Biodiesel Feedstock in Malaysia: A Concise Review. *Processes* **2020**, *8*, 786. [CrossRef]
2. Dzul Rashidi, N.F.; Jamali, N.S.; Mahamad, S.S.; Ibrahim, M.F.; Abdullah, N.; Ismail, S.F.; Siajam, S.I. Effects of Alginate and Chitosan on Activated Carbon as Immobilisation Beads in Biohydrogen Production. *Processes* **2020**, *8*, 1254. [CrossRef]
3. Nguyen-Sy, T.; Yew, G.Y.; Chew, K.W.; Nguyen, T.D.P.; Tran, T.N.T.; Le, T.D.H.; Vo, C.T.; Tran, H.K.P.; Mubashir, M.; Show, P.L. Potential Cultivation of Lactobacillus pentosus from Human Breastmilk with Rapid Monitoring through the Spectrophotometer Method. *Processes* **2020**, *8*, 902. [CrossRef]
4. Wan-Mohtar, W.A.A.Q.I.; Sohedein, M.N.A.; Ibrahim, M.F.; Ab Kadir, S.; Suan, O.P.; Weng Loen, A.W.; Sassi, S.; Ilham, Z. Isolation, Identification, and Optimization of γ-Aminobutyric Acid (GABA)-Producing Bacillus cereus Strain KBC from a Commercial Soy Sauce moromi in Submerged-Liquid Fermentation. *Processes* **2020**, *8*, 652. [CrossRef]
5. Wijaya, H.; Sasaki, K.; Kahar, P.; Quayson, E.; Rachmadona, N.; Amoah, J.; Hama, S.; Ogino, C.; Kondo, A. Concentration of Lipase from Aspergillus oryzae Expressing Fusarium heterosporum by Nanofiltration to Enhance Transesterification. *Processes* **2020**, *8*, 450. [CrossRef]
6. Saw, H.S.; Sankaran, R.; Khoo, K.S.; Chew, K.W.; Phong, W.N.; Tang, M.S.Y.; Lim, S.S.; Mohd Zaid, H.F.; Naushad, M.; Show, P.L. Application of a Liquid Biphasic Flotation (LBF) System for Protein Extraction from Persiscaria Tenulla Leaf. *Processes* **2020**, *8*, 247. [CrossRef]

 *processes* 

Article

The Effects of Biofertilizers on Growth, Soil Fertility, and Nutrients Uptake of Oil Palm (Elaeis Guineensis) under Greenhouse Conditions

Aaronn Avit Ajeng [1], Rosazlin Abdullah [1,*], Marlinda Abdul Malek [2,*], Kit Wayne Chew [3], Yeek-Chia Ho [4,5], Tau Chuan Ling [1], Beng Fye Lau [1] and Pau Loke Show [6,*]

[1] Institute of Biological Sciences, Faculty of Science, University of Malaya, Kuala Lumpur 50603, Malaysia; aaronnavit@gmail.com (A.A.A.); tcling@um.edu.my (T.C.L.); bengfye@um.edu.my (B.F.L.)
[2] Institute of Sustainable Energy (ISE), Universiti Tenaga Nasional, Kajang, Selangor 43000, Malaysia
[3] School of Energy and Chemical Engineering, Xiamen University Malaysia, Jalan Sunsuria, Bandar Sunsuria, Sepang, Selangor 43900, Malaysia; kitwayne.chew@xmu.edu.my
[4] Civil and Environmental Engineering Department, Universiti Teknologi PETRONAS, Seri Iskandar 32610, Malaysia; Yeekchia.ho@utp.edu.my
[5] Centre for Urban Resource Sustainability, Institute of Self-Sustainable Building, Universiti Teknologi PETRONAS, Seri Iskandar 32610, Malaysia
[6] Department of Chemical and Environmental Engineering, Faculty of Engineering, University of Nottingham Malaysia Campus, Semenyih, Selangor 43500, Malaysia
*   Correspondence: rosazlin@um.edu.my (R.A.); marlinda@uniten.edu.my (M.A.M.); PauLoke.Show@nottingham.edu.my (P.L.S.); Tel.: +60-3-7967-4360 (R.A.); +60-3-8921-2020 (M.A.M.); +60-3-8924-8605 (P.L.S.)

Received: 5 August 2020; Accepted: 10 September 2020; Published: 19 December 2020

Abstract: The full dependency on chemical fertilizers in oil palm plantation poses an enormous threat to the ecosystem through the degradation of soil and water quality through leaching to the groundwater and contaminating the river. A greenhouse study was conducted to test the effect of combinations of biofertilizers with chemical fertilizer focusing on the soil fertility, nutrient uptake, and the growth performance of oil palms seedlings. Soils used were histosol, spodosol, oxisol, and ultisol. The three treatments were T1: 100% chemical fertilizer (NPK 12:12:17), T2: 70% chemical fertilizer + 30% biofertilizer A (CF + BFA), and T3: 70% + 30% biofertilizer B (CF + BFB). T2 and T3, respectively increased the growth of oil palm seedlings and soil nutrient status but seedlings in oxisol and ultisol under T3 had the highest in almost all parameters due to the abundance of more efficient PGPR. The height of seedlings in ultisol under T3 was 22% and 17% more than T2 and T1 respectively, with enhanced girth size, chlorophyll content, with improved nutrient uptake by the seedlings. Histosol across all treatments has a high macronutrient content suggesting that the rate of chemical fertilizer application should be revised when planting using the particular soil. With the reduction of chemical fertilizer by 25%, the combined treatment with biofertilizers could enhance the growth of the oil palm seedlings and soil nutrient properties regardless of the soil orders.

Keywords: plant growth promoting rhizobacteria; oil palm seedlings nursery; biofertilizers; chemical fertilizer

---

1. Introduction

The agriculture sector is considered as one of the economy pillars in many developing nations [1]. However, continuous use of agrochemicals such as chemical fertilizers and pesticides in this sector is detrimental to human health such as infant methemoglobinemia [2] and which also cause ecological imbalance [3,4]. The use of chemical fertilizer will also cause air and ground water pollution resulting from eutrophication. This practice also negatively affects the roots of the crops, making them unable to acquire nutrients [5,6]. Therefore, there is a need to replace this conventional agricultural practice by implying a safer alternative to promote the growth of the plants, without affecting the agroecosystem. The effort to reduce the dependence on the chemical fertilizers has been made through the establishment of biological based organic fertilizers (also known as biofertilizer) as an alternative [7]. Biofertilizers are made up from soil bacteria that are beneficial to the plants and it is known as an integrated nutrients system where nutrients required by the plants are provided by the activity of the below-ground microorganisms. This practice of using beneficial microbes in agriculture has started about 60 years ago [8].

The introduction of beneficial microbes in inorganic fertilizers have received a considerable amount of attention in the last decades as the microbes are effective in promoting plant growth by secreting phytohormones and metabolites [9]. The application of bioinoculants containing N-fixing bacteria and P-solubilizing bacteria have proven to improve leaf chlorophyll, plant nutrient uptake, and yield of rice in which the use of N and P fertilizer was able to be minimized by 50% [10]. These beneficial bacteria are termed as the Plant Growth Promoting Rhizobacteria (PGPR) which comprise of nitrogen—fixers, phosphorus (P), and potassium (K) solubilizers, and often combined as consortium with the some beneficial fungi in the production of biofertilizers [7]. The PGPR mechanism of action can be divided into direct and indirect mechanism, with direct mechanism including biofertilization, root stimulation, rhizoremediation, and plant stress control. The indirect mechanism includes the biological control against diseases which includes antibiosis, induction of systemic resistance, and competition for nutrient and niches [11]. Genera belonging to *Rhizobium* spp., *Azospirillum* spp., and *Bacillus* spp. are the symbiotic nitrogen fixing bacteria which efficiently fix the nitrogen in the nodules and roots of the plants, hence reducing the dependence on nitrogenous fertilizers [12,13]. Meanwhile, the production of organic acid such as the gluconic acid, oxalic acid, malic acid, formic acid, 2-ketogluconic acid, propionic acid, lactic acid, D-malic acid, and citric acid by PGPR belonging to the genera *Pseudomonas* spp. [14,15], *Acinetobacter* spp. [16], *Bacillus* spp. [17], *Klebsiella pneumonia* [18], and *Burkholderia fungorum* [19] aid in the P solubilization and making the nutrient accessible to the plants.

Malaysia is currently the world's second largest palm oil producer and the increase in the plantation area from 1.5 million hectares in 1985 to 5.39 million hectares in 2014 has resulted in extensive use of chemical fertilizers [9,20]. Malaysia has tropical soils, most of them are considered as problematic soils such as peat, sandy acid sulfate, and highly weathered soils such as ultisols and oixsols [21]. The acidity of the soils is due to the natural ecosystem through the weathering process, pyrite oxidation in acid sulfate soil, and organic matter deposition and accumulation of forming peat; where these processes are often enhanced by human activities through intensive land-based crop and animal production [22]. The aim of this study is to observe the effects of combined application of chemical fertilizer with the biofertilizer application on oil palm (*Elaeis guineensis*) seedlings using different soil orders under greenhouse conditions as we are looking into reducing the rates of chemical fertilizer application and investigate the effects of common problematic soils in Malaysia on the application of biofertilizers for oil palm seedlings plantation. Different biofertilizers with various compositions of beneficial microbes was used together with the chemical fertilizer and evaluated based on plant growth attributes including the uptake of nutrients by these seedlings.

2. Materials and Methods

2.1. Study Area and Experimental Design

The experiment was carried out at Rimba Ilmu, University of Malaya, Kuala Lumpur. Four soil orders were used in this experiment; histosol, spodosol, oxisol, and ultisol. The chemical properties of the soils used in this study were listed in Table 1. Three quarters of each polybags ($n = 12$) (20 kg for histosol and spodosol, 15 kg for oxisol and ultisol) was filled with each respective soil and 4 g of Christmas Island Phosphate Rock (CIRP) (15% Total P and available P was 58.86) was applied and mixed into each polybag (34 × 45 cm) and incubated for a week before the transplant.

The temperature ambience was 28–33 °C. The experiments were conducted in the Complete Block Design (CBD) with four replicates for each treatment in a single trial. Liquid biofertilizer A (BFA) (effective microorganisms: 1×10^7 CFU/mL) and biofertilizer B (BFB) (effective microorganisms: 1×10^6 CFU/mL) were purchased from local Malaysian manufacturers. BFA consists of *Bacillus* spp. such as *Bacillus cereus* JCM 2152, *Bacillus amyloliquefaciens* strain MPA 1034 and *Bacillus tequilensis* strain 10b *Lactobacillus* spp.; *Azospirillum* spp. and *Rhizobium* spp. Meanwhile, BFB consists of a very diverse group of microbes: Actinomycetes such as *Kocuria rhizophila*, *Arthrobacter methylotrophus*, *Bacillus* spp. such as *B. pumilus*, *B. subtilis* (subspecies Spizizenii), *B. vallismortis*, *B. Thurengiensis*, *B. mycoides*, *B. mucilaginosus*, *Brevibacillus reuszeri*, *Paenibacillus polymax*, and *Paenibacillus azoreducens*. *Azospirillum brasilense* and fungus such as *Aspergillus niger* and *Aspergillus awamori*; yeast such as *Saccharomyces cerevisiae* Hansen were also the beneficial microbes contained in the biofertilizer. The micro and macro nutrient with the organic matter of the biofertilizers were listed in Table 2. NPK blue with the formulation ratio of (12 N:12 P_2O_5:17 K_2O: 2 MgO + TE) was used as the chemical fertilizer. The experiment consists of three treatments: [T1] 100% of CF, [T2] 70% CF + 30% BFA, and [T3] 70% CF + 30% BFB. The amount and dose of fertilizers applied was listed in Table 3. Treatments were done for four rounds (every 30 days).

Table 1. Chemical properties of histosol, spodosol, ultisol, and oxisol.

Soil Properties	Histosol	Spodosol	Ultisol	Oxisol
pH	3.23	5.49	3.83	4.33
Total N (%)	0.61	0.34	0.10	0.12
Availlable P (mg/kg)	75.81	36.66	25.99	32.78
Exchangeable K (mg/kg)	455.2	487.93	358.33	471.1

Table 2. The micro and macro nutrient, and the organic matter of the biofertilizer A and biofertilizer B.

Micro and Macro Nutrients	Biofertilizer A	Biofertilizer B
N	7%	5–6%
P	6%	8–9%
K	9%	10–11%
Ca	2%	-
Mg	1%	0.5–1.0%
Su	1%	-
Bo	0.5%	0.9–1.1%
Fe	50 ppm	282 ppm
Cu	15 ppm	18.4 ppm
Mn	10 ppm	35.8 ppm
Zn	15 ppm	51.4 ppm
Mo	12 ppm	-
Organic matter	Aloe vera Seaweed extract Fulvic acid Amino acid Protein	Aloe vera Seaweed extract Humic acid Amino acid Fish emulsify

Table 3. Chemical fertilizer and biofertilizer application. The biofertilizer was diluted with 200 mL of distilled water before applied to a single seedling.

Month	Control Plot Dosage per Palm (g seedlings^{-1}) (NPK 12-12-17-2 + TE) 100% Chemical Fertilizer	Treatment Plot 75% Chemical Fertilizer	Biofertilizer (mL)
1	15	10	2
2	20	15	2
3	25	20	3
4	30	25	3

2.2. Analysis of Soil Chemical Properties

The soil used was thoroughly mixed and air dried, sieved using a 2 mm mesh sieve and measured for macronutrients (NPK) content. Soil chemical properties such as the pH was accessed using a glass electrode pH meter with 1:2.5 soil to water suspensions (Eutech Instruments, Thermo Fisher Scientific, Woodlands, Singapore). Total N was analyzed using CNS analyzer (LECO TruMac® CNS, St. Joseph, MN, USA) [23] and K was determined using the leaching method with 1 M ammonium acetate buffered at pH 7 [24]. P was analyzed using the method described by Bray and Kurtz [25].

2.3. Measurement of Oil Palm Seedlings Growth

During the treatment period, the growth parameters of the oil palms were observed and taken every two weeks. The height of the fronds (leaflets plus rachis) was measured using measuring tape from the lowest rudimentary to the tip of the rachis. The number of fronds was taken and recorded. The girth size was measured using a digital Vernier caliper at 5 cm from the planting medium. The chlorophyll was measured using a chlorophyll meter (SPAD-502, Minolta Camera Co., Osaka, Japan) of leaf blades from the third frond with a visually green colour at the midrib to maximize the calibration [26]. The readings were taken at three random spots and the amount was averaged throughout the study period. The SPAD-502 was calibrated after and before another reading on different seedlings.

2.4. Plants Analysis

The planting material used in this study was D × P Yangambi (ML 161). The seedlings were harvested after four months of planting. They were carefully removed from the soil and the roots were cleansed of soil particles. The seedlings were cut at the soil level and separated from the roots. The dry mass of both aboveground biomass and root were determined by drying in an oven at 71–75 °C until constant weight was achieved. The aboveground biomass and roots were ground separately using a grinding machine (<2 mm) separately for macronutrients (NPK) analysis. To determine the residual of plant nutrients, the amount of N was determined using a CNS analyzer (LECO TruMac® CNS, St. Joseph, MN, USA) [24] while P was analyzed using the wet digestion method [27] and exchangeable K using displacement of cations with 1.0 N of NH4OAc (pH 7.0) and displacement of absorbed NH_4^+ with 0.1 N K_2SO_4 [28].

2.5. Data Analysis

Tukey's HSD (honest significant difference) test was used to determine significant differences (p-value ≤ 0.05) between different types of soil orders and two-way ANOVA was used to determine significant differences (p-value ≤ 0.05) between the treatments. The data are statistically analyzed using the IBM SPSS Statistics for Windows, version 21.0.

3. Results

3.1. Growth Performance of Fronds

Figure 1 shows the height of the oil palm seedlings at DAT (days after transplant) 131. There is a significant difference between T1 and T3 plots. In general, all seedlings across every soil type show a better growth under the T3. Under the same biofertilizer treatment, seedlings planted in oxisol depicted the highest growth. The application of BFB has a positive effect on the growth of seedlings in oxisol with a total increase of 22% and 17% from T2 and T1, respectively. Most seedlings in the T3 plot are higher and have more developed fronds with leaves as compared to the other treatments. In terms of the number of fronds, no significant difference was observed between the treatments and the type of soils used. However, seedlings in oxisol under the T2 have more frond counts compared to T1 and T3 with an increase of 5% and 12%, respectively. Meanwhile, seedlings in T3 planted using spodosol and ultisol depicted the highest number of fronds where seedlings in T1 show the lowest number of fronds.

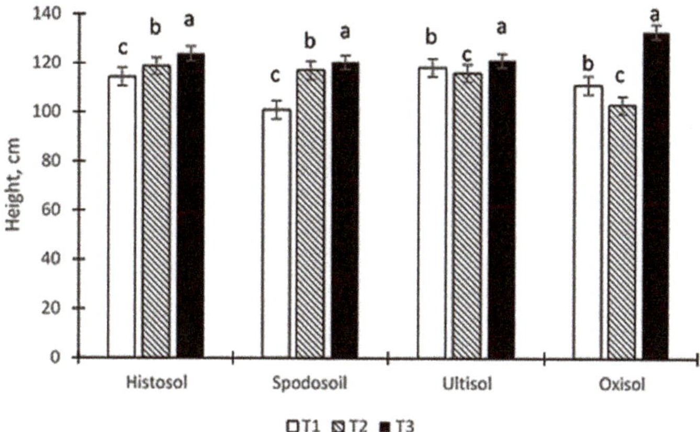

Figure 1. Highest frond height of oil palm seedlings at the end of treatment (DAT 131). Vertical bar represents the standard deviation. Different letters represent significant differences in Tukey's HSD comparison. Means sharing the same letter across treatments do not differ significantly at p-value ≤ 0.05.

3.2. Stem Girth Size

At the time of harvest, all seedlings in histosol, ultisol, and oxisol under T3 plots have the largest girth size (38.38, 38.87, and 38.25 mm, respectively) (Figure 2). The girth size of the seedlings was at least 26–30% larger than seedlings in T1. The stem girth size of seedlings in spodosol under the combined fertilization with BFA was the highest across all soils and treatments. Seedlings in spodosol also have the largest girth size under T1 while the girth size of seedlings in histosol, oxisol, and ultisol was less than 30 mm at 131 DAT. The other soils show an increase in girth size reading under T2 and no girth size below 30 mm was recorded.

3.3. Aboveground Biomass (ABG) and Root Dry Mass Ratio

Table 4 shows the aboveground biomass and root and their dry ratio. The highest ABG dry mass was obtained from oil palms seedlings treated under T3 under oxisol and ultisol while the least dry mass was from seedlings planted under 100% CF. This corresponded with oil palm seedlings height in which ultisol has the highest height compared to other soils. There was a significant difference between T1 and T3 for ABG, however there was no significant difference in ABG dry mass in between the soils used. Heavier root weights can be observed in seedlings under oxisol as compared to other soil types especially in the treatment with BFB. However, there was no significant difference between the treatments in root dry mass, and the root:Aboveground ratio. From this study, the combined fertilization with biofertilizers also showed a positive effect on the proliferation and development of the roots.

3.4. Chlorophyll Content

Figure 3 shows the chlorophyll content of the oil palm seedlings throughout the study period measured by the SPAD meter. The chlorophyll content of the seedlings in spodosol under T1 plots were decreasing starting DAT 41 until the harvesting day and remained the lowest. The combined treatment of chemical fertilizer with the biofertilizers show a positive response on the chlorophyll content of the seedlings. Our results indicate that biofertilizers can substitute chemical fertilizers to sustain N needs by the seedlings to enhance the chlorophyll content even at a reduced rate of fertilizers. There were significant differences among T1 and T3 in all the experimented seedlings across all soil types at p-value ≤ 0.05. Unlike T1, an increasing trend was seen in all seedlings under T2 and T3. Under the combined fertilization with BFA, all seedlings show an increment in the chlorophyll content except for chlorophyll in seedlings planted using oxisol which decreased at 67 DAT but increased again after 131 DAT. Seedlings in histosol depicted the highest chlorophyll reading throughout the last two months of treatment period. The chlorophyll content of seedlings in T3 planted using histosol declined after 30 DAT but increased after 41 DAT and show a slight change from 67 and 131 DAT. Seedlings in ultisol under the same treatment reached the highest peak at 41 DAT with the chlorophyll content reading of 63.18 but decreased to 62.50 at 131 DAT. A steady increase in the chlorophyll content was seen in seedlings under oxisol but it remained the lowest reading throughout the last three months during the treatment period. The addition of biofertilizers seems also to have a positive impact on the chlorophyll reading of the seedlings.

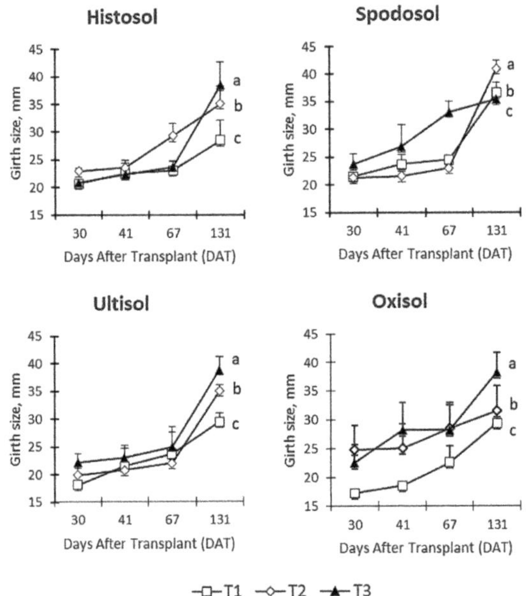

Figure 2. Girth size of the seedlings throughout the treatment period. Vertical bar represents the standard deviation. Different letters represent significant differences in Tukey's HSD comparison. Means sharing the same letter across treatments do not differ significantly at *p*-value ≤ 0.05.

Table 4. Aboveground biomass (ABG) and root dry weight with ABG:root. Different letters represent significant differences in Tukey's HSD comparison. Means sharing the same letter across treatments do not differ significantly at *p*-value ≤ 0.05.

Soil	Treatment	ABG	Root	Root:ABG
Histosol	T1	57.97 ± 9.92b	16.22 ± 4.46a	0.28 ± 0.03a
	T2	62.07 ± 3.47a	16.10 ± 3.31a	0.26 ± 0.04ab
	T3	59.50 ± 17.47b	15.09 ± 3.61a	0.26 ± 0.04b
Spodosol	T1	49.62 ± 14.32b	14.25 ± 4.21a	0.29 ± 0.02a
	T2	63.48 ± 7.08ab	16.52 ± 0.92a	0.26 ± 0.02ab
	T3	64.53 ± 4.99a	15.84 ± 1.17a	0.25 ± 0.02b
Ultisol	T1	53.61 ± 3.80b	11.70 ± 0.68a	0.22 ± 0.01ab
	T2	66.34 ± 2.50ab	15.20 ± 1.26a	0.23 ± 0.02a
	T3	70.39 ± 7.98a	13.92 ± 1.60a	0.20 ± 0.00b
Oxisol	T1	65.97 ± 4.61b	15.55 ± 2.95a	0.24 ± 0.06ab
	T2	58.70 ± 11.13ab	14.30 ± 1.03a	0.25 ± 0.02a
	T3	78.21 ± 14.91a	16.44 ± 0.95a	0.22 ± 0.00b

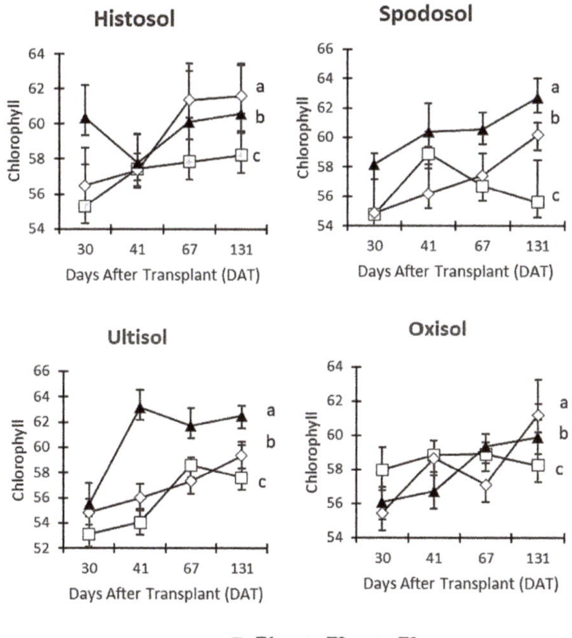

Figure 3. Chlorophyll index of the seedlings throughout the treatment period. Vertical bar represents the standard deviation. Different letters represent significant differences in Tukey's HSD comparison. Means sharing the same letter across treatments do not differ significantly at p-value ($p \leq 0.05$).

3.5. Soil Macronutrient Status

The nitrogen content in histosol was significantly higher across all treatments especially in the T3 plots as compared to spodosol, oxisol, and ultisol at the time of harvest p-value ($p \leq 0.05$) (Figure 4). This indicates that a lower N fertilizer rate is substantial for oil palms seedlings grown in the nursery prior to transplanting especially in tropical peat soil or the histosol. In the T3 treatment, the P content in the soil was higher than T2 followed by the T1 treatment especially in the histosol. However, no significant differences were found in all treatments and soils for P, as well as K soil content.

3.6. NPK Uptake by the Oil Palm Seedlings

The nutrient uptake by the oil palm seedlings is shown in Figure 5. The majority of the seedlings depicted an improved nutrient uptake in treatments with biofertilizers especially BFB under oxisol and ultisol. Overall, seedlings planted in ultisol under T3 have the highest NPK uptake, which is 20%, 38%, and 14% more than histosol, spodosol, and oxisol, respectively. However, there was no significant difference between the treatment at p-value ($p \leq 0.05$) but a significant difference was observed between histosol and spodosol, spodosol and ultisol in the uptake of N by the oil palm seedlings. Our results also indicate that there is a correlation between the high N uptake with an enhanced chlorophyll reading. However, there was no significant difference between the treatment at p-value ($p \leq 0.05$) but a significant difference was observed between histosol and spodosol, spodosol and ultisol in the uptake of N by the oil palm seedlings. No significant difference was found in all soil orders for the uptake of K at p-value ($p \leq 0.05$). Improved nutrient uptake is often correlated with root growth [20], however in our study the majority of seedlings depicted better root growth in histosol and spodosol as compared to the seedlings in oxisol (Figure 6) but the NPK uptake was higher in seedlings planted in oxisol and ultisol especially in T3.

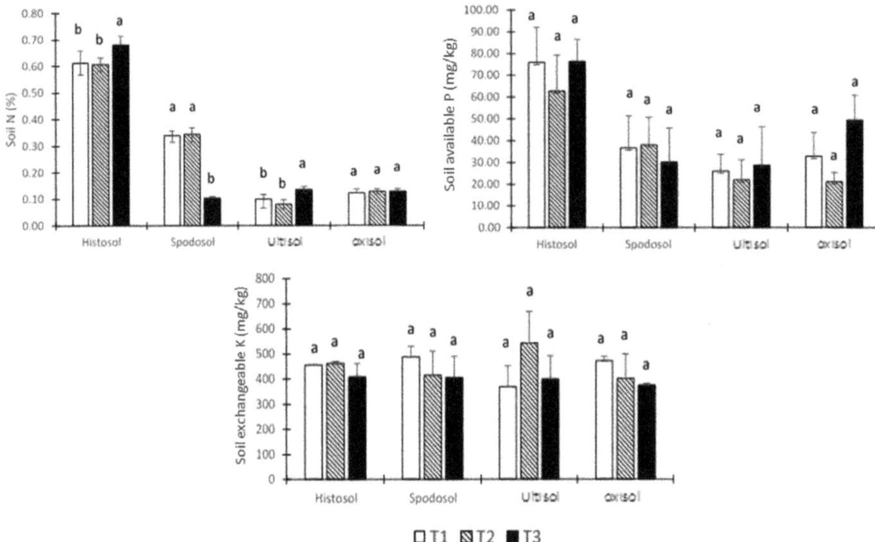

Figure 4. Soil macronutrient content (NPK) after the harvest. Vertical bar represents the standard deviation. Different letters represent significant differences in Tukey's HSD comparison. Means sharing the same letter across treatments do not differ significantly at *p*-value ($p \leq 0.05$).

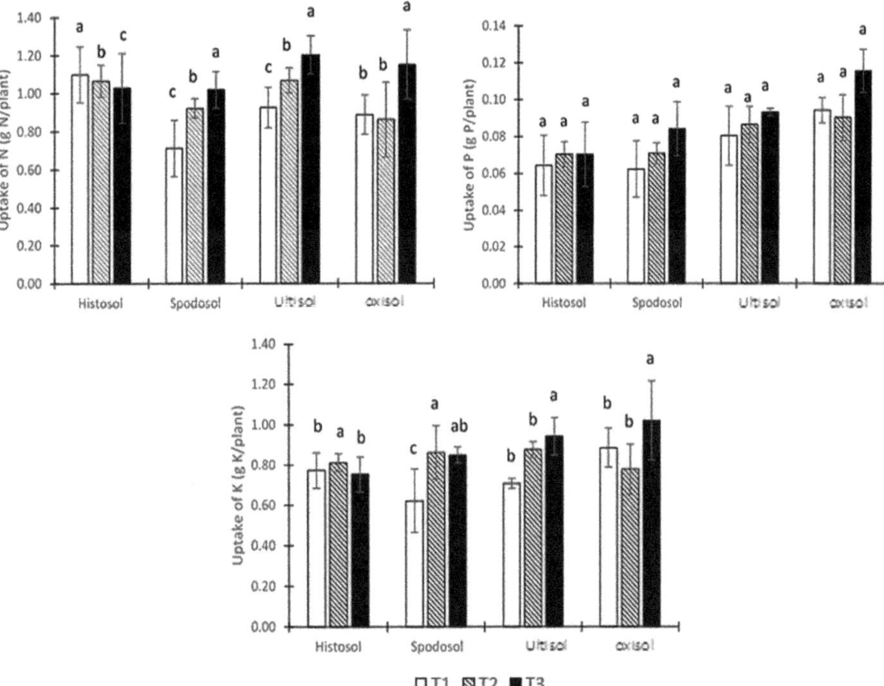

Figure 5. NPK uptake by the seedlings at time of harvest. Vertical bar represents the standard deviation. Different letters represent significant differences in Tukey's HSD comparison. Means sharing the same letter across treatments do not differ significantly at *p*-value ($p \leq 0.05$).

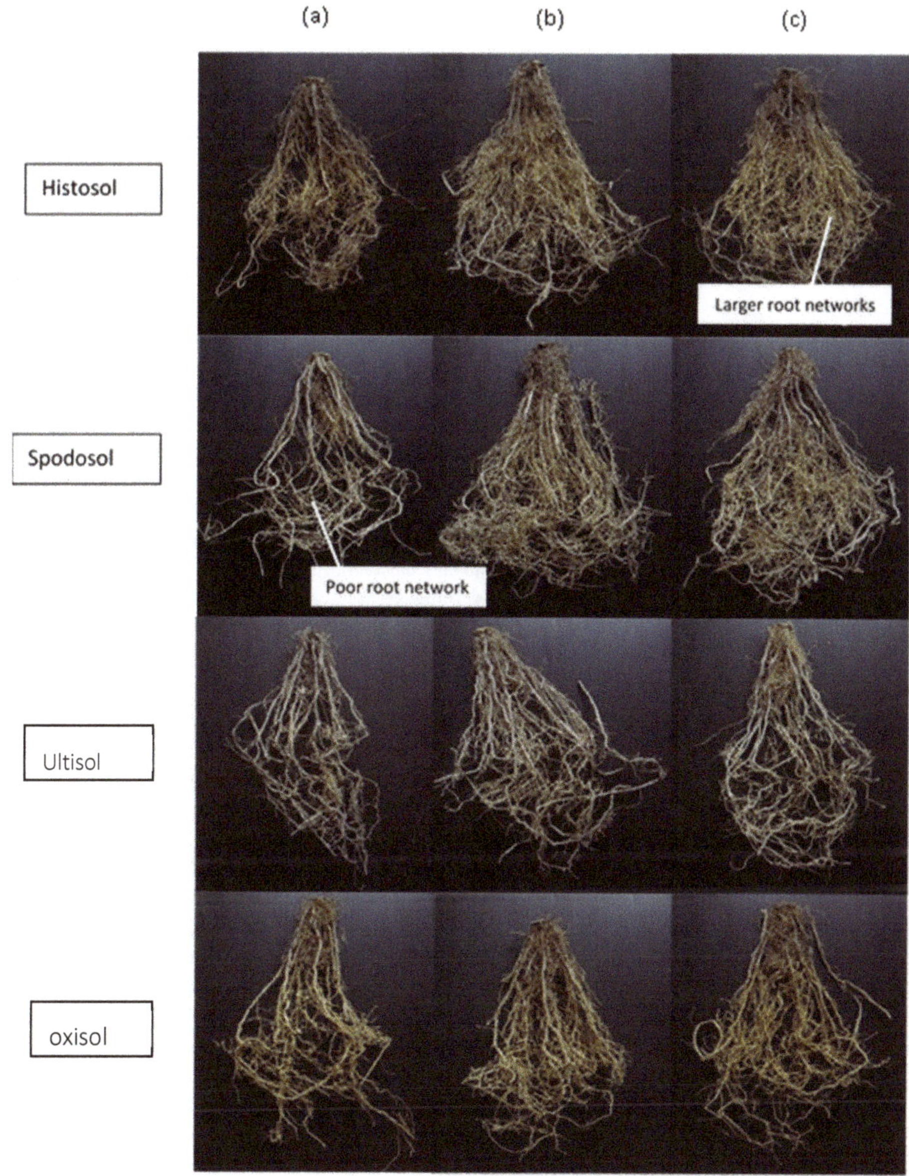

Figure 6. Roots of oil palm seedlings at the end of treatment. (a)T1, (b) T2, and (c) T3. The roots of oil palm seedlings treated with T3 were more in number, longer with more root hairs followed by seedlings in T2 then 100% T1 plots.

4. Discussion

4.1. Growth Performance of Fronds

FAO in 2011 states that about 175.5 million tons of chemical fertilizer is used in agriculture to achieve an optimum crop yield [29]. The enormous amount of chemical fertilizers deposited into soil causes a severe pollution of the river and groundwater which poses serious environmental issues and public health. Thus, the initiative of using organic fertilizers such as bioinoculant has gained immense interest over the years. Although the production of phytohormones, suppression of phytopathogens, activation of phosphate solubilization, and promotion of plant nutrients uptake are the common mechanisms of PGPR in promoting the plant growth but the exact mechanisms are not clearly known [30]. PGPR also influence aerial growth of the inoculated crops which can be observed in increasing plant height, shoot, weight and stem width, as well as increasing the number of leaves per plant [31]. A similar finding was observed by Adiprasetyo et al. [32] where the multi-microbial biofertilizer was able to increase the height, chlorophyll, and the number of leaves of oil palm plants as compared to the sole treatment with chemical fertilizer. Zainuddin et al. [9] also reported the increase of vegetative growth of oil palm seedlings in the treatment containing biofertilizer made with *Bacillus* spp., *Providencia* sp., *Phyllobacterium* sp., *Sphingobacterium* sp., and few fungi species at a lowered rate of CF. In the present study, the bioinoculant is effective in conferring beneficial growth traits to the seedlings under greenhouse condition. Regarding the plant growth promotion mechanisms, *Bacillus* strains in BFA may promote the growth of the oil palm seedlings via the inhibition of pathogens. *B. amyloliquefaciens* is efficient in producing secondary metabolites such as lipopeptides, surfactins, bacillomycin D, and fengycins, which are secondary metabolites mainly with inhibiting pathogens activity [30,33,34]. *B. amyloliquefaciens* produce a mixture of organic acids such as lactic, isovaleric, isobutyric, and acetic acid that were identified as phosphate solubilizers [35]. A study by Yanti et al. [36] reported that the *B. cereus* strain JCM 2152 has the potential to promote the growth of the tomato plant and provide resistance towards *Ralstonia solanacearum* which usually causes bacterial wilt in tomato and other tropical crops. These *Bacillus* strains are ideal to be incorporated into biofertilizer formulation as they will be able to promote the plant growth.

4.2. Stem Girth Size

Abidemi et al. [37] reported that the phosphorus fertilization by the PGPR in most seedlings resulted in a higher number of leaves and stem girth size of oil palm seedlings. In the present study, the PGPR strains used in both biofertilizers A and B were excellent in enhancing the vegetative growth of the seedlings. Although seedlings in spodosol show poor frond growth, it is compensated with the large stem girth size.

4.3. Aboveground Biomass (ABG) and Root Dry Mass Ratio

During the treatment period, the growth of the oil palm seedlings in the greenhouse occurred predominantly aboveground especially the rachis and leaves as similar was observed by Zakry et al. [38]. Batool and Iqbal [39] reported that an increment of 10–95% in root length and shoot length was observed in seedlings treated by phosphate solubilizing bacteria (PSB) inoculum as compared to the control treatment. The *Azospirillum* strains used by Amir et al. [40] was also proven to increase the root growth of oil palm plantlets at the nursery stage. A diverse group of PGPR including few PSB and *Azosprillum brasilense* in BFB might have contributed to the growth and the elongation of the roots alongside the aboveground biomass of oil palm seedlings. The increase in aboveground and root biomass could also be related to the response to PGPR in the production of phytohormones such as auxin, gibberellin, and cytokinin which stimulated the root growth [12,39] and in turn increasing the essential nutrient uptake for an enhanced growth. Seedlings treated under both biofertilizers have a positive impact on the root development and proliferation. The root dry mass was not significantly affected by the treatments but T2 spodosol followed by T3 in ultisol gave the highest root dry mass.

The enhanced root development is likely triggered by the phytohormone production and nitrogen fixation [41]. Seedlings treated using T2 seemed to have a positive impact on the root development and proliferation. This indicates the effectiveness of PGPR to colonize the roots and promote nutrients solubilization and uptake by the seedlings that result in enhanced growth. *B. tequilensis* contained in BFA was reported by Dastager et al. [42] to produce indole-3-acetic acid (IAA) which promoted the proliferation and elongation of the black pepper roots for the uptake of nutrients. Even though the roots development of seedlings in spodosol was much better than seedlings in oxisol and ultisol, the majority of seedlings depicted poor growth. This could be due to poor drainage and soil acidity in association with using spodosol [43]. Furthermore, poor root development was seen in oxisol and ultisol especially in the sole treatment with CF but there seems to be an increase root volume when the seedlings are treated with biofertilizers. Poor root development in soil can be caused by the soil bulk density, with less pore space and soil aeration [20] which gave less space for the roots proliferation. Another possible explanation could be that both oxisol and ultisol have higher bulk density, thus uprooting the seedlings was harder as compared to spodosol and histosol which were less dense. Seedlings in more compacted soils such as oxisol and ultisol might produce less primary and secondary roots, but this might be compensated by the production of longer and thicker tertiary and quaternary roots as observed by Yahya et al. [44]. Our study shows that different soil orders play a different role in oil palm seedlings nutrient uptake especially on the stimulatory efficiency of the bacterial inoculant contained in both BFA and BFB which may be important for successful root inoculation and plant growth promotion. The efficiency of the treatment and inoculation on agricultural crops also depends on few factors such as the ability of the bacteria to colonize roots, compounds exuded by the roots that enhance the colonization, and soil health [45]. Furthermore, the differences in the responses towards fertilization could be due to several factors such as the water regime [46], temperature, wind speed, soil texture, and soil depth [47].

4.4. Chlorophyll Content

The decrease in chlorophyll content in spodosol and ultisol under T1 plots could be due to the fact that CF alone failed to sustain N needs of the oil palm seedlings and this is shown in the N status of the seedlings which was the lowest among other soil orders. N is responsible to enhance the chlorophyll content [10]. Furthermore, the decrease could be due to few factors such as exposure to environmental stresses, herbicides, light irradiance or might be due to a source-sink relationship during the plant growth in later stages [48,49]. Reduction in the leaf chlorophyll content of oil palm seedlings treated with PGPR has also been reported by Amir et al. [12]. The decrease may also be influenced by time of day and underlying changes in solar irradiance where standardization of time and irradiance condition when taking chlorophyll measurements is recommended [50].

4.5. Soil Macronutrient Status

N is responsible for the cellular synthesis of chlorophyll and other components for the plant growth [51]. Soil N status plays a role in determining the effect of the PGPR inoculation especially on the nitrifying and denitrifying communities where the crop could affect the soil N dynamics within the rhizosphere and influence the type and level of mineral N available [52]. Although the addition of biofertilizer alongside with the NPK fertilizer can save up to 48% of the total N requirement in oil palm seedlings [53], from the present study there is a need to study the N rate application on crops to predict the N fertilizer requirement to avoid N rates that exceed the plant requirement [54] as seen in the histosol under the T1. The effect of PGPR also may be varied depending on the N source, N rate, and soil fertility [55]. The low N content in both oxisol and ultisol across all treatments could be an indication that essential macronutrients including P and K are taken up by the seedlings for survival [49]. Phosphate fertilizers applied to soils with lower pH are often precipitated immediately by aluminium and iron after application which makes P not available for the uptake by the crops, thus the practice of using soil pH correctors such as gypsum and liming are common in Malaysia [22]. Another alternative of increasing the P availability for the uptake by the crops is via the use of phosphate-solubilizing bacteria (PSB). Although the soils used in the study were in the acidic range, the addition of biofertilizers was found to improve the available P especially in oxisol and ultisol and histosol but no significant differences were observed. However, a revision on the rates of the applied NPK fertilizers on the histosol must be done due to excessive available N and P in the soil as the chemical fertilizer added was 500 mg P/kg in T1, 420 mg/kg soil in T2 and T3, respectively when the oil palm seedlings growth can be optimized from as low as 90 mg L^{-1} of P$_2$O$_5$ [56]. Through the reduction of NPK rate, the farmers or the oil palm plantation companies can optimize the production cost of oil palm seedlings, and reduce the negative environmental impacts due to the excessive use of chemical fertilizer [57]. Therefore, further studies on the increased rate of biofertilizers and reduced mineral fertilizer must be done to avoid over-fertilization.

4.6. NPK Uptake by the Oil Palm Seedlings

The plant N status correlates with the SPAD chlorophyll readings [58] and in the present study there is a positive correlation between the N status of the seedlings with chlorophyll readings. The inoculation with PGPR can enhance the uptake of N in the early phase of the oil palm cultivation. N-fixer such as *Bacillus sphaericus* could increase the fixed N in the soil for the uptake by the plants especially to the most nutrient demanding part in oil palm such as rachis and leaflets [38]. The microbial inoculant in the biofertilizers consisting of the N-fixer has the ability to increase the nutritional assimilation (total N) and increase the growth and at the same time improve the soil properties [59]. PGPR such as *Azospirillum* contributes about 70% of the total N requirement of the host plant [60]. The low solubilization and precipitation of phosphorus in the soil might be caused by several factors such as the pH and the type of soil used [51]. However, combined treatment with BFB resulted in higher phosphate solubilization as *Bacillus* strains such as *Bacillus brevis* strains, *B. polymyxa*, *B. thurengiensis*, *Paenibacillus*, and *B. subtilis* in BFB were considered as some of the important strains applied to soils to enhance phosphorus solubilization and uptake in plants [61]. *B. tequilensis* is a phosphate solubilizer and it also has shown to improve the macronutrient (NPK) uptake of the black pepper in both acidic and alkaline conditions [42]. The oil palm seedlings growth could be enhanced if the dose of the chemical fertilizer is reduced [62]. Essential nutrients such as the NPK must be present in biofertilizers [9] but it might be in a minute amount for the uptake by the roots. Thus, the major source of K in all treatments might be from the fertilization with chemical fertilizer, not fully by the K solubilization by the PGPR. PGPR contained in the BFB are important to the plant nutrition by increasing the N and P uptake by the plants [63]. This study also shows that although the nutrient supply and fertility are the limiting factors especially in oxisol and ultisol [64], the practice of combining biofertilizers with chemical fertilizers can enhance and optimize the yield for oil palms in these soils.

5. Conclusions

From the present study, the addition of biofertilizers alongside with chemical fertilizers have shown not only enhanced oil palm seedlings growth in terms of the height, girth size, and chlorophyll, it also improves the nutrient uptake of the seedlings and soil nutrient status at a reduced rate of chemical fertilizer. Reduction on the rate of the chemical fertilizer may be needed to avoid over-fertilization of the oil palm seedlings.

Author Contributions: Conceptualization, A.A.A.; methodology, A.A.A., R.A., and T.C.L.; software, R.A.; validation, B.F.L., P.L.S., and K.W.C.; formal analysis, A.A.A.; investigation, A.A.A.; resources, R.A., T.C.L., P.L.S., and K.W.C. ; data curation, A.A.A. and Y.-C.H.; preparation, A.A.A.; writing—review and editing, B.F.L., P.L.S., and K.W.C.; visualization, A.A.A. and Y.-C.H.; supervision, R.A., and T.C.L.; project administration, R.A., M.A.M., and T.C.L.; funding acquisition, R.A., T.C.L., P.L.S., and M.A.M. All authors have read and agreed to the published version of the manuscript.

Funding: This research was funded by the Impact-Oriented Interdisciplinary Research Grant (IIRG) (grant number IIRG004-19IISS). The authors would like to acknowledge the technical and financial support from Universiti Tenaga Nasional UNITEN-RMC Internal Research Grant (RJO 10517919/iRMC/Publication). This work was also supported by the Fundamental Research Grant Scheme, Malaysia (FRGS/1/2019/STG05/UNIM/02/2) and MyPAIR-PHC-Hibiscus grant (MyPAIR/1/2020/STG05/UNIM/1).

Acknowledgments: The authors would like to thank Universiti Malaya (UM) for the facilities provided.

Conflicts of Interest: The authors declare no conflict of interest.

References

1. Hoffmann, U. Assuring food security in developing countries under the challenges of climate change: Key trade and development issues of a fundamental transformation of agriculture. In Proceedings of the United Nations Conference on Trade and Development, Geneva, Switzerland, 22 February 2011; pp. 35–40.
2. Ward, M.H.; Brender, J.D. Drinking water nitrate and human health. In *Encyclopedia of Environmental Health*; Elsevier BV: Amsterdam, The Netherlands, 2019; pp. 173–186.
3. Blankson, G.K.; Osei-Fosu, P.; Adeendze, E.; Ashie, D. Contamination levels of organophosphorus and synthetic pyrethroid pesticides in vegetables marketed in Accra, Ghana. *Food Control* **2016**, *68*, 174–180. [CrossRef]
4. Domínguez, I.; Romero-González, R.; Liébanas, F.J.A.; Vidal, J.L.M.; Frenich, A.G. Automated and semi-automated extraction methods for GC–MS determination of pesticides in environmental samples. *Trends Environ. Anal. Chem.* **2016**, *12*, 1–12. [CrossRef]
5. Wang, B.; Lai, T.; Huang, Q.-W.; Yang, X.-M.; Shen, Q.-R. Effect of N Fertilizers on Root Growth and Endogenous Hormones in Strawberry. *Pedosphere* **2009**, *19*, 86–95. [CrossRef]
6. VoonKheong, L.; Rahman, Z.A.; Musa, M.H.; Hussein, A. Effects of severing oil palm roots on leaf nutrient levels and P uptake. *J. Oil Palm Res.* **2012**, *24*, 1343–1348.
7. Bhardwaj, D.; Ansari, M.W.; Sahoo, R.K.; Tuteja, N. Biofertilizers function as key player in sustainable agriculture by improving soil fertility, plant tolerance and crop productivity. *Microb. Cell Factories* **2014**, *13*, 66. [CrossRef]
8. Itelima, J.U.; Bang, W.J.; Onyimba, I.A.; Sila, M.D.; Egbere, O.J. Bio-fertilizer as key player in enhancing soil fertility and crop productivity: A Review. *Direct Res. J. Agric. Food Sci.* **2018**, *6*, 73–83. [CrossRef]
9. Zainuddin, N. Effect of biofertiliser containing different percentage rates of chemical fertiliser on oil palm seedlings. *J. Oil Palm Res.* **2019**. [CrossRef]
10. Naher, U.A.; Panhwar, Q.A.; Othman, R.; Shamshuddin, J.; Ismail, M.R.; Zhou, E. Proteomic study on growth promotion of PGPR inoculated aerobic rice (*Oryza sativa* L.) cultivar MR219-9. *Pak. J. Bot.* **2018**, *50*, 1843–1852.
11. Vejan, P.; Abdullah, R.; Tumirah, K.; Ismail, S.; Boyce, A.N. Role of plant growth promoting rhizobacteria in agricultural sustainability—A Review. *Molecules* **2016**, *21*, 573. [CrossRef]
12. Amir, H.G.; Shamsuddin, Z.H.; Halimi, M.S.; Marziah, M.; Ramlan, M.F. Enhancement in nutrient accumulation and growth of oil palm seedlings caused by PGPR under field nursery conditions. *Commun. Soil Sci. Plant Anal.* **2005**, *36*, 2059–2066. [CrossRef]
13. Jiménez-Gómez, A.; Flores-Félix, J.D.; García-Fraile, P.; Mateos, P.F.; Menéndez, E.; Velazquez, E.; Rivas, R. Probiotic activities of *Rhizobium laguerreae* on growth and quality of spinach. *Sci. Rep.* **2018**, *8*, 295. [CrossRef] [PubMed]
14. Vyas, P.; Gulati, A. Organic acid production in vitro and plant growth promotion in maize under controlled environment by phosphate-solubilizing fluorescent *Pseudomonas*. *BMC Microbiol.* **2009**, *9*, 174. [CrossRef]
15. Fasim, F.; Ahmed, N.; Parsons, R.; Gadd, G.M. Solubilization of zinc salts by a bacterium isolated from the air environment of a tannery. *FEMS Microbiol. Lett.* **2002**, *213*, 1–6. [CrossRef] [PubMed]
16. Gulati, A.; Sharma, N.; Vyas, P.; Sood, S.; Rahi, P.; Pathania, V.; Prasad, R. Organic acid production and plant growth promotion as a function of phosphate solubilization by *Acinetobacter rhizosphaerae* strain BIHB 723 isolated from the cold deserts of the trans-Himalayas. *Arch. Microbiol.* **2010**, *192*, 975–983. [CrossRef]
17. Saeid, A.; Prochownik, E.; Dobrowolska-Iwanek, J. Phosphorus solubilization by *Bacillus* species. *Molecules* **2018**, *23*, 2897. [CrossRef]
18. Rajput, M.S.; Kumar, G.N.; Rajkumar, S. Repression of oxalic acid-mediated mineral phosphate solubilization in rhizospheric isolates of *Klebsiella pneumoniae* by succinate. *Arch. Microbiol.* **2012**, *195*, 81–88. [CrossRef] [PubMed]
19. Leite, A.D.A.; Cardoso, A.A.D.S.; Leite, R.D.A.; De Oliveira-Longatti, S.M.; Filho, J.F.L.; Moreira, F.M.D.S.; Melo, L.C.A. Selected bacterial strains enhance phosphorus availability from biochar-based rock phosphate fertilizer. *Ann. Microbiol.* **2020**, *70*, 1–13. [CrossRef]
20. Rosenani, A.B.; Rovica, R.; Cheah, P.M.; Lim, C.T. Growth performance and nutrient uptake of oil palm seedling in prenursery stage as influenced by oil palm waste compost in growing media. *Int. J. Agron.* **2016**, *2016*, 1–8. [CrossRef]

21. Halim, N.S.A.; Abdullah, R.; Karsani, S.A.; Osman, N.; Panhwar, Q.A.; Ishak, C.F. Influence of soil amendments on the growth and yield of rice in acidic soil. *Agronomy* **2018**, *8*, 165. [CrossRef]

22. Sung, C.T.; Ishak, C.F.; Abdullah, R.; Othman, R.; Qurban; Panhwar, A.O. Soil properties (physical, chemical, biological, mechanical). In *Soils of Malaysia*; Ashraf, M.A., Othman, R., Ishak, C.F., Eds.; CRC Press, Taylor & Francis Group: New York, NY, USA, 2017.

23. Nelson, D.W.; Sommers, L.E.; Sparks, D.L.; Page, A.; Helmke, P.; Loeppert, R. Total carbon, organic carbon, and organic matter. In *Soil Science Society of America*; Soil Science Society of America and American Society of Agronomy: Madison, WI, USA, 2018; pp. 961–1010.

24. Bremner, J.M.; Mulvaney, C.S. Nitrogen-Total. In *Methods of Soil Analysis. Part 2 Chemical and Microbiological Properties*, 2nd ed.; Page, A.L., Ed.; Soil Science Society of America, Inc.: Madison, WI, USA, 1982.

25. Bray, R.H.; Kurtz, L.T. Determination of total, organic, and available forms of phosphorus in soils. *Soil Sci.* **1945**, *59*, 39–46. [CrossRef]

26. Sim, C.C.; Zaharah, A.R.; Tan, M.S.; Goh, K.J. Rapid determination of leaf chlorophyll concentration, photosynthetic activity and NK concentration of *Elaies guineensis* via correlated SPAD-502 chlorophyll index. *Asian J. Agric. Res.* **2015**, *9*, 132–138. [CrossRef]

27. Havlin, J.L.; Soltanpour, P.N. A nitric acid plant tissue digest method for use with inductively coupled plasma spectrometry 1. *Commun. Soil Sci. Plant Anal.* **1980**, *11*, 969–980. [CrossRef]

28. Hendershot, W.; Lalande, H.; Duquette, M. Ion Exchange and Exchangeable Cations. In *Soil Sampling and Methods of Analysis*, 2nd ed.; Carter, M.R., Gregorich, E.G., Eds.; CRC Press, Taylor & Francis Group: Boca Raton, FL, USA, 2006.

29. FAO; IFAD; IMF; OECD; UNCTAD; WFP; World Bank; WTO; IFPRI; UNHLTF. *Price Volatility in Food and Agricultural Markets: Policy Responses*; Policy Report; FAO: Roma, Italy, 2 June 2011.

30. Almaghrabi, O.A.; Massoud, S.I.; Abdelmoneim, T.S. Influence of inoculation with plant growth promoting rhizobacteria (PGPR) on tomato plant growth and nematode reproduction under greenhouse conditions. *Saudi J. Boil. Sci.* **2013**, *20*, 57–61. [CrossRef] [PubMed]

31. Park, Y.-S.; Park, K.; Kloepper, J.W.; Ryu, C.-M. Plant Growth-Promoting Rhizobacteria stimulate vegetative growth and asexual reproduction of *Kalanchoe daigremontiana*. *Plant Pathol. J.* **2015**, *31*, 310–315. [CrossRef] [PubMed]

32. Adiprasetyo, T.; Purnomo, B.; Handajaningsih, M.; Hidayat, H. The usage of BIOM3G-Biofertilizer to improve and support sustainability of land system of independent oil palm smallholders. *Int. J. Adv. Sci. Eng. Inf. Technol.* **2014**, *4*, 345–348. [CrossRef]

33. Chen, X.-H.; Vater, J.; Piel, J.; Franke, P.; Scholz, R.; Schneider, K.; Koumoutsi, A.; Hitzeroth, G.; Grammel, N.; Strittmatter, A.W.; et al. Structural and functional characterization of three polyketide synthase gene clusters in *Bacillus amyloliquefaciens* FZB 42. *J. Bacteriol.* **2006**, *188*, 4024–4036. [CrossRef]

34. Hossain, M.J.; Ran, C.; Liu, K.; Ryu, C.-M.; Rasmussen-Ivey, C.R.; Williams, M.A.; Hassan, M.K.; Choi, S.-K.; Jeong, H.; Newman, M.; et al. Deciphering the conserved genetic loci implicated in plant disease control through comparative genomics of *Bacillus amyloliquefaciens* subsp. plantarum. *Front. Plant Sci.* **2015**, *6*, 631. [CrossRef]

35. Illmer, P.; Schinner, F. Solubilization of inorganic phosphates by microorganisms isolated from forest soils. *Soil Boil. Biochem.* **1992**, *24*, 389–395. [CrossRef]

36. Yanti, Y.; Warnita, W.; Reflin, R.; Nasution, C.R. Characterizations of endophytic *Bacillus* strains from tomato roots as growth promoter and biocontrol of *Ralstonia solanacearum*. *Biodiversitas J. Boil. Divers.* **2018**, *19*, 906–911. [CrossRef]

37. Abidemi, A.A.; Akinrinde, E.A.; Obigbesan, G.O. Oil Palm (*Elaeis guineensis*) seedling performance in response to phosphorus fertilization in two benchmark soils of Nigeria. *Asian J. Plant Sci.* **2006**, *5*, 767–775. [CrossRef]

38. Zakry, F.A.A.; Shamsuddin, Z.H.; Khairuddin, A.R.; Zin, Z.Z.; Anuar, A.R. Inoculation of *Bacillus sphaericus* UPMB-10 to young oil palm and measurement of its uptake of fixed nitrogen using the 15N isotope dilution technique. *Microbes Environ.* **2012**, *27*, 257–262. [CrossRef] [PubMed]

39. Batool, S.; Iqbal, A. Phosphate solubilizing rhizobacteria as alternative of chemical fertilizer for growth and yield of Triticum aestivum (Var. Galaxy 2013). *Saudi J. Boil. Sci.* **2019**, *26*, 1400–1410. [CrossRef] [PubMed]

40. Amir, G.H.; Shamsuddin, Z.H.; Saud, M.M.H.; Ramlan, M.F.; Marziah, M. Effects of *Azosprillum* inoculation on N2 fixation and growth of oil palm plantlets at nursery stage. *J. Oil Palm Res.* **2001**, *13*, 42–49.

41. Om, A.C.; Ghazali, A.H.A.; Keng, C.L.; Ishak, Z. Microbial inoculation improves growth of oil palm plants (*Elaeis guineensis* Jacq.). *Trop. Life Sci. Res.* **2009**, *20*, 71–77. [PubMed]

42. Dastager, S.G.; Deepa, C.K.; Pandey, A. Growth enhancement of black pepper (*Piper nigrum*) by a newly isolated *Bacillus tequilensis* NII-0943. *Biologia* **2011**, *66*. [CrossRef]

43. Safitri, L.; Hermantoro, H.; Purboseno, S.; Kautsar, V.; Saptomo, S.K.; Kurniawan, A. water footprint and crop water usage of oil palm (*Eleasis guineensis*) in Central Kalimantan: Environmental sustainability indicators for different crop age and soil conditions. *Water* **2018**, *11*, 35. [CrossRef]

44. Yahya, Z.; Husin, A.; Talib, J.; Othman, J.; Ahmed, O.H.; Jalloh, M.B.; Yahya, O.H. Oil Palm (*Elaeis guineensis*) roots response to mechanization in Bernam Series soil. *Am. J. Appl. Sci.* **2010**, *7*, 343–348. [CrossRef]

45. De Souza, R.; Ambrosini, A.; Passaglia, L.M.P. Plant growth-promoting bacteria as inoculants in agricultural soils. *Genet. Mol. Boil.* **2015**, *38*, 401–419. [CrossRef]

46. Auxtero, E.A.; Shamshuddin, J. Growth of oil palm (Elaeis guineensis) seedlings on acid sulfate soils as affected by water regime and aluminium. *Plant Soil* **1991**, *137*, 243–257. [CrossRef]

47. Woittiez, L.S.; Van Wijk, M.T.; Slingerland, M.; Van Noordwijk, M.; Giller, K.E. Yield gaps in oil palm: A quantitative review of contributing factors. *Eur. J. Agron.* **2017**, *83*, 57–77. [CrossRef]

48. Dudeja, S.S.; Chaudhary, P. Fast chlorophyll fluorescence transient and nitrogen fixing ability of chickpea nodulation variants. *Photosynthetica* **2005**, *43*, 253–259. [CrossRef]

49. Nursu'aidah, H.; Motior, M.R.; Nazia, A.; Islam, M.A. Growth and photosynthetic responses of long bean (*Vigna unguiculata*) and mung bean (*Vigna radiata*) response to fertilization. *J. Anim. Plant Sci.* **2014**, *24*, 573–578.

50. Padilla, F.; De Souza, R.; Peña-Fleitas, M.T.; Grasso, R.; Gallardo, M.; Thompson, R. Influence of time of day on measurement with chlorophyll meters and canopy reflectance sensors of different crop N status. *Precis. Agric.* **2019**, *20*, 1087–1106. [CrossRef]

51. Hayat, R.; Ali, S.; Amara, U.; Khalid, R.; Ahmed, I. Soil beneficial bacteria and their role in plant growth promotion: A review. *Ann. Microbiol.* **2010**, *60*, 579–598. [CrossRef]

52. Florio, A.; Pommier, T.; Gervaix, J.; Bérard, A.; Le Roux, X. Soil C and N statuses determine the effect of maize inoculation by plant growth-promoting rhizobacteria on nitrifying and denitrifying communities. *Sci. Rep.* **2017**, *7*, 1–12. [CrossRef] [PubMed]

53. Mia, M.A.B.; Shamsuddin, Z.H.; Wahab, Z.; Marziah, M. High-yielding and quality banana production through plant growth-promoting rhizobacterial (PGPR) inoculation. *Fruits* **2005**, *60*, 179–185. [CrossRef]

54. Zhao, W.; Liang, B.; Yang, X.; Zhou, J. Fate of residual 15 N-labeled fertilizer in dryland farming systems on soils of contrasting fertility. *Soil Sci. Plant Nutr.* **2015**, *61*, 1–10. [CrossRef]

55. Fan, X.-H.; Zhang, S.; Mo, X.-D.; Li, Y.; Fu, Y.-Q.; Liu, Z. Effects of Plant Growth-Promoting Rhizobacteria and N source on Plant Growth and N and P uptake by Tomato Grown on Calcareous Soils. *Pedosphere* **2017**, *27*, 1027–1036. [CrossRef]

56. Mohidin, H.; Rafii, M.Y.; Man, S.; Idris, J.; Hanafi, M.M.; Abdullah, S.N.A.; Idris, A.S.; Sahebi, M. Determination of optimum levels of nitrogen, phosphorus and potassium of oil palm seedlings in solution culture. *Bragantia* **2015**, *74*, 247–254. [CrossRef]

57. Radin, R.; Abu Bakar, R.; Ishak, C.F.; Ahmad, S.H.; Tsong, L.C. Biochar-compost mixture as amendment for improvement of polybag-growing media and oil palm seedlings at main nursery stage. *Int. J. Recycl. Org. Waste Agric.* **2017**, *7*, 11–23. [CrossRef]

58. Gholizadeh, A.; Saberioon, M.; Boruvka, L.; Wayayok, A.; Soom, M.A.M. Leaf chlorophyll and nitrogen dynamics and their relationship to lowland rice yield for site-specific paddy management. *Inf. Process. Agric.* **2017**, *4*, 259–268. [CrossRef]

59. Lim, S.-L.; Subramaniam, S.; Zamzuri, I.; Amir, H.G. Growth and biochemical profiling of artificially associated micropropagated oil palm plantlets with *Herbaspirillum seropedicae*. *J. Plant Interact.* **2018**, *13*, 173–181. [CrossRef]

60. Mia, M.A.B.; Shamsuddin, Z.H.; Wahab, Z.; Marziah, M. Effect of plant growth promoting rhizobacterial (PGPR) inoculation on growth and nitrogen incorporation of tissue-cultured Musa plantlets under nitrogen-free hydroponics condition. *Aust. J. Crop Sci.* **2010**, *4*, 85–90.

61. A Van Veen, J.; Van Overbeek, L.S.; Van Elsas, J.D. Fate and activity of microorganisms introduced into soil. *Microbiol. Mol. Boil. Rev.* **1997**, *61*. [CrossRef]

62. Veeramachaneni, S.; Ramachandrudu, K. Changes in growth, microbial and enzyme activities in oil palm nursery in response to bioinoculants and chemical fertilizers. *Arch. Agron. Soil Sci.* **2019**, *66*, 545–558. [CrossRef]

63. Egamberdieva, D. The effect of plant growth promoting bacteria on growth and nutrient uptake of maize in two different soils. *Appl. Soil Ecol.* **2007**, *36*, 184–189. [CrossRef]

64. Teo, C.B.; Chew, P.S.; Goh, K.J.; Kee, K.K. Optimising return from fertilizer for oil palms: An integrated agronomic approach. In Proceedings of the IFA Regional Conference for Asia and the Pacific, Hong Kong, China, 7–10 December 1998.

Publisher's Note: MDPI stays neutral with regard to jurisdictional claims in published maps and institutional affiliations.

Article

Determination of Dissolved CO_2 Concentration in Culture Media: Evaluation of pH Value and Mathematical Data

Amir Izzuddin Adnan [1], Mei Yin Ong [1], Saifuddin Nomanbhay [1,*] and Pau Loke Show [2]

[1]   Institute of Sustainable Energy, Universiti Tenaga Nasional, Kajang 43000, Selangor, Malaysia;
     amir.izzuddin@uniten.edu.my (A.I.A.); me089475@hotmail.com or Mei.Yin@uniten.edu.my (M.Y.O.)
[2]   Department of Chemical and Environmental Engineering, Faculty of Science and Engineering,
     University of Nottingham Malaysia, Jalan Broga, Semenyih 43500, Selangor, Malaysia;
     pauloke.show@nottingham.edu.my
*   Correspondence: saifuddin@uniten.edu.my; Tel.: +60-3-8921-7285

Received: 23 September 2020; Accepted: 26 October 2020; Published: 29 October 2020

Abstract: Carbon dioxide is the most influential gas in greenhouse gasses and its amount in the atmosphere reached 412 μmol/mol in August 2020, which increased rapidly, by 48%, from preindustrial levels. A brand-new chemical industry, namely organic chemistry and catalysis science, must be developed with carbon dioxide (CO_2) as the source of carbon. Nowadays, many techniques are available for controlling and removing carbon dioxide in different chemical processes. Since the utilization of CO_2 as feedstock for a chemical commodity is of relevance today, this study will focus on how to increase CO_2 solubility in culture media used for growing microbes. In this work, the CO_2 solubility in a different medium was investigated. Sodium hydroxide (NaOH) and monoethanolamine (MEA) were added to the culture media (3.0 g/L dipotassium phosphate (K_2HPO_4), 0.2 g/L magnesium chloride ($MgCl_2$), 0.2 g/L calcium chloride ($CaCl_2$), and 1.0 g/L sodium chloride (NaCl)) for growing microbes in order to observe the difference in CO_2 solubility. Factors of temperature and pressure were also studied. The determination of CO_2 concentration in the solution was measured by gas analyzer. The result obtained from optimization revealed a maximum CO_2 concentration of 19.029 mol/L in the culture media with MEA, at a pressure of 136.728 kPa, operating at 20.483 °C.

Keywords: carbon dioxide; culture media; microorganism; optimization

---

1. Introduction

Fossil fuels are broadly acknowledged as being the principal source of energy, and since the First Industrial Revolution the amount of carbon dioxide (CO_2) in the atmosphere has risen from 280 μmol/mol to 412 μmol/mol. The resulting CO_2 emissions contribute significantly to worldwide climate change [1]. Up to now, the deployment of cutting-edge low-carbon fossil-energy technologies was considered to be the ultimate solution. Preventing worldwide climate change can be achieved by taking two long-term emission objectives into account. First, CO_2 emissions have had to reach their highest point, and then in the second half of the century, the goal has had to be to strive to achieve net greenhouse gas neutrality, by balancing anthropogenic emissions by the sources with the removal by sinks [2]. Hence, it is crucial to decrease such anthropogenic emissions. Second, CO_2 can be captured and used as a significant feedstock to produce valuable commodities. As the world population increases, the need for energy supply rises at an exponential rate. Subsequently, to meet this demand, new and renewable energy sources are required. Along this line, treating CO_2 as a feedstock to many value-added chemicals and fuels addresses both emission-control and energy supply challenges [3].

The concepts mentioned above are commonly used in carbon management from a climate change perspective. The term used is CO_2 capture, utilization, and sequestration (CCUS). The carbon capture and storage (CCS) approach in reducing CO_2 emissions is particularly common nowadays [4]. It refers to technologies that emphasize the selective removal of waste CO_2 from a large point source, its compression into a liquified gas, and finally its transportation and sequestration to a storage site where it will not enter the atmosphere such as underground geologic formations, including depleted oil and gas reservoirs or oceans [5]. Meanwhile, carbon capture and utilization (CCU) technologies capture CO_2 to be recycled for an additional application. It differs from CCS in that CCU does not permanently sequester the CO_2 waste, but rather, treats it as a renewable carbon feedstock to complement the conventional petrochemical feedstocks for conversion into other substances or products with higher economic value [6].

However, due to the thermodynamically stable nature of CO_2, utilizing it in chemical reactions is challenging. High energy input is required to breakdown carbon atoms in CO_2 molecules, which is one of the reasons why CO_2 is not extensively used in current chemical industries. Nevertheless, autotrophic microorganisms are well-known for their ability to utilize light to fix atmospheric CO_2 during the process of photosynthesis. These microorganisms can capture energy in the light cycle and store it for converting adenosine diphosphate (ADP) and nicotinamide adenine dinucleotide phosphate (NADP) into adenosine triphosphate (ATP) and nicotinamide adenine dinucleotide phosphate hydrogen (NADPH) respectively. They then utilize these energy molecules during a dark cycle for transforming CO_2 into valuable organic compounds [7]. Research has been done recently on altering the molecules of autotrophic cyanobacteria and algae through metabolic engineering to take advantage of their abilities to treat CO_2 [8]. Microorganisms require macronutrients, micronutrients, and vitamins to grow [9]. Based on these requirements, a culture broth is required as a growth medium in a closed system, such as a bioreactor, for the production of biomass or organic compounds [10]. Therefore, increasing the CO_2 solubility in a culture broth is an important step toward CO_2 utilization by microorganisms.

Al-Anezi et al. studied the effect of temperature, salinity, and pressure on CO_2 solubility in different aqueous solutions [11]. The relationship between these parameters on CO_2 solubility was presented. Where gas solubility reduced between the temperatures of 25 °C and 60 °C, the effect was less evident at a higher temperature. Meanwhile, higher pressure (one to two bars) resulted in higher gas solubility. Additionally, the study stated that gas solubility decreased as salt increased in the solution. Yincheng et al. compared the CO_2 removal efficiency of sodium hydroxide (NaOH) and aqueous ammonia [12]. The study involved the capture of CO_2 in a spray column, and a fine spray of ammonia and NaOH was used. The key finding of the study was the value of mole ratios of NaOH and ammonia to CO_2 suitable for the spray column, which is 4.43 and 9.68 respectively. Additionally, Martins da Rosa et al. researched the CO_2 fixation of *Chlorella* using monoethanolamine (MEA) [13]. Through this research, it was found that the CO_2 intake was higher for the growth of algae using a certain mass concentration of MEA; 50 mg/L and 100 mg/L for this particular strain of algae. However, the treatment of CO_2 decreased if the MEA concentration used was very low or very high. Thus, the concentration of MEA used was dependent on the microorganism used.

The present paper will investigate various features that determine the concentration of CO_2 dissolved in a culture solution. First, to determine the maximum CO_2 concentration in the NaOH aqueous solution as a comparison, a steady rate of CO_2 was supplied for a certain time. NaOH was chosen because the CO_2 absorption capacity of NaOH solution is high, with a mass ratio of capture, $w(NaOH/CO_2)$ equal to 0.9 [14]. Second, MEA and culture media were used as the absorbent to determine the capability of both solutions in capturing CO_2. Next, different kinds of culture media solutions were prepared; with the addition of either NaOH or MEA, and the absorption was carried out under the same conditions as the previous run in a batch reactor. From the experimental results, the absorption behavior is presented according to pH and time. Then optimization was run by software to determine the optimized condition for CO_2 absorption.

2. Materials and Methods

2.1. Carbon Dioxide (CO₂) Delivery System

A batch-typed glass (borosilicate) cylindrical reactor (Bio Gene®, Australia) with a built-in motor, the total volume capacity of 5 L (D = 140 mm; h = 325 mm), equipped with a pressure gauge, (RS Components, Johor Bahru, Johor, Malaysia), pH (BOQU®, Shanghai, China) and temperature probe (DPSTAR Manufacturing Sdn. Bhd., Kuala Lumpur, Malaysia) was employed for the carbon dioxide (CO_2) absorption as shown in Figure 1. The reactor was connected to a pressurized gas mixture tank (Gaslink Industrial Gases Sdn. Bhd., Puchong, Selangor, Malaysia) through a flowmeter (HERO TECH®, Puchong, Selangor, Malaysia) with a valve for controlling the flow rate of the mixture. The compositions of the gas mixture were 90% CO_2 and 10% Nitrogen (N_2). For providing a vacuum space inside the reactor, a vacuum pump (vacuubrand®, Wertheim am Main, Baden-Württemberg, Germany) was mounted with a valve. The gas analyzer (Geotech Environmental Equipment Inc., Denver, CO, USA) was connected through one of the openings to determine the amount of CO_2 in the headspace. The unwanted opening at the surface of the reactor was closed by a stopper. Additional CO_2 diffuser (FunPetAqua.my, Kuala Lumpur, Malaysia) was employed to increase the CO_2 absorption rate. All dimensions and components of the diffuser are described in Figure 2. All components were made of borosilicate except for the distributor head that was made of porous rubber.

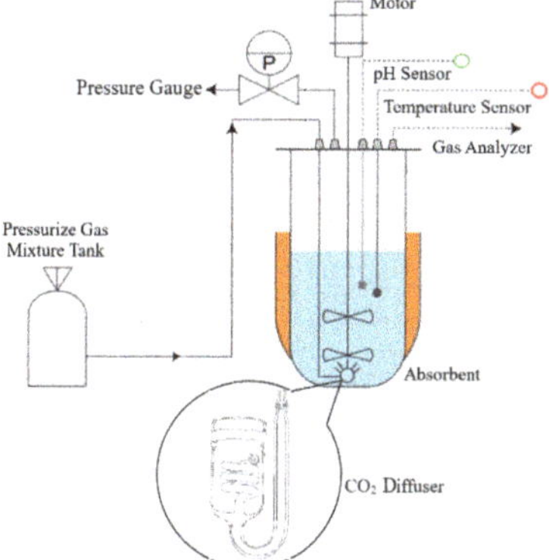

Figure 1. Carbon dioxide absorption system.

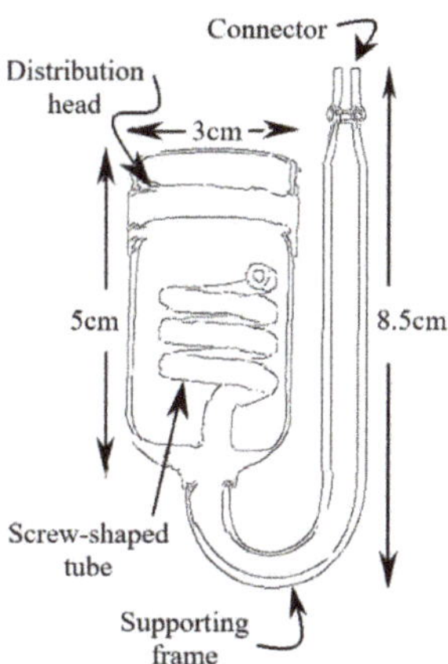

Figure 2. Carbon dioxide diffuser.

2.2. Time Measurement of The Maximum Carbon Dioxide (CO_2) Dissolved

An aqueous solution with the concentration of sodium hydroxide (NaOH) equal to 0.1 mol/L was prepared by dissolving sodium hydroxide pellets (Sigma-Aldrich, St. Louis, MO, USA) in 2 L of distilled water. Before absorption was conducted, the absorbent (NaOH) temperature was maintained at room temperature, 100 kPa pressure and pH = 11.0 by adding hydrochloric acid (HCl) (Sigma-Aldrich, St. Louis, MO, USA) or NaOH to adjust the pH value to a desirable number. All equipment, including tubes, fittings, and the headspace of the reactor, were sufficiently washed by a vacuum pump. After the conditions are met, the absorption was carried out by injecting the gas mixture into the absorbent via a sparger with a flow rate of 4.5 L/min controlled by a mass flow controller. The solution was mixed using a mechanical stirrer at a speed of 180 rpm for uniform reaction in an absorber. The variations of pH during the reaction were measured every 10 s. The gas mixture was bubbled into the absorbent until the pH value stops dropping. The time for the CO_2 component in the mixture to dissolved in the medium was taken and a graph for pH drop against time was drawn. The same experiment was repeated while using 0.1 mol/L monoethanolamine (MEA) (Sigma-Aldrich, St. Louis, MO, USA) solution to determine the pH drop for both absorbents.

Then the experiment was repeated by replacing the absorbent with culture media. Culture media composed of 3.0 g/L dipotassium phosphate (K_2HPO_4), 0.2 g/L magnesium chloride ($MgCl_2$), 0.2 g/L calcium chloride ($CaCl_2$), and 1.0 g/L sodium chloride (NaCl) was prepared (Sigma-Aldrich, St. Louis, MO, USA). A total of 0.1 mol/L aqueous NaOH solution or 0.1 mol/L MEA solution was added to culture media to promote CO_2 absorption. It is important to determine media capability as CO_2 absorbent as culture media will help microbes to utilize CO_2 for producing the chemical commodity. The parameter is set as the previous run, until the CO_2 concentration is maximized, and then the gas mixture supply will be stopped. The graph of pH drop against time for both compositions was plotted to determine the significance of each composition. Then, all the experiments were repeated with the deployment of a CO_2 diffuser at the sparger to observe the change in time for pH drop.

2.3. Optimization through Surface Response Methodology

Additionally, three factors that affect CO_2 solubility (s), composition (X), pressure (p), and temperature (T) were also studied using three-factor, three-level Box-Behnken design (BBD). Each of these independent factors divided into three different levels as shown in Table 1. To describe the relationship between a set of parameters and output, the regression model was developed in BBD design and can be defined by Equation (1):

$$Y = \beta_0 + \sum_{i=1}^{3} \beta_i X_i + \sum_{i=1}^{3} \beta_{ii} X_{ii}^2 + \sum_{i=1}^{2} \sum_{j=i+1}^{3} \beta_{ij} X_i X_j \tag{1}$$

where Y: response; β_0: constant-coefficient; β_i and β_{ii}: linear and quadratic coefficients for the terms X_i and X_{ii}, respectively; β_{ij}: coefficients which represent the interactions of X_i and X_j. Then, an analysis of variance (ANOVA) was used to determine whether the models are acceptable for analysis.

Table 1. Levels and ranges of independent input parameters in the Box-Behnken design.

Parameters	Symbols	Level and Range		
		Low	Centre	High
Absorbent	X	Media + NaOH	NaOH	Media + MEA
Pressure, (kPa)	p	100	125	150
Temperature, (°C)	T	20	30	40

2.4. Analytical Methods

Dissolved carbon dioxide (CO_2) in culture media was determined by calculation based on readings obtained from a gas analyzer (Geotech Environmental Equipment Inc., Denver, CO, USA). Readings of pH and temperature were obtained by direct measurement using a portable pH (BOQU®, Shanghai, China) and temperature probe (DPSTAR, Kuala Lumpur, Malaysia). Time was recorded using a stopwatch.

2.5. Model Description and Calculation

Dissolved carbon dioxide (CO_2) in culture media can be calculated using the information given by the gas analyzer. The reading given by the analyzer consists of the composition of air in the headspace. Since the value given by the analyzer is in percentage, some calculations need to be done. The percentage of CO_2 was multiplied by the volume of headspace to get the volume CO_2 in the headspace in m^3. The volume of headspace was constant throughout the experiment. It was set at 2.5×10^{-3} m^3. By subtracting the total value of CO_2 supplied with the one in the headspace, the value of carbon dioxide aqueous, CO_2(aq) can be obtained.

3. Results and Discussion

3.1. Relationship between Carbon Dioxide (CO_2) Concentration and pH

The pH of the absorbent decreased with increasing CO_2 concentration. A slight difference in the lowest value of pH was observed between the absorbent used; sodium hydroxide (NaOH), culture media, and culture media with either NaOH or monoethanolamine (MEA), but the difference was not significant except for MEA which was observed to take a longer time to achieve minimum pH. A significant difference was only in the time taken to achieve the minimum value of pH. The graph of pH against time was plotted for NaOH, MEA, and culture media in Figure 3. The pH decreased with time until the 18 min mark for NaOH and the final pH was 6.91. Meanwhile for MEA and culture media were $t = 60$ min; pH = 8.52 and $t = 8$ min; pH = 6.46 respectively. Finally, culture media with NaOH and MEA were $t = 9$ min; pH = 6.45 and $t = 21$ min; pH = 7.0 respectively.

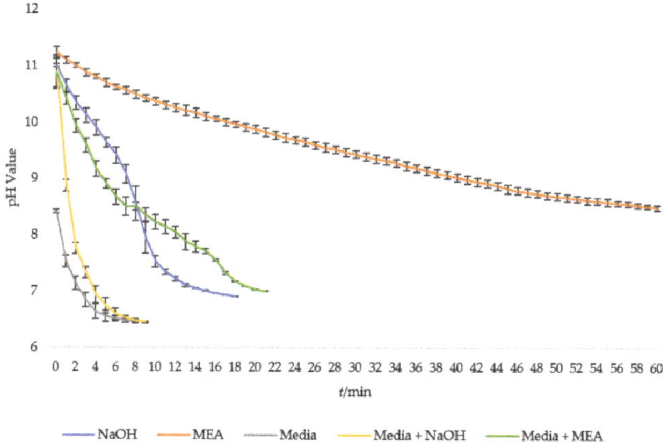

Figure 3. Effect of pH change for five absorbents at $p = 100$ kPa, $T = 20\ °C$, with an initial pH of 11 with exception of culture media with an initial pH of 8.5.

Since the behavior of culture media and MEA absorbents do not follow the trend of the other three absorbents, it not suitable to be included in the optimization process (Table 1). The value of pH for culture media cannot be increased without the addition of other substances, thus, the comparison cannot be done as the initial pH value was way lower than other absorbents. For the case of MEA, the optimization process is used to find the best solution for a range of data, a large difference in data value can resulting in invalid optimization. Because of that, these two absorbents were not included in the optimization analysis. Additionally, the focus was always on culture media with the addition of NaOH and MEA.

Many factors can affect pH in the absorbent. The decrease in pH value in all absorbents was due to the increase of the concentration of dissolved inorganic carbon such as $CO_2(aq)$, HCO_3^-, and CO_3^{2-} present in CO_2 that been dissolved in said absorbents [15]. Theoretically, NaOH will yield a higher CO_2 solubility than culture media present because the total alkalinity of NaOH is higher due to NaOH is a strong alkaline. The mechanism of CO_2 absorption in NaOH aqueous solution is summarized in the following [16]. All Na^+ and OH^- was ionized in pure water, then aqueous CO_2 reacts with OH^- as expressed in the following equation:

$$CO_2(aq) + OH^-(aq) \rightleftharpoons HCO_3^-(aq) \tag{2}$$

$$HCO_3^-(aq) + OH^-(aq) \rightleftharpoons H_2O(l) + CO_3^{2-}(aq) \tag{3}$$

Both equations are reversible reactions with a high effect on pH changes. The reaction is continuous so every $CO_2(aq)$ that was present in the medium will instantaneously be consumed. Due to high alkalinity, reaction Equation (3) was the main reaction occur in the absorbent early on resulting in increasing of CO_3^{2-}, while OH^- is rapidly consumed via both reactions. This explained the reason for the sudden change in pH at the early stage of the experiment. As CO_2 aerated through the medium, OH^- will keep decreasing and CO_3^{2-} keep accumulating. This phenomenon will force a backward reaction of Equation (3) which will accelerate the forward reaction of Equation (2). The pH will keep dropping in this stage. At a certain point in the experiment, pH will stop dropping and remain constant due to the reaction at equilibrium. After all the reactions are at equilibrium, the overall reaction of NaOH with CO_2 can be written as Equation (4).

$$NaOH(aq) + CO_2(g) \rightarrow NaHCO_3(aq) \tag{4}$$

Meanwhile, for the culture media, the phenomenon of CO_2 dissolved can be explained by the reaction between CO_2 and water (because H_2O is present in media). Firstly, aqueous CO_2 will react with water to form carbonic acid [17]:

$$CO_2(aq) + H_2O(l) \rightarrow H_2CO_3(aq) \tag{5}$$

Then the H_2CO_3 can lose one or both of its H^+ to form:

$$H_2CO_3(aq) \rightleftharpoons HCO_3^-(aq) + H^+(aq) \tag{6}$$

$$HCO_3^-(aq) \rightleftharpoons CO_3^{2-}(aq) + H^+(aq) \tag{7}$$

The pH drops in media are due to the released hydrogen ions. However, this equation can operate in both directions depending on the current pH level, working as its buffering system. At a higher pH, this bicarbonate system will shift to the left, and CO_3^{2-} will pick up a free hydrogen ion [18]. But in the system, CO_2 was constantly added to the media, then increasing the value of dissolved CO_2 causing the reaction Equations (6) and (7) forced to be carried out from left to right. This increases H_2CO_3, which decreases pH.

Additionally, based on Figure 2, it shows that culture media with NaOH was the fastest to reach the minimum or equilibrium pH, following by pure NaOH and media with MEA. When the pH reaches its equilibrium point, it indicates that no more CO_2 is being absorbed. With this information, it is concluded that the longer time is taken to reach equilibrium, the higher the CO_2 concentration in absorbent. As expected, culture media capability to absorb CO_2 is weaker than NaOH. However, when MEA was added to culture media, a higher amount of CO_2 was absorbed.

3.2. System Performance with Deployment of Carbon Dioxide (CO₂) Diffuser

One of the main factors in the CO_2 absorption rate is the surface area of the gas–liquid boundary [19]. In the next set of experiments, the factor was been investigated. CO_2 diffuser was mounted at the sparger at the bottom of the reactor where the CO_2 is aerated. This diffuser functions to turn the CO_2 bubble into a more refined form. Parameters used were based on the previous experiments. The result is shown in Figure 4.

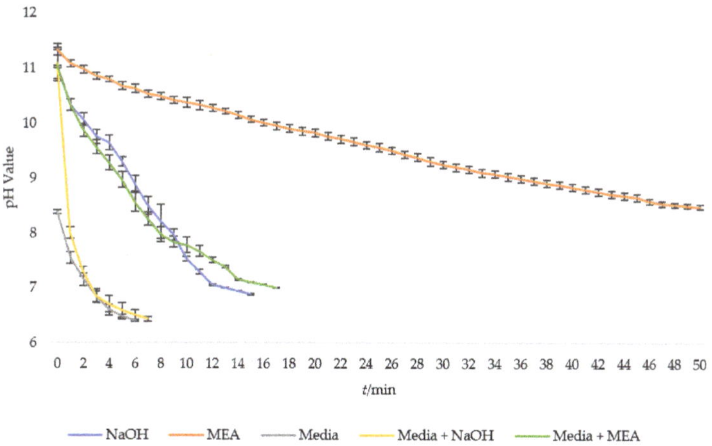

Figure 4. Effect of diffuser on pH change for five absorbents at $p = 100$ kPa, $T = 20\ ^\circ$C, with an initial pH of 11 with exception of culture media with an initial pH of 8.5.

From Figure 4, the addition of the diffuser favorably affected the CO_2 absorption rate by all absorbents. The final pH value were almost identical with previous run, while the time taken for pH to drop for NaOH (pH = 6.88, t = 15 min), MEA (pH = 8.49, t = 50 min), culture media (pH = 6,4, t = 6 min), culture media with NaOH (pH = 6.43, t = 7 min) and culture media with MEA (pH = 7, t = 17 min) were approximately decrease by 20% when using diffuser compared to the previous. By using a diffuser, CO_2 is dissipated into countless small bubbles, which flows in the culture media in the form of a bubbling stream. This process is called atomization. The large number of bubbles scattered in culture media will increase the contact area of gas–liquid that will increase the absorption rate.

A simple relation between the contact surface area and mass transfer rate can be expressed in Equation (8) [20]:

$$m = \int_S j_m dS = S j_m \tag{8}$$

where m, mass transfer, j_m, mass flow, and S, interfacial area, and it is shown that the relationship between mass transfer is directly proportional to the contact area. Additionally, a more detailed mathematical model of absorption rate with other parameters had been derived by Martinez I. et al. [21] using Fick's Law shown in Equation (9):

$$j_g = \frac{q_m}{4\pi r^2} = -D_g \frac{d(w\rho_1)}{dr} \tag{9}$$

where j_g: diffusion mass flux; q_m: mass flow rate; D_g: gas diffusivity; w: mass fraction; ρ_1: density of the liquid used. The assumption is as follows: symmetry is spherical, time-independent, constant liquid and gas density, and residence time of bubble much smaller than the dissolution time, the factor that affects the absorption rate can be defined by Equation (10):

$$w_\infty(t_v) = w_0[1 - \exp(-Kt_v)], \quad \text{where } K \equiv \frac{27v_1 D_g u_g}{2gr_0^4} \tag{10}$$

w_∞: mass fraction before venting; w: gas mass fraction; v: kinematic viscosity of liquid; D_g: diffusion coefficient of CO_2 gas in water; u_g: gas injection speed; g: gravity acceleration; r_0: radius of the bubble. The benefit of this simple mathematical formula is that the relations between different parameters are explicit. Gas diffusion from bubbles to the aqueous phase is measured by w_∞, hence based on this equation, the mass of gas in a liquid is inversely proportional to the fourth power of bubble radius. This experiment only portrays the benefit of a simple diffuser. More complications, especially with made CO_2 diffusers for CO_2 absorption in the reactor can further increase the absorption rate of CO_2.

3.3. Factors Affecting Carbon Dioxide (CO$_2$) Solubilities in Culture Media

CO_2 is one of the key factors in organic acid fermentation [22,23]. Organic acid had a wide application ranging from preservative agents for food to lab application [24]. Due to its benefit, the production of organic acid is widely studied. As mention earlier, one of the three factors that affect CO_2 solubilities is the composition of media. Many studies focusing on organic acid production have increased the CO_2 availability by the addition of chemicals such as magnesium carbonate ($MgCO_3$), sodium bicarbonate ($NaHCO_3$), or calcium carbonate ($CaCO_3$). Through the addition of such chemicals, the CO_2 solubility is greatly increased. Thus, studies on different chemicals such as sodium hydroxide (NaOH) and monoethanolamine (MEA) are also important.

The theoretical access to CO_2 solubilities in culture media is very limited. The effect of organic solutes on gas solubility can be rather complex. But parameters affecting the solubilities were well-studied. Three parameters affecting CO_2 solubilities are pressure, temperature, and media composition [25]. The effects of these parameters on the solubility of a gas in a pure solvent, expressed in a mathematical model were simplified in Table 2. Based on Table 2, composition (X), pressure (p), and temperature (T) had been identified as the three most important independent parameters affecting CO_2 solubility (s) in absorbent and thus are chosen as the inputs for the design.

Table 2. Mathematical model of effects of pressure, temperature, and composition on gas solubility.

Parameters	Model	Remark	Ref.
Composition	Empirical model of Henry's Law $$\log\left(\frac{H}{H_0}\right) = \log\left(\frac{s_{G,o}}{s_G}\right) = \sum_i (h_i + h_G)c_i + \sum_j K_{n,j}c_{n,j}$$ whereas H: Henry's constant in solvent mixtures; s_G: CO_2 solubility in solvent mixtures; subscript 0: value at pure solvent (e.g., water); h: ion-specific parameter; $K_{n,j}$: substance specific model parameter; c_i: concentration of ion; $c_{n,j}$: concentration of an organic substance	Solubility depends on ion and organic substance appearing in solvent	[26]
Pressure	Henry's Law: $$p_{CO_2} = H_0 c_{CO_2}$$ whereas p_{CO_2}: CO_2 partial pressure; H_0: Henry's constant; c_{CO_2}: concentration of dissolved CO_2.	Solubility is directly proportional to the pressure	[26]
Temperature	Henry's Law and Van't Hoff equation: $$c_{CO_2} = k_H p_{CO_2}$$ $$k_H(T) = k_{298K} \exp\left[-\frac{\Delta H_{diss}}{R}\left(\frac{1}{T} - \frac{1}{298}\right)\right]$$ whereas k_h: temperature-dependent Henry's constant; k_{298K}: Henry's constant at 298K; ΔH_{diss}: dissolution enthalpy; R: ideal gas constant	Solubility is inversely proportional to temperature.	[21,22]

3.3.1. Design of Experiment Analysis

In this study, a three-factor, three-level Box-Behnken design (BBD) was used to investigate the effects of composition (X), pressure (p), and temperature (T) the interactions of these factors on the CO_2 solubility (s) in absorbent measured by its concentration. A total of fifteen experimental samples were required for the BBD including three replicated experimental runs using the processing parameters at the center points (Table 3).

Table 3. Actual response of CO_2 concentration at experimental design points.

	Independent Input Variables [X]			Response [Y]
Exp.	Absorbent	T/°C	p/kPa	s/(mol/L)
1	NaOH	20	100	14.82
2	NaOH	40	100	9.7
3	Culture media with MEA	20	125	18.42
4	Culture media with NaOH	20	125	10.44
5 *	NaOH	30	125	10.04
6	Culture media with MEA	30	100	12.11
7	Culture media with MEA	40	125	12.12
8	NaOH	20	150	16.52
9	Culture media with MEA	30	150	15.76
10	NaOH	40	150	10.4
11	Culture media with NaOH	30	100	8.48
12 *	NaOH	30	125	10.29
13	Culture media with NaOH	30	150	9.56
14 *	NaOH	30	125	9.93
15	Culture media with NaOH	40	125	9.57

* Replicated experimental runs.

The ANOVA analysis is performed as shown in Table 4 to determine the significance and adequacy of the regression models. The Model F-value of 135.76 implies the model is significant. There is only a 0.01% chance that an F-value this large could occur due to noise. p-values less than 0.05 indicate model terms are significant and all insignificant model terms had been reduced. Additionally, a graph in Figure 5 shows a good agreement between the actual data and the predicted values from the regression models. The Predicted R^2 of 0.9375 is in reasonable agreement with the Adjusted R^2 of 0.9897; i.e., the difference is less than 0.2. For adequate precision measures of the signal to noise ratio, a ratio greater than 4 is desirable. The ratio of 37.693 indicates an adequate signal. This model can be used to navigate the design space. This result shows that the regression model is statistically significant and adequate for the prediction and optimization of the CO_2 absorption process.

Table 4. Results of ANOVA for the quadratic model of the CO_2 concentration in absorbent.

Source	Sum of Squares	Degrees of Freedom	Mean Square	F-Value	p-Value	
Model	128.74	10	12.87	135.76	0.0001	significant
X-Medium	53.08	2	26.54	279.85	<0.0001	
T-Temperature	42.367	1	42.37	446.76	<0.0001	
p-Pressure	6.35	1	6.35	67.01	0.0012	
XT	9.44	2	4.72	49.78	0.0015	
Xp	2.33	2	1.16	12.28	0.0196	
T^2	14.28	1	14.28	150.60	0.0003	
p^2	2.40	1	2.40	25.34	0.0073	
Residual	0.3793	4	0.0948			
Lack of Fit	0.3112	2	0.1556	4.57	0.1794	not significant
Pure Error	0.0681	2	0.0340			
Corrected Total Sum of Squares	129.12	14				

R^2: 0.9971; adjusted R^2: 0.9897; predicted R^2: 0.9375; Adequate Precision: 37.6933.

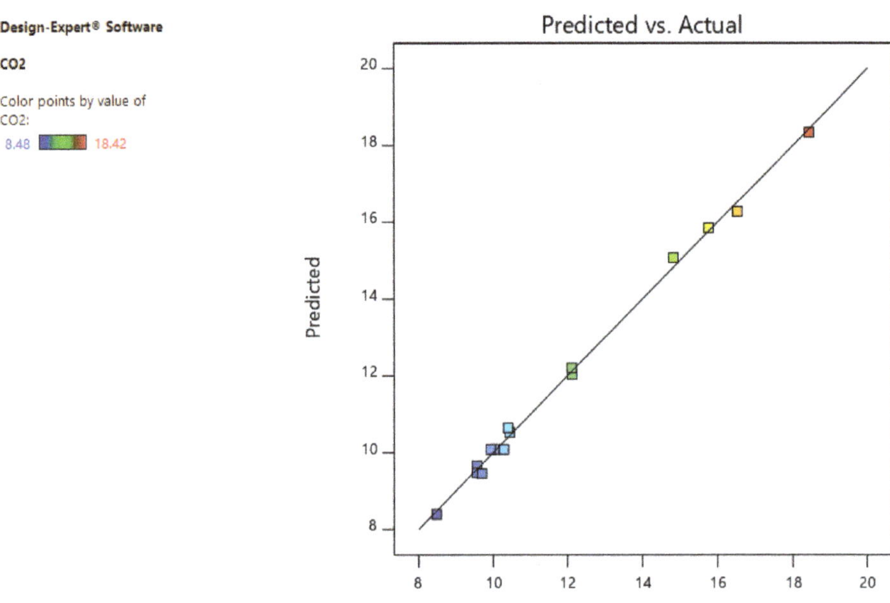

Figure 5. Comparison of actual and predicted results of CO_2 absorption.

Quadratic equations were derived to describe the relationships between the parameters and CO_2 solubility (s) as shown in Equations (11)–(13).

NaOH:

$$s = 0.001291(p)^2 + 0.019667(T)^2 - 0.298667p - 1.46100T + 53.38333 \tag{11}$$

Culture media with NaOH:

$$s = 0.001291(p)^2 + 0.019667(T)^2 - 0.301067p - 1.2235T + 44.5975 \tag{12}$$

Culture media with MEA:

$$s = 0.001291(p)^2 + 0.019667(T)^2 - 0.249667p - 1.495T + 51.4075 \tag{13}$$

3.3.2. Individual Effect of Experiment Parameters on Carbon Dioxide (CO_2) Absorption

The significance of each parameter on CO_2 solubility can be illustrated by the perturbation plot in Figure 6, where steep slope and curvature were obtained for both pressure and temperature indicates that CO_2 solubility is sensitive to the parameters. The maximum value of CO_2 dissolved achieved with the addition of NaOH for culture media was lower than what been achieved for MEA. This is due to the MEA's rapid reaction with CO_2 in low partial pressure [27].

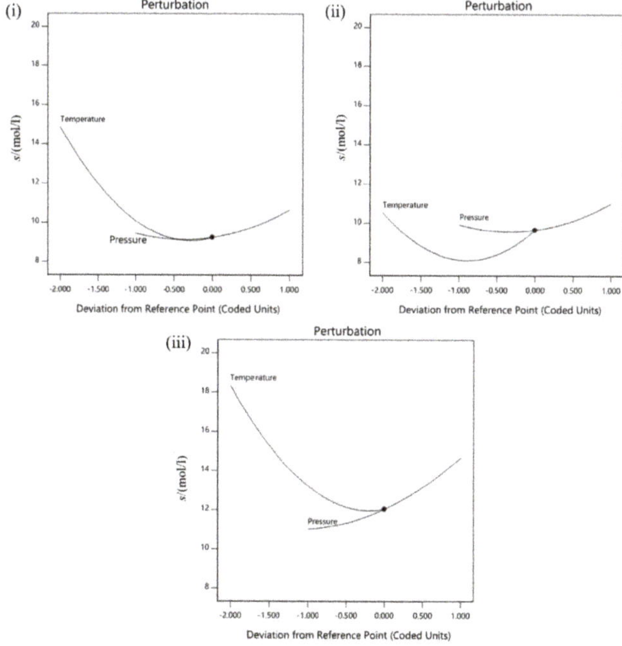

Figure 6. Perturbation plot comparing the response of CO_2 solubility to changes in temperature and pressure in (**i**) pure NaOH, (**ii**) culture media with NaOH, and (**iii**) culture media with monoethanolamine (MEA).

Figure 6 shows the relationship between pressure and final CO_2 concentration which is directly proportional which was described quantitatively by Henry's law shown in Table 2; pressure. External pressure affects the concentration of gas molecules in space. When the partial pressure of the gas above the solution increases, it forces the gas molecules to solute in solution to maintain dynamic equilibrium [28]. Additionally, high temperature demotes CO_2 absorption as opposed to high pressure. Also, the effect of temperature is more significant than the pressure (highest F-value). A slight increase in temperature will greatly reduce CO_2 absorption. While to achieve more significant change by pressure, extremely high pressure needs to be applied [29]. The gas dissolves in liquid because of the interactions between its molecules and absorbent. This interaction will release heat when these new attractive interactions form in an exothermic process. Thus, additional heat will produce thermal energy that overwhelms the attractive forces between the gas and the absorbent molecules resulting in less CO_2 dissolved in solution [28].

3.3.3. Collective Effect of Experiment Parameters on Carbon Dioxide (CO_2) Absorption

The F-value determine the relative importance of a parameter. From Table 4, the temperature has the highest F-value, which is 446.76, showing its importance in CO_2 absorption in this case. Figures 7–9 are three-dimensional response surface and project contour for NaOH, culture media with NaOH, and culture media with MEA, respectively, which show the different experimental parameters and their effects on CO_2 concentration in the absorbent. All graphs derived had curved rather than straight lines, indicating the strong interaction between parameters and the output.

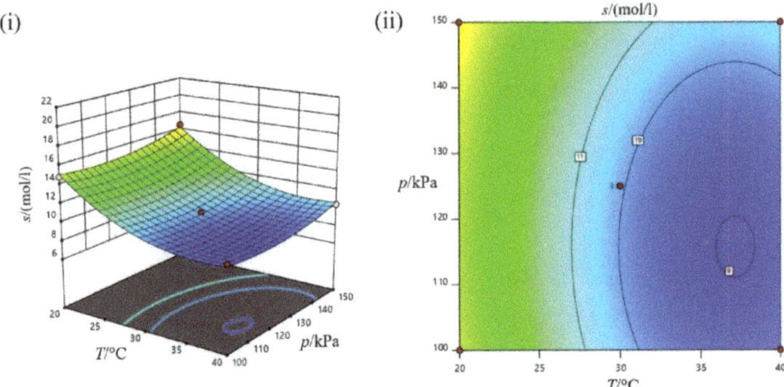

Figure 7. Effect of pressure, temperature and their interaction on CO_2 concentration in NaOH (**i**) three-dimensional surface plot; (**ii**) projected contour plot.

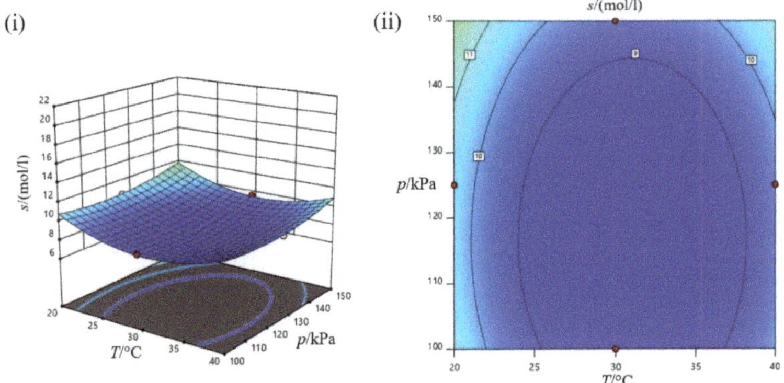

Figure 8. Effect of pressure, temperature and their interaction on CO_2 concentration in culture media with NaOH (**i**) three-dimensional surface plot; (**ii**) projected contour plot.

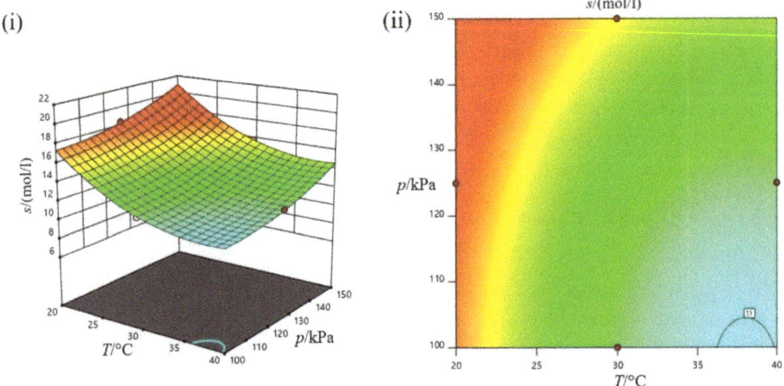

Figure 9. Effect of pressure, temperature and their interaction on CO_2 concentration in culture media with MEA (**i**) three-dimensional surface plot; (**ii**) projected contour plot.

3.3.4. Process Optimization

It was observed that carbon dioxide (CO_2) solubility (s) reaches the maximum values under low temperature operating in high pressure. However, there was a certain point in these parameters where a further change in their value will not affect the final CO_2 solubility (s). Therefore, a balance value needs to be established to achieve the maximum result while optimizing the parameters. The purpose of this step is to observe the combination of the independent variable (i.e., absorbent, temperature, and pressure) to get the maximum CO_2 solubility (s) simultaneously. The overall performance of the CO_2 absorption strongly depends on a wide range of experimental conditions. It is crucial to optimize the CO_2 absorption process with multiple inputs and multiple responses. Thus, a series of optimizing results proposed by response surface methodology (RSM) with multiple inputs and multiple responses are listed in Table 5. The optimal CO_2 solubility (s) can be obtained in the following conditions: culture media with MEA as absorbent; 20.483 °C of temperature; 136.728 kPa of pressure. The optimal CO_2 solubility (s) obtained at this condition was 19.029 mol/L. Additionally, a favorable value for each parameter can be chosen from the RSM. For example, the desired temperature is ≈ 22 °C, pressure must be around 140 kPa to obtain the optimal CO_2 solubility (s). Though from the optimization process, absorbents other than culture media with MEA was found to be undesirable.

Table 5. Process optimization for CO_2 absorption by RSM.

No.	Absorbent	$T/°C$	p/kPa	$s/(mol/L)$
1	Culture media with MEA	20.483	136.728	19.029
2	Culture media with MEA	20.702	133.270	18.537
3	Culture media with MEA	20.557	132.113	18.529
4	Culture media with MEA	21.951	143.633	18.834
5	Culture media with MEA	20.125	132.344	18.850

4. Conclusions

The present work has investigated the features of sodium hydroxide (NaOH) aqueous solution and culture media to capture carbon dioxide (CO_2) at different compositions, pressure, and temperature. It was observed that the CO_2 solubility increases using monoethanolamine (MEA) compared to NaOH. CO_2 absorption was also favorable at high pressure and low temperature. Improvement of the absorption rate can be achieved by deploying a CO_2 diffuser. Using optimization software, the most optimized condition for the CO_2 absorption process was by using culture media added with MEA, at a pressure of 136.728 kPa operating at 20.483 °C. Besides, capturing CO_2 by using culture media will lead to the production of chemical commodities (such as succinic acid, formic acid, and acetic acid) by microbes that can be useful for industrial usage. This work serves as a base for further research on CO_2 absorption by culture media. Further studies such as the feasibility of this method by using gas mixture instead of pure CO_2 are necessary to demonstrate a more flexible process. The ultimate aim of this work is to produce value-added commodities utilizing CO_2 as the main carbon source with the help of suitable microbes.

Author Contributions: Conceptualization—S.N. and A.I.A.; writing, original draft preparation—S.N. and A.I.A.; writing, review, and editing—M.Y.O. and S.N.; writing, proofreading—A.I.A. and P.L.S.; funding acquisition—S.N. All authors have read and agreed to the published version of the manuscript.

Funding: The authors (SN and AIA) would like to thank the Ministry of Education (MOE) Malaysia for the financial support through the Fundamental Research Grant Scheme (MOHE Project Ref. No.: FRGS/1/2018/STG01/UNITEN/01/1) and BOLD 2025 (10436494/B/2019095). AIA also wishes to express his gratitude to UNITEN for providing the UNITEN Postgraduate Excellent Scholarship 2019.

Acknowledgments: The authors would like to express their special thanks to Universiti Tenaga Nasional for providing facilities and equipment to ensure the accomplishment of this project.

Conflicts of Interest: The authors declare that they have no competing interests.

References

1. John, J.C.; Paul, D.H.; James, R.D.; Sam, A.N.; AMichael, S.; James, T.T.; Lynn, D.W. U.S. Energy Information, an Annual Energy Outlook 2015 with projections to 2040. *Off. Integr. Int. Energy Anal.* **2015**, *1*, 1–244.
2. Al-mamoori, A.; Krishnamurthy, A.; Rownaghi, A.A.; Rezaei, F. Carbon Capture and Utilization Update. *Energy Technol.* **2017**, *5*, 834–849. [CrossRef]
3. Markewitz, P.; Kuckshinrichs, W.; Leitner, W.; Linssen, J.; Zapp, P.; Bongartz, R.; Schreiber, A.; Müller, T.E. Worldwide innovations in the development of carbon capture technologies and the utilization of CO_2. *Energy Environ. Sci.* **2012**, *5*, 7281–7305. [CrossRef]
4. Gerard, D.; Wilson, E.J. Environmental bonds and the challenge of long-term carbon sequestration. *J. Environ. Manag.* **2009**, *90*, 1097–1105. [CrossRef] [PubMed]
5. Boot-Handford, M.E.; Abanades, J.C.; Anthony, E.J.; Blunt, M.J.; Brandani, S.; Mac Dowell, N.; Fernández, J.R.; Ferrari, M.; Gross, R.; Hallett, J.P.; et al. Carbon capture and storage update. *Energy Environ. Sci.* **2014**, *7*, 130. [CrossRef]
6. Styring, P.; Jansen, D.; de Coninck, H.; Reith, H.; Armstrong, K. *Carbon Capture and Utilisation in the Green Economy*; The Centre for Low Carbon Futures 2011 and CO2Chem Publishing 2012: Sheffield, UK, 2011; ISBN 9780957258815.
7. Zhao, B.; Su, Y. Process effect of microalgal-carbon dioxide fixation and biomass production: A review. *Renew. Sustain. Energy Rev.* **2014**, *31*, 121–132. [CrossRef]
8. Gong, F.; Cai, Z.; Li, Y. Synthetic biology for CO_2 fixation. *Sci. China Life Sci.* **2016**, *59*, 1106–1114. [CrossRef] [PubMed]
9. Razzak, S.A.; Hossain, M.M.; Lucky, R.A.; Bassi, A.S.; de Lasa, H. Integrated CO_2 capture, wastewater treatment and biofuel production by microalgae culturing—A review. *Renew. Sustain. Energy Rev.* **2013**, *27*, 622–653. [CrossRef]
10. Singh, S.P.; Singh, P. Effect of CO_2 concentration on algal growth: A review. *Renew. Sustain. Energy Rev.* **2014**, *38*, 172–179. [CrossRef]
11. Al-Anezi, K.; Somerfield, C.; Mee, D.; Hilal, N. Parameters affecting the solubility of carbon dioxide in seawater at the conditions encountered in MSF desalination plants. *Desalination* **2008**, *222*, 548–571. [CrossRef]
12. Yincheng, G.; Zhenqi, N.; Wenyi, L. Energy Procedia Comparison of removal efficiencies of carbon dioxide between aqueous ammonia and NaOH solution in a fine spray column. *Energy Procedia* **2011**, *4*, 512–518. [CrossRef]
13. Da Rosa, G.M.; de Morais, M.G.; Costa, J.A. Green alga cultivation with monoethanolamine: Evaluation of CO_2 fixation and macromolecule production. *Bioresour. Technol.* **2018**, *261*, 206–212. [CrossRef] [PubMed]
14. Yoo, M.; Han, S.; Wee, J. Carbon dioxide capture capacity of sodium hydroxide aqueous solution. *J. Environ. Manag.* **2013**, *114*, 512–519. [CrossRef] [PubMed]
15. Kurihara, H.; Shirayama, Y. Effects of increased atmospheric CO_2 on sea urchin early development. *Mar. Ecol. Prog. Ser.* **2004**, *274*, 161–169. [CrossRef]
16. Fleischer, C.; Becker, S.; Eigenberger, G. Detailed modeling of the chemisorption of CO_2 into NaOH in a bubble column. *Chem. Eng. Sci.* **1996**, *51*, 1715–1724. [CrossRef]
17. Knoche, W. *Chemical Reactions of CO_2 in Water BT—Biophysics and Physiology of Carbon Dioxide*; Bauer, C., Gros, G., Bartels, H., Eds.; Springer: Berlin/Heidelberg, Germany, 1980; pp. 3–11.
18. Fundamentals of Environmental Measurements. pH of Water. Available online: https://www.fondriest.com/environmental-measurements/parameters/water-quality/ph/ (accessed on 23 April 2020).
19. Bureau, I. Method and Apparatus for Rapid Carbonanation of a Fluid. WO Patent 2016/122718, 4 August 2016.
20. Geng, H.; Chen, Q.; Zhao, G. The Experiment Study of Biogas Atomization Upgrading with Water Scrubbing at Atmospheric Pressure. In *2015 International Conference on Industrial Technology and Management Science*; Atlantis Press: Paris, France, 2015.
21. Martínez, I.; Casas, P.A. Simple model for CO_2 absorption in a bubbling water column. *Braz. J. Chem. Eng.* **2012**, *29*, 107–111.
22. Song, H.; Lee, J.W.; Choi, S.; You, J.K.; Hong, W.H.; Lee, S.Y. Effects of dissolved CO_2 levels on the growth of Mannheimia succiniciproducens and succinic acid production. *Biotechnol. Bioeng.* **2007**, *98*, 1296–1304. [CrossRef]

23. Xi, Y.; Chen, K.; Li, J.; Fang, X.; Zheng, X.; Sui, S.; Jiang, M.; Wei, P. Optimization of culture conditions in CO_2 fixation for succinic acid production using Actinobacillus succinogenes. *J. Ind. Microbiol. Biotechnol.* **2011**, *38*, 1605–1612. [CrossRef]

24. Hauser, C.; Thielmann, J.; Muranyi, P. *Chapter 46—Organic Acids: Usage and Potential in Antimicrobial Packaging*; Academic Press: San Diego, CA, USA, 2016; pp. 563–580. ISBN 978-0-12-800723-5.

25. Schumpe, A.; Quicker, G. Gas Solubilities in Microbial Culture Media. In *Reaction Engineering*; Springer: Berlin/Heidelberg, Germany, 1982.

26. Gunnarsson, I.B.; Alvarado-Morales, M.; Angelidaki, I. Utilization of CO_2 fixating bacterium Actinobacillus succinogenes 130Z for simultaneous biogas upgrading and biosuccinic acid production. *Environ. Sci. Technol.* **2014**, *48*, 12464–12468. [CrossRef]

27. Wang, M.; Lawal, A.; Stephenson, P.; Sidders, J.; Ramshaw, C. Post-combustion CO_2 capture with chemical absorption: A state-of-the-art review. *Chem. Eng. Res. Des.* **2011**, *89*, 1609–1624. [CrossRef]

28. *LibreTexts 13.4: Effects of Temperature and Pressure on Solubility—Chemistry LibreTexts*; Department of Education Open Textbook Pilot Project: Washington, DC, USA, 2019; pp. 1–5.

29. Wiebe, R.; Gaddy, V.L. The Solubility of Carbon Dioxide in Water at Various Temperatures from 12 to 40° and at Pressures to 500 Atmospheres. Critical Phenomena. *J. Am. Chem. Soc.* **1940**, *62*, 815–817. [CrossRef]

Publisher's Note: MDPI stays neutral with regard to jurisdictional claims in published maps and institutional affiliations.

 *processes*

Article

Effects of Alginate and Chitosan on Activated Carbon as Immobilisation Beads in Biohydrogen Production

Nur Farahana Dzul Rashidi [1], Nur Syakina Jamali [1,*], Siti Syazwani Mahamad [1], Mohamad Faizal Ibrahim [2], Norhafizah Abdullah [1], Siti Fatimah Ismail [1] and Shamsul Izhar Siajam [1]

[1]   Department of Chemical and Environmental Engineering, Faculty of Engineering, Universiti Putra Malaysia, Serdang 43400, Selangor Darul Ehsan, Malaysia; anafarahana19@gmail.com (N.F.D.R.); syazwani.mahamad@gmail.com (S.S.M.); nhafizah@upm.edu.my (N.A.); sitifismail@gmail.com (S.F.I.); shamizhar@upm.edu.my (S.I.S.)

[2]   Department of Bioprocess Technology, Faculty of Biotechnology and Biomolecular Sciences, Universiti Putra Malaysia, Serdang 43400, Selangor Darul Ehsan, Malaysia; faizal_ibrahim@upm.edu.my

*   Correspondence: syakina@upm.edu.my; Tel.: +60-3-9769-4464

Received: 28 July 2020; Accepted: 14 September 2020; Published: 6 October 2020

Abstract: In this study, the effects of alginate and chitosan as entrapped materials in the biofilm formation of microbial attachment on activated carbon was determined for biohydrogen production. Five different batch fermentations, consisting of mixed concentration alginate (Alg), were carried out in a bioreactor at temperature of 60 °C and pH 6.0, using granular activated carbon (GAC) as a primer for cell attachment and colonisation. It was found that the highest hydrogen production rate (HPR) of the GAC–Alg beads was 2.47 ± 0.47 mmol H_2/l.h, and the H_2 yield of 2.09 ± 0.22 mol H_2/mol sugar was obtained at the ratio of 2 g/L of Alg concentration. Next, the effect of chitosan (C) as an external polymer layer of the GAC–Alg beads was investigated as an alternative approach to protecting the microbial population in the biofilm in a robust environment. The formation of GAC with Alg and chitosan (GAC–AlgC) beads gave the highest HPR of 0.93 ± 0.05 mmol H_2/l.h, and H_2 yield of 1.11 ± 0.35 mol H_2/mol sugar was found at 2 g/L of C concentration. Hydrogen production using GAC-attached biofilm seems promising to achieve consistent HPRs at higher temperatures, using Alg as immobilised bead material, which has indicated a positive response in promoting the growth of hydrogen-producing bacteria and providing excellent conditions for microorganisms to grow and colonise high bacterial loads in a bioreactor.

Keywords: biohydrogen production; immobilised cells; entrapment; alginate; chitosan; activated carbon

1. Introduction

Biohydrogen is a popular energy carrier as its production promises clean energy that only generates water upon combustion, with higher energy content per unit weight (122 kJ/g) than any other fuel [1]. Biological methods have been studied to ensure the hydrogen production is safer and more economical than thermochemical methods. Dark fermentation is being recognised as an excellent biological method of hydrogen production because of its ability to perform without light energy and oxygen source [2]. Fermentation is the process of using specific microorganisms to convert organic substrates into hydrogen, carbon dioxide, and other solutes such as acetate, butanol, and ethanol [3].

The production of biohydrogen using a suspended cultivation system through high-temperature operation (50–60 °C) has gained attention due to its higher yield and hydrogen productivity (HPR) capability. This operation is preferable in pathogenic destruction as it can restrain the growth of hydrogen consumers such as homoacetogens and methanogens [4,5]. However, the lower microbial cell density at this temperature is a disadvantage for the fermentation. Difficulties in the retention of biomass in the suspended system and cell washout are regularly experienced inside the reactor, which usually happens during the short hydraulic retention time (HRT) [6,7]. As a consequence, attached cell immobilisation is an approach to maximising and maintaining biomass, such that it can work at a higher rate of dilution without biomass washout from the reactor. Immobilisation technology has been developed to increase hydrogen production by providing a favourable environment of support for microbial cells during fermentation. Hence, the selection of the supporting material is imperative because it affects the overall performance of biohydrogen production.

Alginate (Alg) is an excellent support material, making it a practical choice in immobilisation. It has been reported that biohydrogen production increases three-fold when using alginate beads supplemented with aluminum oxide and titanium oxide [8,9]. Meanwhile, Wu et al. [10] said that biohydrogen production is two-times greater when alginate beads are enhanced with activated carbon. Nonetheless, even though alginate beads have been widely used in immobilisation, they are reported to still suffer from certain limitations like weak mechanical strength and reduced porosity [11,12]. Therefore, several approaches have been studied to improve the permeability and mechanical stability of alginate matrices, such as incorporating other materials like cellulose, metal, and carbon sources.

A previous study reported that the entrapment between chitosan and alginate forms a strong ionic interaction between carboxyl groups of alginate and amino groups of chitosan, thus resulting in an improvement in the mechanical properties of the matrix support [13,14]. In other studies, the formation of high, crosslinked, porous beads, with better mechanical and chemical stability of the support matrix in the buffered medium, is produced from the ionotropic gelation of chitosan, leading to low rates of cell leakage even at higher cell loading [15]. It was also reported that the effectiveness of chitosan coating enables the physical isolation of bacteria from the outer environment and reduces cell detachment during fermentation, besides improving the mechanical strength of alginate bead carriers during storage [16–18].

The entrapment technique is widely used, which can be done in a simple procedure. This technique involves an entrapment process in which the enzyme is crosslinked within polymeric materials such as calcium alginate, polyacrylamide (PAM) gel, and agar [19]. The research reported that the stability of immobilised cells can be enhanced via the fusion of microbial cells into a rigid network of the polymer due to the mechanical firmness and good porosity of the carrier, thus providing the right anaerobic conditions to microorganisms during hydrogen production [20].

In this study, the effect of alginate and chitosan was investigated as one of the potential entrapped-material approaches in cell immobilisation and cell encapsulation. In the first part of this research, the ability of microbial culture to attach and maintain itself on a granular activated carbon (GAC) surface as their support material was performed on mixtures of glucose and xylose as the carbon source. GAC has a high surface area, low toxicity, and excellent mechanical properties, which are ideal for fermentation at high temperatures. Moreover, its characteristic of having highly porous structure helps to preserve cell viability, which serves an excellent purpose in the field of microbial colonisation where fermentative bacteria can expand freely on the surface of the supporting material and form a biofilm [21]. Efforts have been made to improve the high cell density that facilitates the good production of hydrogen. Furthermore, entrapment is part of immobilisation methods that have been used to improve the productivity of enzyme or microbial cells. This study investigated the variations of GAC–alginate (GAC–Alg)- and GAC–alginate–chitosan (GAC–AlgC)-immobilised beads, which acts as a support carrier during biofilm development in batch fermentation of biohydrogen production.

2. Materials and Methods

2.1. Microorganism, GAC Carrier, Alginate, and Chitosan Carriers

The microorganism source was collected from a sludge pit of palm oil mill effluent (POME) that was located at Sime Darby Plantation, Selangor, Malaysia. The sludge containing mixed culture underwent a heat-treated process at 80 to 90 °C for 60 min to prevent the development of the methanogenic population prior to use. The GAC carrier originated from shells of coconuts that were supplied by KI Carbon Solutions Sdn Bhd. GAC was sieved to attain 2–3 mm of particle size. Sodium alginate powder and chitosan flakes originated from crab shells were supplied by BT Science Sdn Bhd. Sodium alginate powder was dissolved into 1 L of distilled water and stirred using a hot plate magnetic stirrer for 30 min to attain homogeneity prior to use.

2.2. Biofilm Formation on Activated Carbon

Biofilm was primarily developed on the surface of GAC using the surface attachment method as one of the immobilisation approaches. A similar ratio of GAC to sludge, 10:10 (*w/v* g/L), was acclimatised in the synthetic medium inside a 1 L modified Schott bottle using the sequencing batch operation mode of 2 days HRT. The biofilm was continuously developed until biogas production was consistently obtained. The synthetic medium was used in sequencing batch fermentation. The medium contained (per liter of deionised water): KH_2PO_4 0.75 g L^{-1}, NH_4Cl 1 g L^{-1}, $K_2HPO_4.3H_2O$ 1.5 g L^{-1}, $NaCl$ 2 g L^{-1}, $NaHCO_3$ 2.6 g L^{-1}, $MgCl_2.6H_2O$ 0.5 g L^{-1}, $CaCl2.2H_2O$ 0.05 g L^{-1}, yeast extract 2 g L^{-1}, xylose 10 g L^{-1}, and glucose 10 g L^{-1}. The fermentation system was cultivated for 48 h in a water bath shaker at 60 °C and 120 rpm, with the pH of the culture medium adjusted to pH 6.0 [21,22].

The gas produced was monitored using the water displacement method. The measuring cylinder was put invertedly in the hydrochloric acid solution (with pH 2) to avoid the gases from being released into the environment. The volume of biogas was recorded in every cycle and collected once the stationary phase was achieved. The experimental setup of this study is shown in Figure 1.

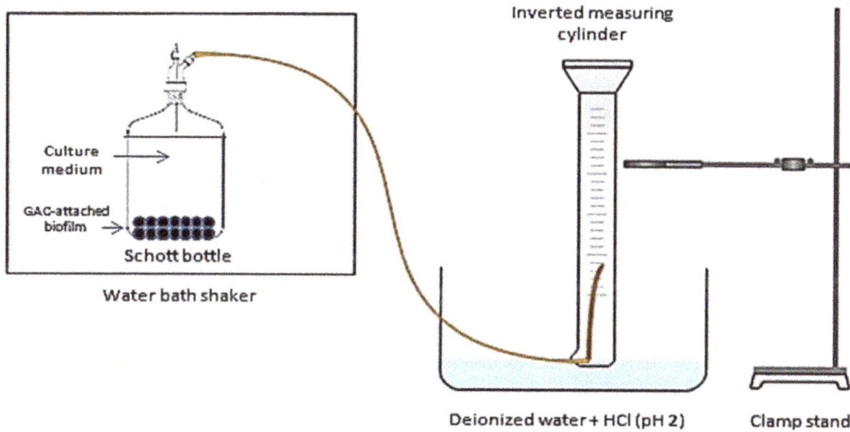

Figure 1. Experimental setup for cell acclimatisation

2.3. Development of GAC—Attached Biofilm Entrapped in Alginate Beads (GAC–Alg)

The different concentrations of alginate were prepared by dissolving 0.5, 1, 2, 3, and 4 g of sodium alginate powder into 1 L of distilled water, as shown in Table 1. About 40 g of GAC-attached biofilm were put into the alginate solution and mildly stirred until well mixed. The mixed granules were then dropped into a 2% (*w/v*) solution of 100 mL of calcium chloride ($CaCl_2$) in a separate beaker to form and harden the beads and left for 30 min. The hardened beads, obtained with a diameter range of approximately 4–5 mm, were filtered and rinsed with sterile water before use.

Table 1. Samples of granular activated carbon–alginate (GAC–Alg)- and granular activated carbon–alginate–chitosan (GAC–AlgC)-immobilised beads *v/w*.

Sample Labelling	GAC-Alg	GAC-AlgC
A	1:0.5	1:0.5
B	1:1	1:1
C	1:2	1:2
D	1:3	1:3
E	1:4	1:4

2.4. Entrapment of GAC–Alg Beads with Chitosan (GAC–AlgC)

The previous method of GAC-attached biofilm with alginate was repeated. The different concentrations of chitosan were prepared by dissolving 0.5, 1, 2, 3, and 4 g of chitosan flakes into 5% (*v/v*) of acetic acid (5 mL) in a separated beaker, as shown in Table 1. The beads formed were added into chitosan solution until they were well immersed. The beads were then sunk into 40 g of NaOH for 30 min to make it hardened and fully coated before being filtered and rinsed with sterile water and used in fermentation [23,24].

2.5. Batch Fermentation of Biohydrogen Production

The fermentation process was carried out in a 250 mL Schott Duran bottle, as shown in Figure 1 Nitrogen gas was pumped into the bottle for 2 min before fermentation to eliminate the oxygen inside the bottle. The fermentation process was carried out for 12 h, with an initial medium of pH 6.0, as well as temperature and shaking speed at 60 °C and 120 rpm, respectively. The process was repeated periodically for two batches with different types of substrates [25,26]. The gas samples generated during the fermentation were collected when the biogas amount was consistently achieved.

2.6. Analysis of Gaseous, Hydrogen Yield, and Productivity

The hydrogen yield (HY) was determined based on the amount of hydrogen produced over the amount of sugar consumed. Hydrogen yield was represented as hydrogen moles per mole of sugar consumed. The percentage of biogas composition was examined using gas chromatography (GC) (Model HP6890N, Agilent Technology, USA) consisting of two detectors: a thermal conductivity detector (TCD) andflame-ionization detector (FID). The internal diameter and film thickness of the column was 0.53 mm and 0.5 mL, respectively. The oven temperature was set at 75 °C, and the carrier gas flow rate (argon) was 6 mL/min. Then, 0.5-mL samples of gas were taken using a 1-mL gas-tight syringe, injected into the GC immediately. The TCD was calibrated with standard gas (Air Product, Malaysia) mixtures, consisting of H_2, CH_4, CO, and CO_2 in nitrogen, at periodic intervals.

Modified Gompertz was presented to correlate the cumulative hydrogen gas production using the Solver add-in in Excel. Theoretically, the Gompertz equation was modified [26]

$$H_t = H_m.exp\left\{-exp\left[\frac{R_m.e}{H_m}(\lambda - t) + 1\right]\right\}. \tag{1}$$

where H_t is the cumulative hydrogen production (mL), H_m is the maximum hydrogen production (mL), R_m is the maximum hydrogen production rate (mL.h^{-1}), e is the Euler number (e = 2.73), λ is the lag phase time (h), and t is the incubation time (h).

2.7. Analysis of Volatile Fatty Acid and Sugar

HPLC analysis was used to determine the number of monosaccharides mainly found as xylose and glucose, and also the amount of volatile fatty acids (TVFAs) that was present in a sample. The liquid samples were filtered into vials via a 0.22-μm syringe. The soluble microbial product (SMP) and monomeric sugar concentrations were quantified by HPLC analysis fitted with a refractive index detector (RID) with a column (Phenomenex, RPM Pb2+). The mobile phase used was 5 mM water at a constant flow rate of 0.6 mL/min at room temperature. The column temperature was maintained at 80 °C, and the HPLC sample injection volume was 20 µL. The intended compounds were identified by conducting standard curves of different concentrations of SMPs and sugar concentrations.

2.8. Scanning Electron Microscope (SEM)

The formation of cell attachments on the immobilisation beads was observed by using scanning electron microscopy (SEM) [27]. The appearance of beads was seen before and after the fermentation process. The size of both types of beads was measured as 4–6 mm per bead. The physical stability of the beads was observed by putting the bead samples separately into test tubes with a medium of synthetic solution at pH 6.0 and keeping them in the same water bath shaker for fermentation. The state of the physical changes of the beads was recorded after they began to degrade. For further analysis, the beads were taken right after the fermentation process and left at −20 °C before the morphology test. The experiment was conducted using a scanning electron microscope (SEM; model Q250, Thermo Scientific, Waltham, MA, USA). The beads were cut into half with a knife to inspect the structure inside. The gel beads were then mounted on metal stubs, and the inside layer underwent sputter-coating with gold for 6 min. Then, the surfaces were examined and captured.

3. Results and Discussion

3.1. Microbial Cells Self-Attached to GAC for Hydrogen Production

The microbes were cultivated in a synthetic medium containing glucose and xylose mixtures as the sole carbon and energy source until biogas production was consistently achieved. The ability of the cells to bind themselves (self-attach) to the GAC surface had been thoroughly evaluated. Biogas production (mL) was plotted over fermentation time (day), as shown in Figure 2. It can be seen that the biogas fluctuated over the fermentation period and started to be consistently produced at Day 30, towards the end of 40 days of fermentation, with cumulative biogas production being 2115.75 ± 413.03 mL and 2274.75 ± 411.83 mL for 10 and 20 g/L sugar loading, respectively. The results of the production of biogas via dark fermentation immobilised with GAC are shown in Table 2.

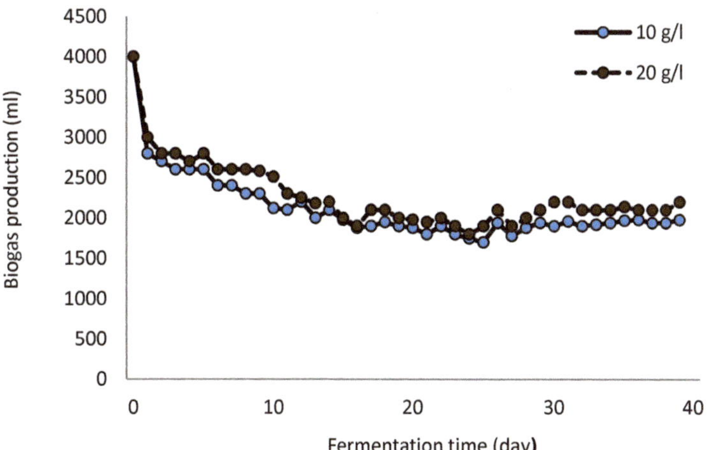

Figure 2. Biogas production (mL) of 10 g/L and 20 g/L sugar loading over fermentation time (day) with 2 days HRT in sequencing mode reactor.

Table 2. Hydrogen productivity and H_2 yield obtained from each of the different sugar loadings.

Sugar Loading	Glucose	Xylose	H_2	H_2 Productivity Rate (HPR)		H_2 Yield
g/L	g/L	g/L	mL	mmol H_2/l.d	mmol H_2/l.h	mol H_2/mol Total Sugar
10	5	5	1743.42 ± 42.16	3.71 ± 0.09	2.56 ± 0.17	3.27 ± 0.08
20	10	10	1840.42 ± 97.76	3.92 ± 0.21	2.70 ± 0.29	3.38 ± 0.18

The process continued until the biogas was stable and ready for gaseous analysis by using GC. The average hydrogen production rate was recorded as 3.71 ± 0.09 mmol H_2/l.d at 10 g/L sugar used and 3.92 ± 0.21 mmol H_2/l.d at 20 g/L sugar used. This indicates that 20 g/L is the optimal amount of sugar to be used for immobilisation beads and, thus, as the optimum substrate for future experiments. In parallel, our previous work also suggested that 20 g/L sugar was the optimal amount of sugar to use [21]. From the data obtained, it was found that the granular activated carbon could provide a suitable matrix to become a primer for cell attachment and colonisation before entrapment for hydrogen production. This is due to the mechanical stability of the biofilms formed on the activated carbon, which have a high propensity in binding capacity, providing a nutrient-rich environment, and thus promoting microbial adhesion [28,29]. The attachment-formed biofilms also help to sustain cell viability and prevent cell washout from the reactor, thus increase cell density [30].

3.2. Biohydrogen Production of Immobilised Beads GAC–Alg

The different concentrations of alginate (Alg) ratios were added to GAC to determine an optimum amount of alginate for biohydrogen production. The results of hydrogen gas produced in each run were plotted against time. From Section 3.1, the optimum result was obtained when 20 g/L amount of sugar was used as a substrate in this experiment, which was subjected to 20% *w/v* of GAC–Alg as immobilised beads in 200 mL working volume.

Hydrogen production using the entrapment technique as immobilised beads was evaluated. Figure 3 shows the comparison of biogas trend production for GAC–Alg beads during the acclimatisation period using a synthetic medium as a substrate. Hydrogen production started to increase at 4 h of fermentation for five different concentrations of alginate, dominantly by GAC–Alg beads at C with a ratio of 1:2.

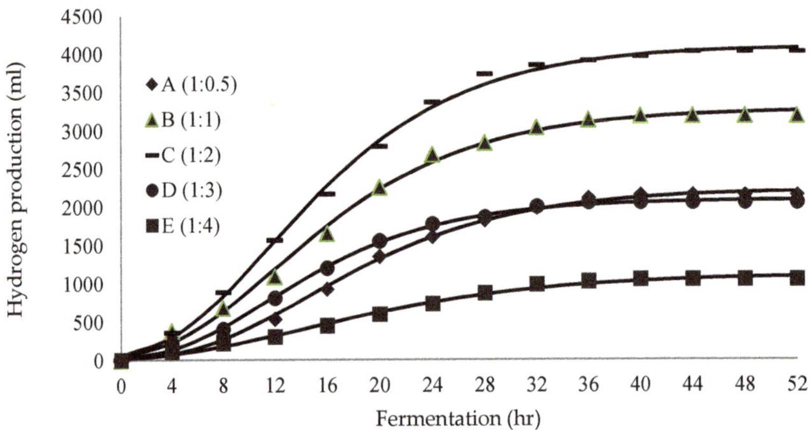

Figure 3. Hydrogen production (mL) at different concentrations of alginate in 200 mL of a 250-mL modified bioreactor in batch fermentation.

The results of HPR (mmol H_2/l.h) and hydrogen yield (mol H_2/mol sugar) were plotted against different concentrations of GAC–Alg, as shown in Figure 4. It can be seen that the highest hydrogen production was found for C at the GAC–Alg ratio of 1:2, with HPR of 2.47 ± 0.47 mmol H_2/l.h.

The highest hydrogen yield was 2.09 ± 0.22 mol H_2/mol total sugar run at C. The cell density (in VSS) of the GAC–Alg was produced at the highest value of 1.65 g/L, as compared to the density of the lowest concentration of GAC–Alg at Run A, which was only 0.48 g/L.

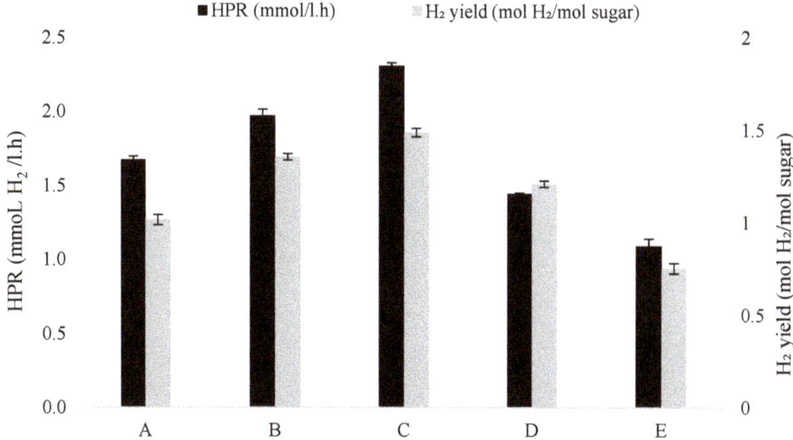

Figure 4. Hydrogen productivity rate (mmol H_2/l.h) and hydrogen yield (mol H_2/mol sugar) in the different concentrations of alginate for hydrogen production.

The trends of HPR and hydrogen yield obtained in Table 3 were comparable to the other runs that contained GAC–Alg at different ratios of B, D, E, and F. It was slightly different in terms of hydrogen yield under different concentrations of alginate, comparing the higher results of Runs B and C to Run D. It can be determined that the irregular space and porous structure present on the carrier's surface provided the microbes with ample space to develop well, in agreement with the results obtained by [31,32]. However, the trends decreased with the increment of added alginate in Runs E and F. It shows that when surrounded by a large number of support carriers, the microbial population had some limitations to grow. Increasing the concentration of alginate did not improve the beads' robustness, and the production of hydrogen gas was slower because a higher amount of alginate also acted as a barrier to the substrate and products [33].

Table 3. Hydrogen productivity obtained from different concentrations of alginate in batch fermentation.

Samples	H_2		Modified Gompertz Equation Parameter Values for H_2 Production (Per Working Volume)		
	Yield	HPR	H_m	R_m	λ
	mol H_2/mol Sugar Consumed	mmol H_2/l.h	mL	mL/h	h
A (1:0.5)	1.36 ± 0.13	1.66 ± 0.05	2225.03	95.83	5.89
B (1:1)	1.73 ± 0.06	1.92 ± 0.05	3273.37	145.48	3.84
C (1:2)	2.09 ± 0.22	2.47 ± 0.47	4099.59	187.58	3.72
D (1:3)	1.54 ± 0.06	1.51 ± 0.45	2091.09	104.85	4.11
E (1:4)	1.11 ± 0.34	1.00 ± 0.17	1119.49	39.79	4.10

The optimum ratio of GAC to alginate of 1:2 remarked the occupation of the optimal porous space of GAC–Alg immobilised beads by the microbes to promote stable biological activity for biohydrogen production. Hence, the study revealed that a combination of GAC and alginate as immobilised beads gave a positive response in bacterial immobilisation, especially in promoting the growth of hydrogen-producing bacteria during the fermentation [34]. The positive performance of the GAC–Alg beads was due to the presence of granule activated carbon inside, which acted as a support for the alginate carrier and maintained the stability of beads [21].

3.3. Bacteria Immobilisation in GAC–Alg Entrapped with Chitosan on Hydrogen Production

The development of entrapped GAC–Alg in chitosan was studied to investigate the adherence of GAC–Alg beads with regards to the mucoadhesion behaviour of chitosan. As reported by Szymańska and Winnicka [35], chitosan possesses good mucoadhesion behaviour resulting from the cationic properties, existence of amino groups, and free hydroxyl, which allow the polymers to interact with each other by electrostatic and hydrogen bonding. The capability of chitosan to trap the GAC–Alg beads had been thoroughly evaluated. Different concentrations of chitosan subjected to 20% *w/v* of GAC–AlgC as immobilisation beads in 200 mL working volume (*w/v*) were used. The results of the hydrogen production (mL H_2) were plotted over fermentation (hr), as shown in Figure 5. It showed the comparison of hydrogen production trends for different concentrations of chitosan g/L used during the acclimatisation period, using a synthetic medium as a substrate. Hydrogen production started to increase at 4 h of fermentation and dominantly during Run C, which had a ratio of chitosan of 1:2. The consistency of hydrogen against chitosan concentration g/L was consistent after 40 h of operation.

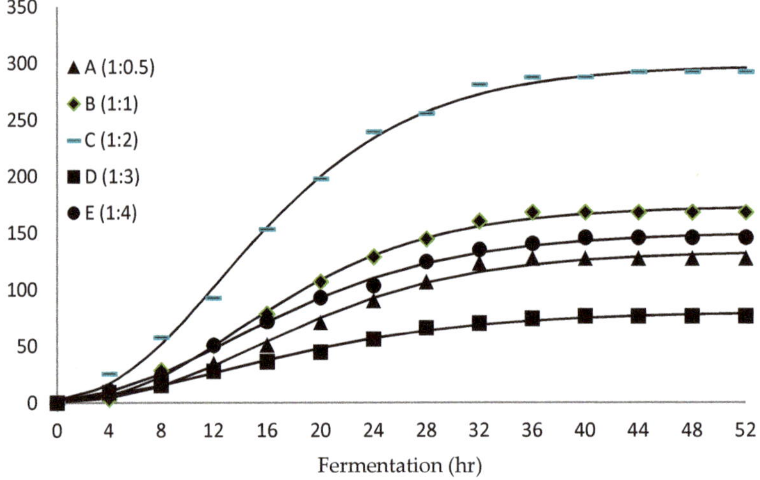

Figure 5. Hydrogen production (mL) at different concentrations of chitosan in 200 mL of a 250 mL modified bioreactor in batch fermentation.

The results of HPR (mmol H_2/l.h) and hydrogen yield (mol H_2/mol sugars consumed) were plotted over different concentrations of chitosan (g/L), as shown in Figure 6. The entrapment of GAC–Alg beads with varying concentrations of chitosan was measured from the evolved gas during the acclimatisation process. The results were analysed and presented in Table 4, which shows that the chitosan concentration at C with 2 g/L reached the highest level for both HPR (0.93 ± 0.05 mmol H_2/l.h) and H_2 yield (1.11 ± 0.35 mol H_2/mol sugar consumed) with 86.63 H_2 %. Meanwhile, at a lower concentration than C, which is concentration atB reached the second highest HPR of 0.85 ± 0.08 mmol H_2/l.h and H_2 yield of 0.97 ± 0.21 mol H_2/mol total sugar with 84.79 H_2 %. The beads of Run D, with 3 g/L, followed as the thirdighest HPR of 0.74 ± 0.15 mmol H_2/l.h and H_2 yield of 0.88 ± 0.12 mol H_2/mol total sugar with 64.54 H_2%. HPR of ratios A 0.5 g (as the lowest concentration) and E (as the highest concentration, with 4 g) was proportionate between those two and was recognised as causing lower hydrogen production than Concentrations B, C, and D after 52 h of operation, individually at 0.58 ± 0.20 and 52.94, and 0.70 ± 0.20 and 76.43 (mmol H_2/l.h; H_2%).

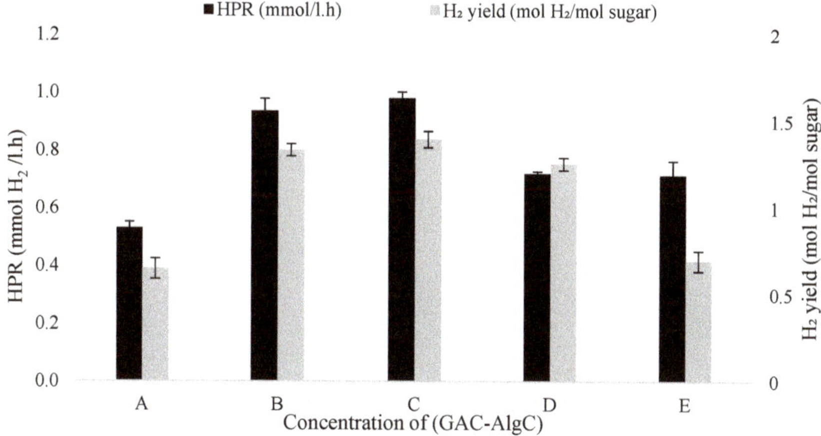

Figure 6. Hydrogen productivity rate (mmol H_2/l.h) and hydrogen yield (mol H_2/mol sugar) in the different concentrations of chitosan.

Table 4. Hydrogen productivity obtained from different concentrations of chitosan in batch fermentation.

Samples	H_2		Modified Gompertz Equation Parameter for H_2 Production (Per Working Volume)		
	Yield	HPR	H_m	R_m	λ
	mol H_2/mol Sugar Consumed	mmol H_2/l.h	mL	mL/h	h
A (1:0.5)	0.46 ± 0.12	0.58 ± 0.20	133.77	5.55	6.26
B (1:1)	0.97 ± 0.21	0.85 ± 0.08	173.19	7.76	5.57
C (1:2)	1.11 ± 0.35	0.93 ± 0.05	297.76	13.53	4.45
D (1:3)	0.88 ± 0.12	0.74 ± 0.15	80.37	2.90	3.01
E (1:4)	0.67 ± 0.30	0.70 ± 0.20	150.61	6.03	4.14

The comparison of these findings revealed that the immobilised beads of Run C reached the highest hydrogen production and they were examined as the optimum concentration for microbial support matrix in immobilisation bead development. The work by Damayanti et. al [36] reported that among a variety of chitosan applications, chitosan in encapsulation technology is widely applied whether as a second-layer coating or in combination with other polymers. It was also reported that chitosan could improve the stability of the capsules. In other studies reported by Žuža et al. [16], it was claimed that the mechanical confidence of alginate beads increased up to seven days when coated with chitosan, which significantly contributes to the preservation of carrier strength during fermentation. The formation of the shape of cell-immobilised GAC, cell-immobilised GAC-Alg and GAC-AlgC were presented in Figure 7. Figure 7a image of cell-immobilised into GAC, (b) the shape of cell-immobilised GAC with alginate and (c) the shape of cell- immobilised GAC-Alginate beads with chitosan. It can be seen both GAC-Alg and GAC-Alg coated with chitosan were not much spherical. The GAC covered with alginate were transparent, while the beads coated with chitosan have slightly cloudy of physical appearance. Generally, the differences in the shape of beads were caused by the gravity and surface tension imbalance when the beads dropped from the syringe. The beads shape formation also were affected by the viscosity of alginate and chitosan, and distance of dropper to gel solution [34]. This is a new method introduced to improve the stability of biofilm formation and adsorption capacity as well as enhanced the mechanical strength of the carrier, thus enhanced the hydrogen yield production. Table 5 summarizes the comparison of a similar study on the efficiency of different types of immobilisation beads on hydrogen production. Table 5 shows that the carbon source and fermentation process were different to produce hydrogen gas. In the present work, the highest hydrogen yield was 2.09 mol H_2 /mol sugar obtained from GAC-Alg immobilised beads. It should be noted that the type of fermentation process and carbon source can always influence the results [36]. Likewise, the selection of the materials to be used as immobilising carriers also playing an important role in biohydrogen production, in term of high resistance towards temperature, mechanical strength as well as the possibility to recycle the immobilised carriers. Hence, in this research, attached-biofilm of hydrogen-producing bacteria on GAC from adsorption approach are stabilised using alginate entrapped with chitosan to form stable cells immobilised beads as a novel approach.

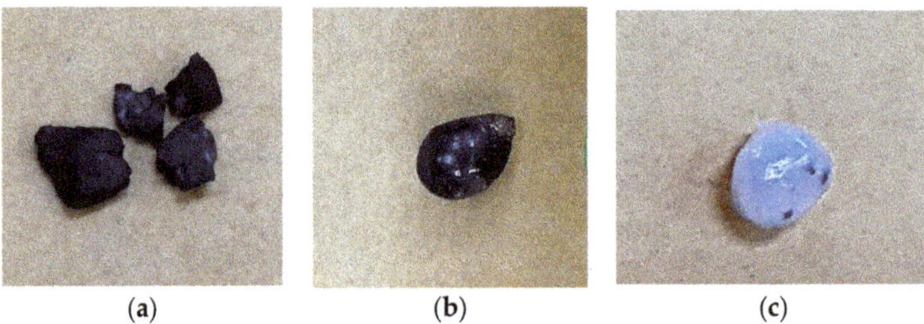

(a)   (b)   (c)

Figure 7. (a) Cell-immobilised GAC; (b) cell-immobilised GAC–Alg; (c) cell-immobilised GAC–AlgC.

Table 5. Comparative study on the efficiency of different types of immobilisation beads on hydrogen production.

Carbon Source	Type of Carrier	Temp	Fermentation Process	Hydrogen Yield	References
Glucose	CA-AC	35	Batch	2.6 mol H_2/mol sucrose	[10]
Sucrose	CA-C-TiO_2	35	Batch	2.60 mol H_2/mol sucrose	[37]
Glucose	CA	60	Batch	1.90 mol H_2/mol glucose	[38]
Glucose	CA	40	Batch	17 L/g mol glucose	[39]
Brewery wastewater	CA	37	Batch	14 g/L COD	[40]
glucose	GAC	37	Continuous	0.4–1.7 mol H_2/mol sugar	[41]
Glucose	CA-AC	36	Batch	0.029 mol H_2/mol glucose	[34]
Sucrose	CA	35	Batch	1.7 mol H_2/mol sucrose	[42]
Xylose	CA	37	Continuous	3.15 mmol H_2/mol xylose	[43]
Sucrose	CA-AC	35	Continuous	2.67 mol H_2/mol sucrose	[8]
Glucose/Xylose	GAC-AlgC	60	Batch	1.11 mol H_2/mol sugar	This study
Glucose/Xylose	GAC-Alg	60	Batch	2.09 mol H_2/mol sugar	This study

Ac = activated carbon; GAC = granular activated carbon; CA = calcium alginate; C = chitosan.

3.4. Effect of Alginate and Chitosan Concentration on Volatile Fatty Acid Production

Hydrogen production performance is usually monitored collectively with the formation of acetic acid (HAc) to butyric acid (HBu) and total volatile fatty acids (TVFAs). In this anaerobic hydrogen production, the concentration of TVFAs and their relative proportions were effectively used as indicators. The plot between HPR (mmol H_2/l.h) and volatile fatty acid (VFA) concentration (mg/L) for GAC-Alg- and GAC-AlgC-immobilised beads in batch fermentation is presented in Figure 8a for GAC-Alg and Figure 8b for GAC-AlgC. The summary of total volatile fatty acids at various concentrations of GAC-Alg and GAC-AlgC shown in Table 6

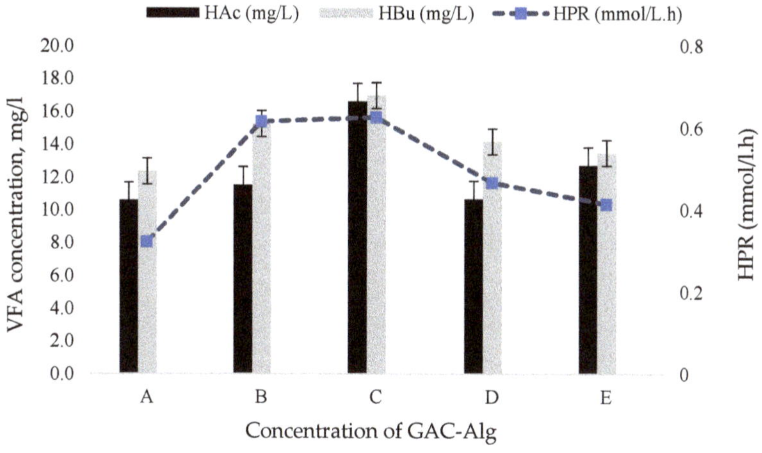

(**a**)

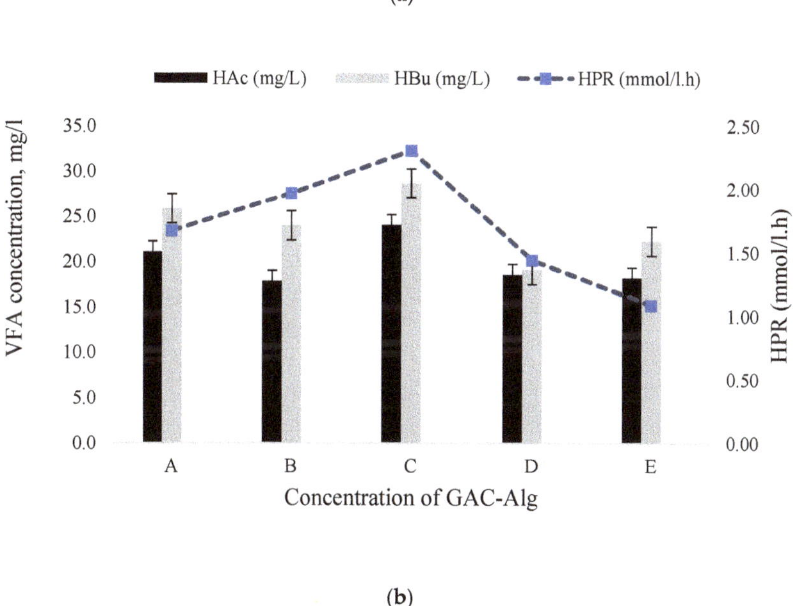

(**b**)

Figure 8. Hydrogen production rate (mmol/l.h) and VFA concentration (HAc-acetate acid and HBu-butyrate acid) in batch fermentation of (**a**) GAC-Alg and (**b**) GAC-lgC beads.

Table 6. Summary of total volatile fatty acids at various concentrations of GAC–Alg and GAC–AlgC, all at 200 mL working volume of a 250 mL modified bioreactor in batch fermentation.

Samples	HPR		Hydrogen	HAc	HBu	TVFAs
	mmol/l.d	mmol/l.h	%	mM	mM	mM
GAC–Alg						
A (1:0.5)	3.63	1.66	84.60	21.08	25.88	46.95
B (1:1)	4.27	1.92	78.97	17.87	24.01	41.87
C (1:2)	5.00	2.47	85.87	24.05	28.66	52.71
D (1:3)	3.13	1.51	65.74	18.58	19.14	37.72
E (1:4)	2.36	1.00	73.18	18.21	22.29	40.50
GAC–AlgC						
A (1:0.5)	1.15	0.58	52.94	10.59	12.36	22.95
B (1:1)	2.03	0.85	84.79	11.53	15.26	26.79
C (1:2)	2.12	0.93	86.63	16.60	16.97	33.57
D (1:3)	1.56	0.74	64.54	10.64	14.18	24.82
E (1:4)	1.55	0.70	76.43	12.71	13.49	26.20

TVFAs (total volatile fatty acids) = HAc + HBu.

Glycolysis is the gateway for the metabolic process of cells that convert glucose into pyruvate as an intermediate metabolite. Pyruvate reacts to acidogenesis and generates VFAs, including butyric acid, acetic acid, and also propionic acid under anaerobic conditions [34]. Theoretically, the maximum amount of H_2 yields when all glucose has been converted to HAc is 4 mol H_2 per mole of glucose in Equation (2), while HBu is 2 mol H_2 per glucose in Equation (3) [44].

$$C_6H_{12}O_6 + 2H_2O \rightarrow 2CH_3COOH + 4H_2 + 2CO_2 \tag{2}$$

$$C_6H_{12}O_6 + 2H_2O \rightarrow CH_2CH_2CH_2COOH + 2H_2 + 2CO_2 \tag{3}$$

It was found that HBu and HAc were the primary volatile fatty acids that were produced, while propionate (HPr) contributed to negligible amounts. The result of GAC–Alg dominated, with 23.99 ± 3.60 mM of HBu and 19.96 ± 2.61 mM of HAc found at Run C. For GAC–AlgC, the result was dominated by the ratio of 1:2, with 14.45 ± 1.76 mM of HBu and 12.41 ± 2.49 mM of HAc. Hydrogen production was slightly increased as the concentration of alginate increased from Runs A to C as Hbu and HAc increased. These results are close to similar metabolic pathways of hydrogen production found by [45,46], who stated that when the composition of HAc and HBu increases, the hydrogen production efficiency should also increase.

Based on these findings, it was found that HBu and HAc were eminently affected by hydrogen production in Equations (2) and (3). The highest range found in butyrate acid proposed that the biohydrogen production from a mixture of xylose and glucose was of a butyrate type in Equation (3). Hence, the investigation suggests that butyrate is significantly affected by glucose and xylose consumption rather than acetate, as reported by [45].

3.5. SEM of Immobilized Beads

Furthermore, this study observed the microbial cell culture on carriers using a scanning electron microscope (SEM). GAC as a primer for cell attachment and colonisation was observed before and after acclimatisation. The image of the micropores of clean GAC is shown in Figure 9a, while the image of Figure 9b shows the microbial cells that have successfully attached to the GAC surface. Both of the images were captured using a field emission scanning electron microscope (FESEM) at 10.00 k magnification [27]. The porosity of GAC was provided with a pleasant environment and conditions for cells to adhere themselves firmly inside the pores, thus helping the cells to grow and form the population. This overcomes the problem of self-detachment of hydrogen-producing cells during repeated batch fermentation in the culture medium [4].

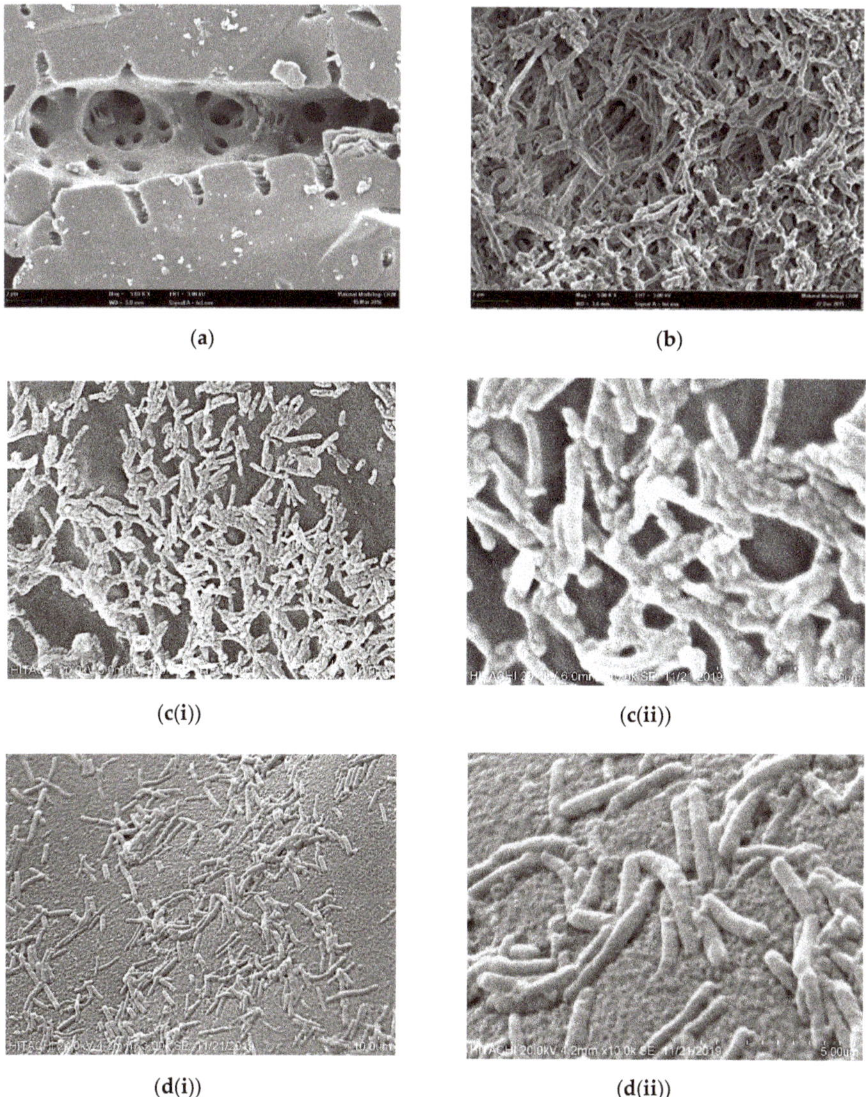

(a)

(b)

(c(i))

(c(ii))

(d(i))

(d(ii))

Figure 9. Images of (**a**) clean GAC; (**b**) GAC-attached biofilm; (**c**) immobilised cells on GAC–Alg at (**i**) 3.00 k magnification and (**ii**) 10.00 k magnification; (**d**) immobilised cells on GAC–AlgC at (**i**) 3.00 k magnification and (**ii**) 10.00 k magnification.

A significant number of microbial cells were observed to have successfully immobilised into the alginate surface, as shown in Figure 9c(i), and agglomerated each other (Figure 9c(ii)). These features indicate that the alginate does not provide toxic and non-nutritive environments towards the microbial cells, but gives them a suitable place to grow dominantly, besides protecting the cells inside the beads. A part of the entrapment of cells into alginate, the image of the predominantly rod-shaped microbial species, was captured after the fermentation process, which can be clearly seen in Figure 9d(ii). The rod-shaped bacterial cells appeared to be the dominant consortium on the GAC based on their morphologic properties. This study is in agreement with a previous study by Jamali et al. [21]. It has been identified that the dominant species of anaerobic hydrogen producers is *Thermoanaerobacterium thermosaccharolyticum*, which is stated to have spores and is rod-shaped and Gram-positive [21,28]. A similar source of inoculum was used in this research as this species has been recognised as having a high propensity to be cultivated at an optimal temperature of 50–60 °C and a pH of approximately 5.5 to 6.5. This bacterium played a significant role in the production of butyric acid, acetic acid, and hydrogen. This research is supported by [21], where Thermoanaerobacterium thermosaccharolyticum has been documented to be effective hydrogen producers of xylose and glucose, with butyrate and acetate as the main byproducts of fermentation in a synthetic medium. Thermoanaerobacterium thermosaccharolyticum has also been documented as capable of fermenting a broad variety of carbohydrates and complex sugars that are present in almost all wastewater [47]. The accumulation of microbial cells around the GAC–AlgC carrier surface membrane, as illustrated in Figure 9d(ii), explains the right conditions of the carrier, allowing the retention of the biological activity of the encapsulated cells. There was a significantly high mortality of cells when using chitosan as an external encapsulation agent. The hydrophobicity behaviour of chitosan was favoured by the undesired protein adsorption and denaturation process. Moreover, the diffusion issues also affected the molecular traffic of substrates and products of the microbial enzymatic process between the outsides and the insides of the carrier [48,49]. Generally, high microbial loadings hosted within the carrier showed that the cells are protected from microbial attack and physical or mechanical damage. The high porosity of the microbial matrix support will provide the right places for cells to grow and immobilise. The suitable chemical nature of carriers also can help the cells to extend as well as their protein can be easily accommodated within the channel. Nevertheless, the unavoidably fragile and prone-to-grinding environment of the carrier needs to be enhanced with optimal entrapment agents.

4. Conclusions

This work has successfully developed the entrapment of immobilised GAC with alginate and chitosan in the batch fermentation system from xylose and glucose fermentation. It was found that concentrations of alginate- and chitosan-immobilised beads at 2 g/L presented the highest amount of hydrogen productivity. The results showed that the immobilised beads maintained their stability in hydrogen production after 40 h; a consistent HPR of 2.47 ± 0.47 mmol H_2/l.h and H_2 yield of 2.09 ± 0.22 mol H_2/mol total sugar was found with GAC–Alg beads. The consistent HPR obtained with GAC–AlgC beads was 0.93 ± 0.05 mmol H_2/l.h, along with H_2 yield of 0.88 ± 0.12 mol H_2/mol total sugar. In accordance with all of the significant results, it is emphasised that the acclimatisation of GAC–Alg and GAC–AlgC beads as support carriers ensures the continuity of HPR and enhances cultural density in the handling of synthetic wastewater for thermophilic hydrogen production.

Author Contributions: Conceptualisation, N.S.J., S.S.M., and N.F.D.R.; methodology, N.F.D.R. and S.S.M.; validation, N.F.D.R.; investigation, N.F.D.R.; resources, N.S.J.; writing—original draft preparation, N.F.D.R.; visualization, S.F.I.; software, S.I.S.; supervision, N.S.J., N.A., and M.F.I.; project administration, N.S.J. All authors have read and agreed to the published version of the manuscript.

Funding: This research was funded by the Ministry of Education Malaysia (MOE) under the Fundamental Research Grant Scheme (FRGS) through the vot code of 5540208 (Ref: FRGS/1/201/STG05/UPM/02/28) towards the success of this study.

Acknowledgments: These authors would like to express their gratitude to all the laboratory staff at the Department of Chemical and Environmental Engineering, Universiti Putra Malaysia, for assisting this research.

Conflicts of Interest: The authors declare no conflict of interest.

References

1. Kim, J.K.; Nhat, L.; Chun, Y.N.; Kim, S.W. Hydrogen production conditions from food waste by dark fermentation with Clostridium beijerinckii KCTC 1785. *Biotechnol. Bioprocess Eng.* **2008**, *13*, 499–504. [CrossRef]
2. Hu, B.; Liu, Y.; Chi, Z.; Chen, S. Biological hydrogen production via bacteria immobilized in calcium alginate gel beads. *Biol. Eng. Trans.* **2007**, *1*, 25–37. [CrossRef]
3. Liu, X.; Yu, X. Enhancement of Butanol Production: From Biocatalysis to Bioelectrocatalysis. *ACS Energy Lett.* **2020**, *5*, 867–878. [CrossRef]
4. Jamali, N.S.; Jahim, J.; Nor, W.; Wan, R.; Abdul, P.M. Particle size variations of activated carbon on biofilm formation in thermophilic biohydrogen production from palm oil mill effluent. *Energy Convers. Manag.* **2016**, *141*, 354–366. [CrossRef]
5. Mohamad, N.R.; Marzuki, N.H.C.; Buang, N.A.; Huyop, F.; Wahab, R.A. An overview of technologies for immobilization of enzymes and surface analysis techniques for immobilized enzymes. *Biotechnol. Biotechnol. Equip.* **2015**, *29*, 205–220. [CrossRef]
6. Mishra, P.; Krishnan, S.; Rana, S.; Singh, L.; Sakinah, M.; Ab Wahid, Z. Outlook of fermentative hydrogen production techniques: An overview of dark, photo and integrated dark-photo fermentative approach to biomass. *Energy Strategy Rev.* **2019**, *24*, 27–37. [CrossRef]
7. Singh, L.; Wahid, Z.A.; Siddiqui, M.F.; Ahmad, A.; Hasbi, M.; Rahim, A.; Sakinah, M. Biohydrogen production from palm oil mill effluent using immobilized Clostridium butyricum EB6 in polyethylene glycol. *Process Biochem.* **2013**, *48*, 294–298. [CrossRef]
8. Wu, S.-Y.; Lin, C.-N.; Chang, J.-S. Hydrogen Production with Immobilized Sewage Sludge in Three-Phase Fluidized-Bed Bioreactors. *Biotechnol. Prog.* **2003**, *19*, 828–832. [CrossRef]
9. Smidsrød, O.; Skjåk-Bræk, G. Alginate as immobilization matrix for cells. *Trends Biotechnol.* **1990**, *8*, 71–78. [CrossRef]
10. Wu, S.-Y.; Lin, C.-N.; Chang, J.-S.; Lee, K.-S.; Lin, P.-J. Microbial Hydrogen Production with Immobilized Sewage Sludge. *Biotechnol. Prog.* **2002**, *18*, 921–926. [CrossRef]
11. Sekoai, P.T.; Awosusi, A.A.; Yoro, K.O.; Oloye, O.; Ayeni, A.O.; Bodunrin, M.; Thabang, P.; Awosusi, A.A.; Yoro, K.O.; Oloye, O.; et al. Critical Reviews in Biotechnology Microbial cell immobilization in biohydrogen production: A short overview. *Crit. Rev. Biotechnol.* **2017**, *38*, 157–171. [CrossRef]
12. Sekoai, P.T.; Yoro, K.O.; Bodunrin, M.O.; Ayeni, A.O.; Daramola, M.O. Integrated system approach to dark fermentative biohydrogen production for enhanced yield, energy efficiency and substrate recovery. *Rev. Environ. Sci. Bio Technol.* **2018**, *17*, 501–529. [CrossRef]
13. Ngah, W.S.W.; Fatinathan, S. Adsorption of Cu (II) ions in aqueous solution using chitosan beads, chitosan-GLA beads and chitosan-alginate beads. *Chem. Eng. J.* **2008**, *143*, 62–72. [CrossRef]
14. Shu, X.; Zhu, K. The release behavior of brilliant blue from calcium-alginate gel beads coated by chitosan: The preparation method effect. *Eur. J. Pharm. Biopharm.* **2002**, *53*, 193–201. [CrossRef]
15. Overgaard, S.; Scharer, J.M.; Moo-Young, M.; Bols, N.C. Immobilization of hybridoma cells in chitosan alginate beads. *Can. J. Chem. Eng.* **1991**, *69*, 439–443. [CrossRef]
16. Žuža, M.G.; Obradović, B.M.; Knežević-Jugović, Z.D. Hydrolysis of Penicillin G by Penicillin G Acylase Immobilized on Chitosan Microbeads in Different Reactor Systems. *Chem. Eng. Technol.* **2011**, *34*, 1706–1714. [CrossRef]
17. Stojkovska, J.; Kostić, D.; Jovanović, Ž.; Vukašinović-Sekulić, M.; Mišković-Stanković, V.; Obradović, B. A comprehensive approach to in vitro functional evaluation of Ag/alginate nanocomposite hydrogels. *Carbohydr. Polym.* **2014**, *111*, 305–314. [CrossRef]
18. Obradović, N.S.; Krunić, T.Ž.; Damnjanović, I.D.; Vukašinović-Sekulić, M.S.; Rakin, M.B.; Rakin, M.P.; Bugarski, B.M. Influence of whey proteins addition on mechanical stability of biopolymer beads with immobilized probiotics. *Tehnika* **2015**, *70*, 397–400. [CrossRef]
19. Mahajan, R.; Gupta, V.K.; Sharma, J. Comparison and suitability of gel matrix for entrapping higher content of enzymes for commercial applications. *Indian J. Pharm. Sci.* **2010**, *72*, 223–228. [CrossRef]
20. Kumar, G.; Mudhoo, A.; Sivagurunathan, P.; Nagarajan, D.; Ghimire, A.; Lay, C.-H.; Lin, C.-Y.; Lee, D.-J.; Chang, J.-S. Recent insights into the cell immobilization technology applied for dark fermentative hydrogen production. *Bioresour. Technol.* **2016**, *219*, 725–737. [CrossRef]

21. Jamali, N.S.; Farahana, N.; Rashidi, D.; Jahim, J.; O-thong, S.; Jehlee, A.; Engliman, N.S. Thermophilic biohydrogen production from palm oil mill effluent: Effect of immobilized cells on granular activated carbon in fluidized bed reactor. *Food Bioprod. Process.* **2019**, *117*, 231–240. [CrossRef]

22. Angelidaki, I.; Sanders, W. Assessment of the anaerobic biodegradability of macropollutants. *Rev. Environ. Sci. Biotechnol.* **2004**, *3*, 117–129. [CrossRef]

23. Duarte, J.C.; Rodrigues, J.A.R.; Moran, P.J.S.; Valença, G.P.; Nunhez, J.R. Effect of immobilized cells in calcium alginate beads in alcoholic fermentation. *AMB Express* **2013**, *3*, 1–8. [CrossRef]

24. Khanna, N.; Das, D. Biohydrogen production by dark fermentation. *Wiley Interdiscip. Rev. Energy Environ.* **2013**, *2*, 401–421. [CrossRef]

25. Wang, C.C.; Chang, C.W.; Chu, C.P.; Lee, D.J.; Chang, B.V.; Liao, C.S. Producing hydrogen from wastewater sludge by Clostridium bifermentans. *J. Biotechnol.* **2003**, *102*, 83–92. [CrossRef]

26. Sekoai, P.T.; Yoro, K.O.; Daramola, M.O. Batch Fermentative Biohydrogen Production Process Using Immobilized Anaerobic Sludge from Organic Solid Waste. *Environments* **2016**, *3*, 38. [CrossRef]

27. Chen, W.-H.; Chen, S.-Y.; Kumar Khanal, S.; Sung, S. Kinetic study of biological hydrogen production by anaerobic fermentation. *Int. J. Hydrog. Energy* **2006**, *31*, 2170–2178. [CrossRef]

28. Syakina, N.; Jahim, J.; O-thong, S. ScienceDirect Hydrodynamic characteristics and model of fluidized bed reactor with immobilised cells on activated carbon for biohydrogen production. *Int. J. Hydrog. Energy* **2019**. [CrossRef]

29. Chu, Y.; Wei, Y.; Yuan, X.; Shi, X. Bioconversion of wheat stalk to hydrogen by dark fermentation: Effect of different mixed microflora on hydrogen yield and cellulose solubilisation. *Bioresour. Technol.* **2011**, *102*, 3805–3809. [CrossRef]

30. Syakina, N.; Jahim, J. ScienceDirect Biofilm formation on granular activated carbon in xylose and glucose mixture for thermophilic biohydrogen production. *Int. J. Hydrog. Energy* **2016**, 1–11. [CrossRef]

31. Seifan, M.; Samani, A.K.; Hewitt, S.; Berenjian, A. The Effect of Cell Immobilization by Calcium Alginate on Bacterially Induced Calcium Carbonate Precipitation. *Fermentation* **2017**, *3*, 57. [CrossRef]

32. Lutpi, N.A.; Jahim, J.; Mumtaz, T. RSC Advances Physicochemical characteristics of attached bio fi lm on granular activated carbon for thermophilic biohydrogen production. *RSC Adv.* **2015**, 19382–19392. [CrossRef]

33. Mesran, M.H.; Mamat, S.; Pang, Y.R.; Tan, Y.H.; Muneera, Z.; Ghazali, N.F.; Ali, M.A.; Mahmood, N.A. Preliminary Studies on Immobilized Cells-Based Microbial Fuel Cell System on Its Power Generation Performance. *J. Asian Sci. Res.* **2014**, *4*, 428–435.

34. Damayanti, A.; Sediawan, W.B.; Syamsiah, S. Performance analysis of immobilized and co-immobilized enriched-mixed culture for hydrogen production. *J. Mech. Eng. Sci.* **2018**, *12*, 3515–3528. [CrossRef]

35. Szymańska, E.; Winnicka, K. Stability of chitosan—A challenge for pharmaceutical and biomedical applications. *Mar. Drugs* **2015**, *13*, 1819–1846. [CrossRef]

36. Talebnia, F.; Taherzadeh, M.J. In situ detoxification and continuous cultivation of dilute-acid hydrolyzate to ethanol by encapsulated S. cerevisiae. *J. Biotechnol.* **2006**, *125*, 377–384. [CrossRef]

37. Wu, K.; Chang, J. Biohydrogen Production Using Suspended and Immobilized Mixed Microflora. *J. Chin. Inst. Chem. Eng.* **2006**, *37*, 545–550.

38. Basile, M.A.; Dipasquale, L.; Gambacorta, A.; Vella, M.F.; Calarco, A.; Cerruti, P.; Malinconico, M.; Gomez d'Ayala, G. The effect of the surface charge of hydrogel supports on thermophilic biohydrogen production. *Bioresour. Technol.* **2010**, *101*, 4386–4394. [CrossRef]

39. Kumar, A.; Jain, S.R.; Sharma, C.B.; Joshi, A.P.; Kalia, V.C. Increased H2 production by immobilized microorganisms. *World J. Microbiol. Biotechnol.* **1995**, *11*, 156–159. [CrossRef]

40. Zhu, K.; Arnold, W.A.; Sakkos, J.; Davis, C.W.; Novak, P.J. Achieving high-rate hydrogen recovery from wastewater using customizable alginate polymer gel matrices encapsulating biomass. *Environ. Sci. Water Res. Technol.* **2018**, *4*, 1867–1876. [CrossRef]

41. Zhang, Z.P.; Show, K.Y.; Tay, J.H.; Liang, D.T.; Lee, D.J. Enhanced Continuous Biohydrogen Production by Immobilized Anaerobic Microflora. *Energy Fuels* **2008**, *22*, 87–92. [CrossRef]

42. Wu, S.Y.; Lin, C.N.; Chang, J.S.; Chang, J.S. Biohydrogen production with anaerobic sludge immobilized by ethylene-vinyl acetate copolymer. *Int. J. Hydrog. Energy* **2005**, *30*, 1375–1381. [CrossRef]

43. Zhao, L.; Cao, G.; Wang, A.; Guo, W.; Liu, B.; Ren, H.; Ren, N.; Ma, F. Enhanced bio-hydrogen production by immobilized Clostridium sp. T2 on a new biological carrier. *Int. J. Hydrog. Energy* **2011**, *37*, 162–166. [CrossRef]

44. Karimi, K. *Lignocellulose-Based Bioproducts*; Springer: Berlin/Heidelberg, Germany, 2015; ISBN 9783319350349.

45. Zhang, S.; Kim, T.-H.; Lee, Y.; Hwang, S.-J. Effects of VFAs Concentration on Bio-hydrogen Production with Clostridium Bifermentans 3AT-ma. *Energy Procedia* **2012**, *14*, 518–523. [CrossRef]

46. Jamali, N.S.; Jamaliah, J. Optimization of Thermophilic Biohydrogen Production by Microflora of Palm Oil Mill Effluent: Cell Attachment on Granular Activated Carbon as Support Media Zirconia ceramic membrane as a separator in microbial fuel cell View project. *Artic. Malays. J. Anal. Sci.* **2016**. [CrossRef]

47. Ueno, Y.; Sasaki, D.; Fukui, H.; Haruta, S.; Ishii, M.; Igarashi, Y. Changes in bacterial community during fermentative hydrogen and acid production from organic waste by thermophilic anaerobic microflora. *J. Appl. Microbiol.* **2006**, *101*, 331–343. [CrossRef]

48. Fu, Y.-L.; Xiong, Y.; Liu, X.-D.; Yu, W.-T.; Wang, Y.-L.; Yu, X.-J.; Ma, X.-J. Study of alginate/chitosan microcapsules for immobilization of Escherichia coli DH5 alpha. *Sheng Wu Gong Cheng Xue Bao* **2002**, *18*, 239–241.

49. Zucca, P.; Sanjust, E. Inorganic Materials as Supports for Covalent Enzyme Immobilization: Methods and Mechanisms. *Molecules* **2014**, *19*, 14139–14194. [CrossRef]

 *processes*

Article

Assessing Supply Chain Performance from the Perspective of Pakistan's Manufacturing Industry Through Social Sustainability

Maryam Khokhar [1], Wasim Iqbal [2], Yumei Hou [1,*], Majed Abbas [1] and Arooj Fatima [1]

[1] College of Economics and Management, Yanshan University, Qinhuangdao 066004, China; maryamkhokhar@stumail.ysu.edu.cn (M.K.); abbasmajed@gmail.com (M.A.); Aroojfatima132@yahoo.com (A.F.)

[2] Department of Management Science, College of Management, Shenzhen University, Shenzhen 518060, China; wasimiqbal01@yahoo.com

* Correspondence: hym@ysu.edu.cn

Received: 20 June 2020; Accepted: 4 August 2020; Published: 1 September 2020

Abstract: The industry is gradually forced to integrate socially sustainable development practices and cross-social issues. Although researchers and practitioners emphasize environmental and economic sustainability in supply chain management (SCM). This is unfortunate because not only social sustainable development plays an important role in promoting other sustainable development programs, but social injustice at one level in the supply chain may also cause significant losses to companies throughout the chain. This article aimed to consolidate the literature on the responsibilities of suppliers, manufacturers, and customers and to adopt sustainable supply chain management (SSSCM) practices in the Pakistani industry to identify all possible aspects of sustainable social development in the supply chain by investigating the relationship between survey variables and structure. This work went beyond the limits of regulations and showed the status of maintaining sustainable social issues. Based on semi-structured interviews, a comprehensive questionnaire was developed. The data was collected through a survey of the head of the supply chain in Karachi, Pakistan. The results of this paper showed that organizational learning was the most important dimension of supplier social sustainability with a value of 40.5% as compared to the effectiveness of the supply chain and the supplier performance with values 37.7 and 9.6%, respectively. In terms of the manufacturer's social responsibility, the highest score for operational performance was 47%, while productivity was 20%, and corporate social demonstration was 20%. Finally, for the customers' social sustainability, two dimensions were determined, namely, customer satisfaction and customer commitment with scores of 47 and 40%, respectively. We also solved sustainable social problems from the perspective of suppliers, manufacturers, and customers. The study would help professionals anywhere to emphasize their considerations and would improve the management of social sustainability in their supply chain.

Keywords: sustainable supply chain management (SSSCM); social sustainability; qualitative research; Pakistan

---

1. Introduction

Due to the rigidity of the industrial environment, communities, Non-Government Organizations (NGOs), and consumer awareness policies are under pressure. Organizations must implement sustainable implementation in supply chain management. Sustainable development combines economic, social, and environmental characteristics and transcends limits within and between industries. Therefore, sustainable business is directly related to the sustainable supply chain management (SSSCM) proposition [1]. Sustainability is defined by contextualization [2], "the ability to meet the needs of the next generation without compromising today's needs". Besides, SSSCM can be defined as "how to manage social issues that can sustain long-term strategies of the organization".

The above aspects are not limited to the internal operations of suppliers, manufacturers, and customers, but also can be extended to external social issues of the organization. The environmental and economic sustainability has a considerable impact on the literature and practice. However, social sustainability has not received enough attention. In order to enhance sustainability and other aspects of practice, a socially sustainable industrial scale needs to be accurately achieved. Most researchers focus on how the industry develops social sustainability when working with upstream or downstream companies [3]. Despite much research on South Asian countries, there is less evidence of Pakistani industries with different social norms. Some advocates emerging and protection of labor rights and how these measures support claims for greater efficiency [4]. To date, few studies have incorporated social factors into their SSSCM framework and have not incorporated sustainable management practices. Several cases have attempted to address the sustainability of the business sector with a short-term focus on social initiatives undertaken by the organization. These attempts are not conducive to improving social measures and building the capabilities and resources needed to comprehensively and systematically manage the social impact of the organization's supply chain management. Some studies have taken the first step in identifying and examining some useful issues and aspects related to social sustainability.

The purpose of this study was to investigate social sustainability in Pakistan from the perspective of major companies, first-tier suppliers, and first-line customers. This research made a significant contribution to the existing literature. First, we extended the current literature by testing and validating models to enhance Pakistan's sustainable supply chain management practices through social commerce motivations, mechanisms, and performance results. Second, our data analysis revealed significant differences between regulatory constraints that indicated a position to maintain sustainable social issues. Third, this research raised awareness that in the context of social commerce, socially sustainable management in its supply chain is still crucial. Fourth, our research provided empirical evidence that standard adaptation practices of manufacturing companies would positively affect customers' willingness to buy in social enterprises.

2. Theoretical Background

2.1. Sustainable Supply Chain Management (SSSCM)

Sustainable supply chain management can often be described as the processes and practices that are carried out within and across organizations to achieve emotional synchronization to increase output, reduce costs, maximize asset utilization, and maximize customer service. The transactions involve its activities, resources, information funds, and the impact of the supply chain on the social well-being of its employees, society, and customers. It also minimizes environmental impacts [5]. Increasingly, companies are held accountable for the social, economic, and ecological decisions that arise from their internal and supplier operations [6]. For the past two decades, SSSCM has worked to integrate social, economic, and environmental goals through focused business processes. It has become a practice for companies to achieve sustainable development results in their supply chains [7]. However, the management register is not able to inspire global SSSCM. SSSCM is recommended to increase stakeholder attention to the impact of social enterprise internal supply chain operations on social systems [8].

The business sector performance can be improved by the sustainability of the supply chain. This directly affects the competitiveness of the industry and the performance of the supply chain. Koberg and Longoni [9] adopted a plan to solve social problems at multiple levels. According to Wan Ahmad et al. [10] in order to minimize waste and save costs, the plan must be implemented with sustainable motivation, and suppliers and manufacturers should be strengthened through the refinement between the capabilities and assistance of internal representatives. In order to implement SSSCM, the social, economic, and environmental requirements of industry models and practices have been proposed. According to Yuen et al. [11], in order to improve the sustainability of the business sector, its long-term goal is to manage the well-being of ordinary people who must control social and economic management operations. This is why many industries use sustainability indices to assess their social sustainability capabilities [12].

2.2. Social Sustainable Supply Chain Management (SSSCM)

With regard to the environmental and economic sustainability of industrial practices, SSSCM should also be considered when companies are seeking successful and sustainable growth. Social sustainability should be managed through social issues to improve the long-term life of the industry [13]. These issues are important characteristics of sustainable corporate development that are being considered and evaluated [14,15]. Former scholars use the same report to address social issues in the supply chain. Determining the scope and measures of universal and global social sustainability is challenging because of the lack of conceptual clarification, especially in South Asian countries like Pakistan, where it involves social issues. It is, therefore, clear that supply chain managers do not have a sufficient understanding of the social issues involved and how to evaluate and manage [16].

A review of the existing literature shows that there are huge challenges in addressing SSSCM and social issues related to the business sector; few studies have explicitly and broadly focused on the dimensions of social issues and sustainability [17]. According to research by Moroke et al. [18] further research is necessary to observe the social sustainability dimensions of Pakistan. Therefore, this article focused on the social sustainability of supply chain management of the business sector. Figure 1 illustrates the conceptual model of the study.

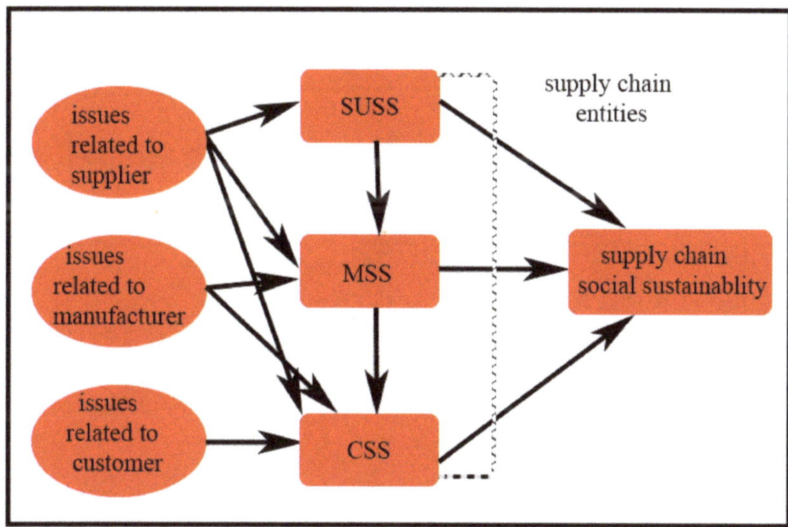

Figure 1. Conceptual model of this study. Note: SUSS: supplier social sustainability, MSS: manufacturer social sustainability, and CSS: customer social sustainability.

2.3. Social Sustainable Supply Chain Issues and Dimensions

Socially sustainable supply chain practices include managing social issues in all three stages of the supply chain. Social issues fall into two categories: basic issues (good for health and safety) and progress issues (reporting goods and practices) [19]. Ollila and Macy [20] suggested that improved organizational social issues should include simple basic questions: What do you say about social issues? Whom are you fighting for? Is it a mechanism? Our previous arguments refute the previous query needs for basic and progressive social issues. The response to the next query comes from the reorganization of complex individuals in the supply chain. Stakeholder perceptions interpret customers, manufacturers, suppliers, governments, and society as stakeholders in the industry and the right of a company's activities to affect their health and safety benefits. As such, stakeholders play a vital role in leveraging the company's social and sustainable management practices [21]. Analysis changes social issues at various stages of the supply chain and stakeholders [22]. For example, supplier social sustainability issues and dimensions, manufacturer social sustainability issues and dimensions, and customer social sustainability issues and dimensions are very powerful. Besides, these dimensions are directly related to society.

2.3.1. Supplier Social Sustainability

Sustainable supply chain management (SSCM)audits usually carefully consider suppliers, social issues, and dimensions, which always marks the complexity of individuals in supply chain management [23]. Parsa, Roper, Muller-Camen, and Szigetvari [24] used case studies to find improved labor rights and their impact on the supply chain from a supplier's perspective. In contrast, Sendlhofer and Lernborg [25] classified labor rights based on the perception of suppliers in South Asian countries. Others have demonstrated the organization's commitment to management, occupational safety, health management, wages, education and training, labor relations, worker welfare, research and development, ethics, and children and slavery and their role in sustainable supply chains [26]. Taking into account the above issues of suppliers, it can help to achieve the social sustainability of upstream supply chains [27].

Others believe that the implementation of social sustainability means many influential services in the upstream supply chain and their relationships with suppliers, jobs, and customer representatives. Merad et al. [28] validated the positive link between middle management and customer pressure and the sustainable development of society. However, their research did not establish a positive correlation between management pressure and the implementation of social sustainability. Similarly, Fortunati and O'Sullivan [29] found that imitation, regulation, and daunting pressures exist in the implementation of sustainable social development. Once a supporter of the principles of sustainable development, it is deeply entrenched that companies implement socially sustainable development policies. Further the supply chain social issues identified through different researcher are mentioned in Table 1.

Table 1. List of supply chain social issues identified through literature.

Social Issues	Suppliers	Manufacturer	Customer	Literature
Child labor and forced labor	X	X	X	[30,31]
Diversity	X	X	X	[32,33]
Discrimination	X	X	X	[34,35]
Health and Safety	X		X	[36,37]
Unethical practice	X	X	X	[38,39]
Philanthropy	X	X	X	[40,41]
Labor practices	X	X		[39,42]
Human rights	X	X		[43,44]
Wages	X	X		[45,46]
Education	X	X	X	[46]
Sustainable sourcing	X	X		[32,47]
Local sourcing	X	X		[48,49]
Product responsibility	X	X	X	[50,51]
Employee welfare	X	X	X	[32]
Employment creation		X		[20]
Poverty alleviation		X		[29]
Local economic development		X		[32,33]
Stakeholders engagement		X		[50,51]

2.3.2. Manufacturer's Social Sustainability

In addition, considering social issues at the manufacturing level strengthens the overall sustainability of key businesses. Manufacturers' social sustainability requires the management of social issues that theoretically plague individuals, workers, society, and customers. These issues are complex in manufacturing practice. Through the case study process, based on stakeholder awareness, Ikram et al. [52] identified organizations' management, occupational safety and health management, wages, labor rights, education and training, labor relations, children and slavery, organizational commitment, worker welfare, research and development, altruism, stakeholders, product liability, and social issues in the Pakistani Industry and their role in the SSSCM. It is also believed that true statements of sustainable development practices for stakeholders and consumers tend to take a gradual approach and expand consideration of the business sector [53,54].

The value of social sustainability promotes manufacturing management's responsibility to provide equal opportunities, positive scope, and refinement of natural life, which is spread throughout the community. Some opinions favor the development of appropriate education and training resources, fair policies, and worker welfare. Quality management in manufacturing can better manage social issues that deserve improvement. MARTIN [55] discussed 14 points on quality development, work environment, self-improvement plan, fearless job training, and fair income, which are essential to increase production value. Other practical studies have established a link between quality improvement, employee satisfaction, knowledge improvement, and inclusion programs. The organization's commitment to management, occupational safety and health management, and wages and labor rights is positively related to the strategic presentation of companies, which demonstrates what is observed among stakeholder's status and reputation. Subsequently, based on speculative support provided by stakeholders, Mani et al. [56] discussed child labor and forced labor practices, diversity, altruism, employee safety, welfare, and ethical issues and how to focus on solving these problems in your business. The convenience and other performance of social representatives are considered welfare.

2.3.3. Customer's Social Sustainability

Customer social sustainability points to social issues in the downstream indicators of the supply association, focusing primarily on the personal issues of customers and suppliers. Generally, it falls into two categories. The former contract is concluded by social issues that prevail at the sellers, which mark the health and safety benefits of the working class, which, in turn, affects trade administration and forms a concern for the entire supply chain [57]. The issue of the next contract to use manufactured items establishes the foundation for end-consumer health and safety issues and empowers the company's sustainable implementation. Among the shared actual customers, downstream associations play a leading role in their dominant position as they stimulate manufacturers and suppliers.

In previous work, customer-related social sustainability issues are broadly related to health and safety management, social issues, customer rights, and education [58]. These issues are gradually affecting executive representatives [59]. In a study of Pakistan (Atef et al.) [60] vendors considered various social issues, including health management, social issues, customer rights, education, corruption and bribery, job creation, ethical labels, and respect for customer privacy that affects SSSCM. Through previous work, the characteristics of social sustainability point to diversity, ethical issues, health and safety, human rights, and job creation. However, the semi-characteristics and events of companies are different, and their attitudes to these issues in business activities are also different [61]. Regardless of the increasing literature on suppliers, sustainable social development plays a role, but this information system is quietly developing and becoming a prerequisite for further consideration in emerging economies [62].

This theoretical background recognizes the need to study social sustainability standards in South Asian countries, especially Pakistani industries, from the perspectives of key companies, front-line traders and consumers. The different Dimensions of Social Sustainability in Supply Chain according to the different researcher point of view are presented in Table 2.

Table 2. Dimensions of social sustainability in the supply chain according to literature (scale items and measures for social sustainability) [63].

Measures	Items Measures	Source
Organizational commitment to management (suppliers and manufacturers)	OCM1	[64,65]
Occupational safety and health management (suppliers, manufacturers, and customers)	OCH1	[66]
Wages (suppliers, manufacturers, and customers)	WS1	[67]
Labor rights (suppliers and manufacturers)	LR1	[63]
Customer issues (customers)	CI1	[64]
Educational training (customers)	ET1	[68]
Labor relations (suppliers and manufacturers)	IR1	[69]
Stakeholder (manufacturers)	ST1	[70]
Child and bonded (suppliers and manufacturers)	CB1	[71]
Worker welfares (suppliers and manufacturers)	WW1	[32]
Research and development (suppliers and manufacturers)	RD1	[72]

This theoretical background recognizes the need to study social sustainability standards in South Asian countries, especially Pakistani industries, from the perspectives of key companies, front-line traders, and consumers.

3. Methodology

3.1. Measurement Development

In this study, we tested 41 projects on a pilot scale with 19 supply chain manufacturing managers. In developing and completing the questionnaire, we followed the generally accepted recommendations on wording issues [73]. Appendix A lists the measurement items and their related sources.

We used a five-point Likert scale (1 totally disagrees, 5 completely agrees). The Likert scale has been used in several sustainability measurement studies [74]. To ensure the content is valid, we conducted expert reviews to improve the tool. All construction projects were originally developed in English.

It is a Ph.D. student who is proficient in English who participated in the translation process. The original Urdu questionnaire was tried by some of our colleagues and online friends. Before being accepted as the final version, 40 useful responses were returned. Our model included several control variables to ensure that empirical results were not biased due to covariance between variables.

3.2. Survey Design

We designed a survey to test our research hypotheses and conduct it on the Pakistani industry. We chose the survey method because this quantitative research predicts behavior and examines the relationship between variables and construction. In line with cheap labor and the government's pursuit of creating an encouraging environment for manufacturers, Pakistan's sustainability assessment ranks six of the most ideal industrial goals. Therefore, the Pakistan Business Council in 2005 recognized various factors that promote industrial effectiveness and sustainability. To collect survey data, a pilot test was conducted for this study.

Our target group included 19 supply chain manufacturing managers and experts in facial expression effectiveness and readability, including general manager, assistant general manager, senior manager, CEO, and VP (sustainability), who participated in the bi-annual supply chain management IIMB Conference held in December 2018. In addition, the Pakistani Corporate and Regulatory Affairs and the Pakistan Security and Transaction Commission (SECP) have ordered fully registered organizations to observe and issue a Business Responsibility Report (BRR) horizontally through its business statement. Figure 2 shows the industries targeted in our survey and the percentage of participating companies in each industry. Appendix A lists the industries and all the related companies.

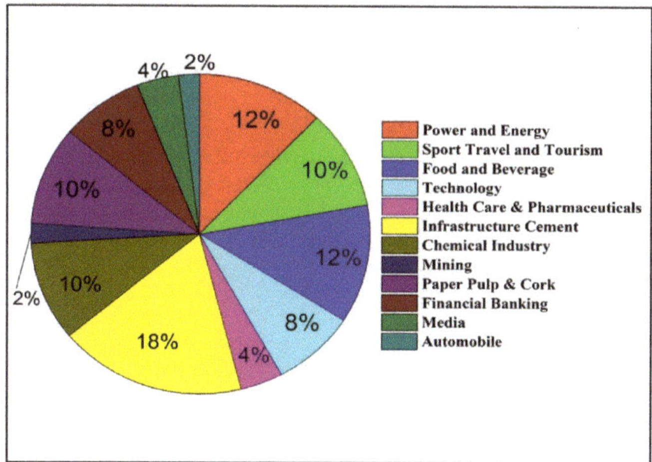

Figure 2. The share of industries targeted in our survey.

3.3. Data Collection

We conducted semi-structured interviews to accumulate data. Closed-ended questioning and semi-structured interviews helped to reach core rationality by confirming that a comparative assessment of the responses of all respondents was made. A pre-test was conducted to assess the authenticity of the interview etiquette, and then a pilot test was conducted with the supply supervisor. The semi-structured form was sent to 12 different industries, and its release address can be found on its official website link.

We invited an administrative representative to confirm that the questionnaire was filled by supply chain operations, procurement, logistics, and production supervisors. In the sample structure, all respondents who received pre-tests belonged to different industries (as described below) and were not sponsors of the main research. The corresponding supply chain executives representing various industries in Pakistan participated in preliminary tests to complete the semi-structured questionnaire. Managers were selected based on their experience; they should have at least five years of experience in the supply chain and sustainable development of society. The data collected was triangulated with supplementary data sources (i.e., print media, company reports, retired members of the company) to define the extent to which respondents in the trial structure were able to truly answer the questions asked. After making minor changes to the interview etiquette, we determined that the protocol (Appendix A) and trial structure were appropriate for this study.

The trial structure consisted of Pakistan's well-known supply chain directors. We sought to contribute to the most familiar and respected supervisors. Therefore, our trial structure involved special envoys or spokespersons who have been requested at the Pakistan Supply Chain Association (SCAP) meetings over the past few years. Respondents' data were collected by the conference leader. The 25 directors were contacted primarily based on the technical level, industry, and corporate income. Overall, 11 members answered preliminary appeals. The interview was scheduled in May 2019. The honorary member carried archived data and other credentials on sustainable social efforts in his supply chain.

3.4. Data Analysis

Data analysis was done by having each respondent mark their opinions within a given dimension. Each questionnaire was based on respondents' industry experience. To improve response times, an alternate email was sent (twice a week). Eventually, 503 responses were collected from 12 industries by ordinary post. From the 503, 412 (81.90%) were selected for post-evaluation, and 91 (18.10%) were rejected due to incomplete data or information. After each conversation, a comprehensive summary was organized, recording key facts identified by each participant. The applicant's true identity was clarified by phone. The interview was then translated; the meeting and archive information were carefully observed to discover the subject. To improve internal coherence, an interview with an expert in supplier sustainability education was conducted individually. Interviewers and other experts were recorded separately and interviewed. Analyzing all conversations could lead to trivial deviations. To improve external effectiveness, this approach was generalized to the research setting [75]. This research used contributors from various fields of the Pakistani industry. In addition, we introduced the members by translating interview information and conclusions in order to obtain a response on the representativeness and authenticity of the data.

4. Results and Discussion

4.1. Dimensions of Supplier Social Sustainability Analysis

We used the induction-deduction method in our analysis. Our analysis yielded a variety of topics related to the adoption of suppliers, focal companies, and customers. If these practices were cited in the report, they were cited as implementations for that particular company. First, various social sustainability topics were identified through a literature review. Based on the literature review, a social sustainability taxonomy for the supply chain was established. This helped us identify the keywords used to run the query. By comparing the sustainable supply chain management with socially sustainable supply chain management practices mentioned in the literature review and the report, we identified patterns and themes related to the practices adopted by Pakistani companies. At the same time, word frequency and potential coding techniques led to the identification of new topics that were not part of the listed taxonomy.

Job satisfaction describes various aspects of supplier social sustainability and related issues. Organizational commitment also has been broadly covered in management research. The members highlighted activities, such as those related to increasing personal productivity, increasing effort, and increasing job satisfaction, which are important features of supplier SSSCM. Other characteristics, such as reduced absenteeism and employment of workers in the organization, were grouped together and labeled under "organizational commitment to management" (OCM). The results in Table 3 indicated that the dimensions related to OCM were 2.4% high individual productivity, 35.3% increased endeavor, 28.7% higher job satisfaction, 13.5% less absenteeism, and 20.1% worker employment in an organization being the most significant. Asrar-ul-Haq et al. [76] stated the importance of job satisfaction and organizational commitment to supply chain sustainability. The issue of corporate responsibility for management is even more common in Pakistan's supply chain. Existing expansions are further straining supply-side supply chain management (tier 2, 3, etc.), as smaller companies tend to use organization-based ideals and rules to reduce absenteeism and worker employment.

Participants also discussed issues related to occupational health and safety management conditions, occupational hazard prevention, and the implementation of 28% safety and health roles (called "health and safety") in supplier workplaces. The safety and security of 28% of female employees were essential for the reporting of cumulative incidents via social media. Social issues related to security were also highlighted in 19.9% of the total responses related to occupational safety and health management (OSHM). Finally, issues related to ensuring compliance with regulations and the measurement of health and hygiene status were highlighted in 20.9% of the responses.

Some contributors have defined the importance of labor rights (13.3%) [77]. Forced labor and human trafficking (35.8%) are recorded in Pakistan. Another important dimension is child labor, i.e., children's engagement in hazardous work and debt bondage. Many executives have suggested banning child labor and restricted labor. One manager explained that the minimum age for employment was observed as 47.9%, child abuse was prohibited, and old-age benefits were paid recorded as 41.9%.

Table 3. Supplier social sustainable supply chain management dimensions.

Constructs	Items	Description	F	P	C
Organizational commitment to management	OCM1	High individual productivity	10	2.4	2.4
	OCM2	Increased endeavor	149	35.3	37.7
	OCM3	Higher job satisfaction	121	28.7	66.4
	OCM4	Less absenteeism	57	13.5	79.9
	OCM5	Worker employment in an organization	85	20.1	100
Occupational Safety and health management	OSH1	Occupational hazards prevention and implementation of safety and health roles	118	28	28.6
	OSH2	Maintaining the safety measurement for women at workplace	122	28.9	58.3
	OSH3	Ensuring that the company complies with the regulations	84	19.9	78.6
	OSH4	Measurement of health and hygiene conditions	88	20.9	100
Wages	W1	Reasonable wages paid to employees	157	37.2	38.1
	W2	No wages period should exceed one month	199	47.2	86.4
	W3	Not violates labor laws	56	13.3	100
Labor rights	LR1	Forced labor and human trafficking is illegal	151	35.8	36.7
	LR2	No child under the age of fourteen years engaged in any hazardous work	230	54.5	92.5
	LR3	Secure the well-being of people	31	3.7	100
Educational training	ET1	Encourage employees to become productive	174	41.2	42.2
	ET2	Reduce workplace injuries and accidents	207	49.1	92.5
	ET3	Produce a sense of responsibility	31	7.3	100
Industrial relations	IR1	Vast access to the whole seller	112	26.5	27.2
	IR2	Local and international exposure	145	34.4	62.4
	IR3	Right of the collective bargaining	62	14.7	77.4
	IR4	Right of the strike or power to go slow	93	22.2	100
Child and bonded	CB1	Follow the rule with the minimum age for employment	202	47.9	49
	CB2	Child abusing is prohibited	210	49.8	100
Worker welfare	WW1	Provision for the old-age benefits grant	177	41.9	43
	WW2	Social security for workers	156	37	80.8
	WW3	Compensation in case of sickness, maternity, injury, or death	79	18.7	100
Research and development	RD1	Adopt R&D culture in future growth prospect	210	49.8	51
	RD2	Valid through products and process innovation	202	47.9	100

Participants also described how to use "sweet shops or violate two or more labor laws". In practice, the labor force in small cities is more disintegrating, in violation of two or more labor laws, sweatshop labor, reasonable wages paid to employees, lower typical substandard working conditions, and wages provided. Managers stressed the wage period must not exceed one month, with minimal compensation to maintain workers and sustainability. The supply chain manager agreed.

Supply chain managers often considered the role of education and skills improvement. These pieces of training included health and safety, hygiene, achievement, and capacity development in new careers. Scholar Conaty and Robbins [78] emphasized the impact of reducing payroll injuries and accidents, creating a sense of responsibility, extensive contact with the entire seller, and workers' educational benefits on supplier performance and supply chain management practices. However, in South Asian countries, funds are still included in education accounts because it means that traders and suppliers can save more money. Here, suppliers are mainly required to invest in employers' health and safety hygiene practices training.

Applicants emphasized on worker welfare to improve social sustainability. Although issues, such as workers' social security and sickness, maternity leave, compensation for injuries or deaths, appear to be similar in South Asian countries and especially in Pakistan, suppliers in these countries differ in engaging in such activities. The suppliers have adopted a culture of R&D, which is anticipated to be 49.8% in future growth prospects. They have also effectively modernized the temple through product and process innovation, and the impact of donated schools and hospitals on elementary schools on social vendor representatives corresponds to 47.9%. Despite the analysis of humanitarian charities, a study has confirmed different philanthropic measures in South Asian countries.

4.2. Dimensions of Manufacturer Social Sustainability

Metrics related to manufacturers' social sustainability were not sufficient for an explicit assessment of the primary business and transient environment. Table 4 provides a list of the manufacturers' social dimensions of sustainability derived from the given data. Workers emphasized on social activities, such as adding local suppliers, buying from local suppliers, buying women-owned plans, supporting different communities in building hospitals, schools, and colleges, skilled training centers, conducting worker training, and seeking employment. Others discussed the importance of building health centers, hospitals, and health camps to improve social health. The importance of building public canters to promote social well-being and expand support for sustainable social agriculture was also discussed as a means to improve social and community sustainability. In addition, managers discussed the importance of establishing portable drinking water facilities in cities because too many workers cannot drink pure water. Workers with higher effective commitment were calculated to be 67.8%, which was very high, as shown in Table 4.

Table 4. Manufacturer social sustainable supply chain management dimensions.

Constructs	Items	Description	F	P	C
Organizational commitment to management	OCM1	Workers with higher effective commitment	286	67.8	70.1
	OCM2	Normative commitment involves a feeling of the moral obligation of workers	126	29.9	100
Occupational safety and health management	OSH1	Integrated affected mechanisms designed	160	37.9	38.8
	OSH2	Control the risk that may affect worker's health and safety	168	39.8	79.6
	OSH3	Define certain policies for women's safety at the workplace	84	19.9	100
Wages	W1	Reasonable wages paying to employees	328	77.7	79.6
	W2	Wages are to be paid within seven days after the end of the wage period	84	19.9	100
Labor rights	LR1	Ensuring appropriate labor working conditions	253	60	61.4
	LR2	Work rights are exceptional from cast, creed, or race	159	37.7	100
Educational training	ET1	For skill enhancement	127	30.1	30.8
	ET2	For development	126	29.9	61.4
	ET3	For accelerating productivity	117	27.7	89.8
	ET4	For high supervision	42	10	100
Industrial relations	IR1	There are critical factors in industrial relations unitary, pluralist, Marxist, and radical	160	37.9	38.8
	IR2	Based on a healthy relationship between employee and employer	252	59.7	100
Child and bonded	CB1	Child under the age of 15 is prohibited at workplace	244	57.8	59.2
	CB2	Employers are incurred not by the children themselves	168	39.8	100
Worker welfare	WW1	Housing (employer-provided or employer-paid)	85	20.1	20.6
	WW2	Furnished or not	84	19.9	41
	WW3	With or without free utilities	84	19.9	61.4
	WW4	Group insurance (health, dental, life, etc.)	126	29.9	92
	WW5	Disability income protection	33	7.8	100
Research and development	RD1	R&D regards the effort increases the pressure to perform in case of failure	286	67.8	69.4
	RD2	Plan developed under R&D more productive than the current plan	126	29.9	100
Altruism	A1	Increases another person's welfare, belief that the others are equally treated	143	33.9	34.7
	A2	Traditional virtue	126	29.9	65.3
	A3	Communities and supplier to behave ethically	143	33.9	100
Stakeholders	S1	Demand for sustainable measures to the society	96	22.7	23.3
	S2	Seek to understand strategically integrated issues	144	34.1	58.3
	S3	Impact the firm sustainability	124	29.4	88.3
	S4	Emerging economies	48	11.4	100

Another activity included disseminating employment opportunities for qualified young people to activists in response to activists' past complaints that major companies did not serve 29.9% of young people. Issues, such as health and hunger, adequate housing, and employment formation, have been explored in developing countries. For developing countries, other issues, such as helping society with sustainable agriculture, drinking water facilities, and establishing primary healthcare centers, are unique characteristics. More discussions about workers are "teaching for the professional development of workers" or "training for the efficiency of industrial institutions", and the main focus is on workers' sustainability education.

A large number of participants (37.9%) drew attention to health and safety procedures. The rest 39.8% advocated the ethical role of manufacturers in protecting contract workers. However, they were not entitled to permanent positions. In addition, some people mentioned the cleanliness of the organization, hoping to provide a healthy atmosphere for employees. For the benefit of cooperation, 19.9% of executives emphasized on women's health and safety issues. Our research found social issues related to the right to health and security in developing countries, especially Pakistan.

Supply chain managers emphasized ethical aspects that did not allow employees to participate, such as environmental pollution, coercion, bribery. Gender discrimination in employment, transfer, and promotion was also highlighted. This research provided organizations with a platform on which they can develop social sustainability by applying human methods, such as donating to cancer hospitals, religious organizations, NGOs, orphanages (sweet homes), schools refurbishment, and cultural heritage donations.

During the interview, many executives highlighted issues related to child labor and labor, as shown in Table 4. The social participation of children in labor activities is sad, which is very high (57.7%) in Pakistan. Finally, the issues that are most reflected at the enterprise level are often ensuring sanitation in the surrounding area, youth unemployment, the construction of primary healthcare centers, health awareness seminars, and drinking water facilities in public places. Companies may investigate to ignore local needs to identify major issues.

4.3. Dimensions of Customer Social Sustainability

Table 5 shows a list of dimensions related to the social sustainability of customers. Unsurprisingly, many customer-related issues were comparable to social issues related to suppliers and manufacturers. From the perspective of SSSCM, customers had mainly contributed to business-to-business (B2B) customers. However, some interviewees could plan carefully when dealing with end consumers. Based on our research data, preventing child-parent relationships and protecting human rights suggested key aspects of research. In developing countries, such as Pakistan, both issues support the entire system.

Respondents emphasized the importance of using non-toxic resources that could harm customers' health, and these were organized by H & S and recorded as just 37.5%. Respondents judged social issues by declaring responsive packaging, using appropriate product labels, non-toxic resources used in packaging, and maintaining customer H & S during product use recorded as 13%. They also discussed the social issues of building customer objection and response tools. Managers worried that healthcare protection for channel workers must be linked to supply chain management representatives. Gender diversity in the appointment and promotion of channel staff was also highlighted. Supply chain managers emphasized and discussed training to adapt workers to capacity growth and business development. Several managers' training of employees was full of attention to employee maintenance and sustainability.

Table 5. Social Sustainable dimensions of customers.

Constructs	Items	Description	F	P	C
Safety and health management	SHM1	The place must be hazards free for visiting customers	179	37.5	43.4
	SHM2	Risk prosecution	62	13.0	58.5
	SHM3	May lose customer tendency	171	35.8	100
	SHM4	Access to fresh drinking water, corruption, native violence, drug use	46	9.6	11.2
Social issue	SI1	Environmental contamination	180	37.7	54.9
	SI2	Inadequate emergency services	87	18.2	76.0
	SI3	Inequality, poverty, racism	99	20.8	100
Customer rights	CR1	The right information about the product	169	35.4	41
	CR2	Right to have accessibility for considering substitutes	90	18.9	62.9
	CR3	Defense from false and distorted rights in advertising labeling and observing	153	32.1	100
Customer issues	CI1	Social sustainable and supply chain management practices impact cooperation in their strategic and operational performance	170	35.6	41.3
	CS2	Customer's social sustainability is positively associated with local businesses' supply chain performance	242	50.7	100
Education	E1	Increase awareness	196	41.1	47.6
	E2	Emphasized for product moves	127	26.6	78.4
	E3	Awareness of product issues that cause safety and health issues	89	18.7	100

4.4. Discussion and Implications

Table 6 contains a description of each dimension and its relative frequency. These frequencies indicated the level of socially sustainable entities and their importance to corporate productivity. Mangers insisted that there was a proportional relationship between supplier performance and time management of goods (delivery time) with less turbulence. This would lead to a healthy and worry-free environment. An executive said that as social sustainability is adopted, all of the above factors would reduce work stress, minimize operational risks, and improve product quality and the ability to meet buyer needs.

Table 6. Dimension wise results and processes (supplier, manufacturer, and customer).

Construct	Items	Results and Related Process	F	P	C
Supplier social sustainability	SSS1	Organizational learning: improved collaboration between suppliers, manufacturers, and customers	190	40.5	46.1
	SSS2	Effective supply chain: manufacture value improvement and timely delivery to buyers	177	37.7	89.1
	SSS3	Supplier performance: capable against buyers, fewer fluctuations in supply	45	9.6	100.0
Manufacturer social sustainability	MSS1	Productivity: raised yield per worker, the adaptation of new technology	94	20.0	22.8
	MSS2	Corporate social demonstration: ethical business execution, productivity, trustworthy suppliers	96	20.5	46.1
	MSS3	Operational performance: competence, quality products, and consistency	222	47.3	100.0
Customer social sustainability	CSS1	Customer association and commitment: enhance attentiveness, improved communication skill	188	40.1	45.6
	CSS2	Customer performance: enlightened patience, increased sale, time management skill development	224	47.8	100.0

The social sustainability of the organization helped to leverage operational performance, quality, and reliability, thereby enhancing the company's social representation and promoting results. An executive added the company's social description. Considering all shareholders' research findings on customer social issues, the company's image was enhanced. New customer acquisition and loyalty related to social sustainability practices. This was a continuous process. The executive concluded that the method originally used in the study was to conduct a preliminary inspection of SSSCM and explore key businesses, front-line traders, and consumers by applying the social sustainability dimension. Secondly, social problems were attributed to the consequences of socially sustainable development and analyzed in terms of dimensions. Next, the aspects and results of social sustainability in different industries in Pakistan were investigated.

Compared to environmental and economic sustainability, our research provided the social characteristics of sustainability, which is not practiced at all [79]. Promoting discussions about social sustainability is a daunting task because it benefits the well-being and personal prestige of society. Even in terms of sustainability, emphasis should be placed on, for example, "organizational commitment to customer management, occupational safety and health management, customer comfort, wages, labor rights for ethical production, education with product accessibility, industrial relations, children and slavery, worker welfare, research, development, and contribution to society".

Considering the social sustainability of suppliers, the results were linked to [79], which establishes a bond and plays a role in the formal recognition and firm conviction between supplier sustainability measures mediating role. Nonetheless, researchers from [78] only focused on the realization of suppliers and purchasing functions, the process of social sustainability, and this research proposed a key business of social sustainability, the front-line trader, and consumer side. In addition, by applying ethical values, the ethical behavior exhibited by suppliers also recognizes the purpose of achieving company sustainability). We emphasized that moral values were related to the social sustainability of developing countries. This finding was in line with the results and the research direction of [78], but we improved their work by suggesting the implementation of social sustainability standards in developing countries. We, therefore, agreed to explore more discoveries about the social sustainability of developing countries.

Considering the social sustainability of manufacturers, most activities are mainly related to the business of the organization and the social responsibility of stakeholders and industry representatives. These findings lay the foundation for cooperation between social and environmental sustainability for the future and recognize the diversity and characteristics of SSSCM integration in developing countries and measures and help export and commercialize social representatives.

This study proposed numerous activities related to joint results and the social sustainability of clients. Our findings were related to [77], who was accustomed to building a link between corporate reputation and the sustainable performance of customer society. This research has brought new insights into the phenomenon of social sustainability and recommended a more comprehensive study of sustainable supply chain management, including suppliers, manufacturers, and customers.

5. Conclusions and Limitations

This study discussed many dimensions of SSSCM in Pakistan's manufacturing industries. These dimensions addressed the social issues in SSSCM and were distinctly different from advanced economies. The study raised many social questions about how companies can retain social assets, theoretically improve sustainability, and differentiate on a participant basis. In addition, the study revealed consequences in social expressions, such as sustainable implementation, how it mimics trade procedures. This study gave serious attention to the SSSCM literature by providing insights into various social issues and their scale, results, and experiments in developing countries. The ensuing social issues and scale of sustainable development were related to the industrial social supply chain and provided decision-making services to supply chain managers, who proposed to establish socially accessible supply chains in developing countries. In addition, the results and processes of socially sustainable supply chain management were also presented.

Nonetheless, the study had some limitations. We used data collected from several trade industries in Pakistan. Although the number of pilots was not too large, the demographics of the applicants (in terms of participation in the senior management industry, number of participants, and percentage of participants) were not very different from the source. This approach needs to improve overall social issues by selecting applicants in certain industries, and it must also be believed that, over time, treating positions diagonally must provide contributors with different altitudes and different groups of skill. However, further research should follow the example settings. We assume that Pakistan has representatives of several emerging countries.

However, conducting this research can test or develop our findings through assessments in other developing countries. Analysis can also explore the links between trade and social sustainability outcomes. In addition, improving the sympathetic relationship between environmental and social dimensions is an important area of research, but it has not responded much to past research. For future consideration and positive quantitative analysis, these findings can be used to examine the strength of the proposed multidimensional social sustainability model. For this reason, it is reasonable to conduct a preliminary study on the effect or importance of each standard size. Finally, results and processes related to the social sustainability dimension are reported. Advanced studies can be added to confirm these and their recommended associations.

Author Contributions: Conceptualization, M.K. and Y.H.; Methodology, M.K.; Software, W.I.; Validation, M.A. and A.F.; Investigation, W.I.; Resources, Y.H.; Data Curation, Y.H.; Writing-Original Draft Preparation, M.K.; Writing-Review & Editing, M.A., M.K.; Visualization, A.F.; Supervision, Y.H.; Project Administration, Y.H.; Funding Acquisition, Y.H. All authors have read and agreed to the published version of the manuscript.

Funding: We highly praise the honorable Professor Hou Yumei for her valuable guidance and great support in this research. This research was supported and funded by the project Joint Optimization of Omni-Channel Retailer Procurement and Pricing Considering Consumer Behavior (G2019203387) under the umbrella of Hebei Province Natural Science Foundation Project in 2019.

Acknowledgments: The author would like to thank the five anonymous reviewers for their helpful comments. This has greatly improved the performance of the paper.

Conflicts of Interest: The authors declare that there is no conflict of interest regarding the publication of this paper.

Interview Decorum: We thank all the interviewees for their valuable opinions (valuable opinions on the supply of goods (raw materials), manufacturing process, and customer behavior). We hope to conclude that there is no right or wrong statement. We have observed positive and negative responses and found useful feedback. We are trying to analyze the social sustainability processes in your organization's supply chain.

Appendix A

Table A1. Industries demographics and companies.

Power and Energy	Code	Sports, Travel, and Tourism	Code	Food and Beverages	Code	Technology	Code
PAEC (Pakistan Atomic Energy Commission)	A1	PKSF (Pakistan Kettlebell Sports Federation)	B1	Dalda Foods Pvt Ltd	C1	Suparco	D1
PEM (Pakistan Energy Mix)	A2	UNWTO (Developing Tourism Industry of Pakistan)	B2	Mair Foods	C2	Al and loT for Pakistan	D2
Ministry of Energy	A3	PTDC (Pakistan Tourism Development Corporation)	B3	Mitchell's Fruit Farms Limited	C3	Reditus	D3
EIA: Electricity	A4	Kashmir Pakistan Tourism	B4	Murree Brewery	C4	TECH Pakistan	D4
AEMC (Atomic Energy Medical Center)	A5	TAAP	B5	National Food	C5		
CREA (Center for Research on Energy and Clean Air)	A6			OMORE	C6		

Healthcare and Pharmaceuticals	Code	Infrastructure and Cement	Code	Chemical Industries	Code	Mining	Code
PPMA	E1	WEP	F1	Solvay	G1	Pakistan International Bulk Terminal Limited	H1
SEARLE	E2	Refhankaisha Hankuri Foundation	F2	CUF	G2		
		PCF	F3	ICL	G3		
		Jeevan Welfare Foundation	F4	ICI Pakistan	G4		
		Lucky Cement	F5	Descon	G5		
		Askari Cement Ltd.	F6				
		Bestway Cement Ltd.	F7				
		Attock Cement Pak Ltd.	F8				
		Kohat Cement Company Ltd.	F9				

Paper, Pulp, and Cork	Code	Financial, Banking, and Insurance	Code	Media	Code	Automobile	Code
FARAH INTERNATIONAL, Karachi	I1	NBP	J1	Dawn News	K1	Suzaki	L1
FARUKI PULP MILL LTD, Lahore	I2	MCB	J2	Geo News	K2		
ASIA CELL INTERNATIONAL, Karachi	I3	Bank Alfalah	J3				
IPO Pakistan	I4	Meezan Bank	J4				
PFVA	I5						

References

1. Angell, L.C.; Klassen, R.D. Integrating environmental issues into the mainstream: An agenda for research in operations management. *J. Oper. Manag.* **1999**, *17*, 575–598. [CrossRef]
2. Everard, M.; Longhurst, J.W.S. Reasserting the primacy of human needs to reclaim the 'lost half' of sustainable development. *Sci. Total. Environ.* **2018**, *621*, 1243–1254. [CrossRef] [PubMed]
3. Fedorova, E.; Pongrácz, E. Cumulative social effect assessment framework to evaluate the accumulation of social sustainability benefits of regional bioenergy value chains. *Renew. Energy* **2019**, *131*, 1073–1088. [CrossRef]
4. Solangi, Y.A.; Solangi, Z.A.; Aarain, S.; Abro, A.; Mallah, G.A.; Shah, A. Review on Natural Language Processing (NLP) and Its Toolkits for Opinion Mining and Sentiment Analysis. In Proceedings of the IEEE 5th International Conference on Engineering Technologies and Applied Sciences, Piscataway, NJ, USA, 22 September 2018. [CrossRef]
5. Cao, M.; Zhang, Q. Supply chain collaboration: Impact on collaborative advantage and firm performance. *J. Oper. Manag.* **2011**, *29*, 163–180. [CrossRef]

6. Huq, F.A.; Chowdhury, I.N.; Klassen, R.D. Social management capabilities of multinational buying firms and their emerging market suppliers: An exploratory study of the clothing industry. *J. Oper. Manag.* **2016**, *46*, 19–37. [CrossRef]

7. Brömer, J.; Brandenburg, M.; Gold, S. Transforming chemical supply chains toward sustainability—A practice-based view. *J. Clean. Prod.* **2019**, *236*, 117701. [CrossRef]

8. Moreno-Camacho, C.A.; Montoya-Torres, J.R.; Jaegler, A.; Gondran, N. Sustainability metrics for real case applications of the supply chain network design problem: A systematic literature review. *J. Clean. Prod.* **2019**, *231*, 600–618. [CrossRef]

9. Koberg, E.; Longoni, A. A systematic review of sustainable supply chain management in global supply chains. *J. Clean. Prod.* **2019**, *207*, 1084–1098. [CrossRef]

10. Rezaei, W.N.K.; Tavasszy, J.L.A.; de Brito, M.P. Commitment to and preparedness for sustainable supply chain management in the oil and gas industry. *J. Environ. Manag.* **2016**, *180*, 202–213. [CrossRef]

11. Yuen, K.F.; Wang, X.; Wong, Y.D.; Ma, F. A contingency view of the effects of sustainable shipping exploitation and exploration on business performance. *Transp. Policy* **2019**, *77*, 90–103. [CrossRef]

12. Totten, M.P. Flourishing Sustainably in the Anthropocene? Known Possibilities and Unknown Probabilities. *Ref. Modul. Earth Syst. Environ. Sci.* **2018**. [CrossRef]

13. Liebetruth, T. Sustainability in Performance Measurement and Management Systems for Supply Chains. *Procedia Eng.* **2017**, *192*, 539–544. [CrossRef]

14. Roy, M.; Roy, A. Nexus of Internet of Things (IoT) and Big Data: Roadmap for Smart Management Systems (SMgS). *IEEE Eng. Manag. Rev.* **2019**, *47*, 53–65. [CrossRef]

15. Iqbal, W.; Fatima, A.; Yumei, H.; Abbas, Q.; Iram, R. Oil supply risk and affecting parameters associated with oil supplementation and disruption. *J. Clean. Prod.* **2020**, *255*, 120187. [CrossRef]

16. Spence, L.; Rinaldi, L. Governmentality in accounting and accountability: A case study of embedding sustainability in a supply chain. *Account. Organ. Soc.* **2014**, *39*, 433–452. [CrossRef]

17. Nawaz, W.; Linke, P.; Koç, M. Safety and sustainability nexus: A review and appraisal. *J. Clean. Prod.* **2019**, *216*, 74–87. [CrossRef]

18. Moroke, T.; Schoeman, C.; Schoeman, I. Developing a neighbourhood sustainability assessment model: An approach to sustainable urban development. *Sustain. Cities Soc.* **2019**, 101433. [CrossRef]

19. Barbee, A.P.; Christensen, D.; Antle, B.; Wandersman, A.; Cahn, K. Successful adoption and implementation of a comprehensive casework practice model in a public child welfare agency: Application of the Getting to Outcomes (GTO) model. *Child. Youth Serv. Rev.* **2011**, *33*, 622–633. [CrossRef]

20. Ollila, J.; Macy, M. Social studies curriculum integration in elementary classrooms: A case study on a Pennsylvania Rural School. *J. Soc. Stud. Res.* **2019**, *43*, 33–45. [CrossRef]

21. Mani, V.; Gunasekaran, A.; Papadopoulos, T.; Hazen, B.; Dubey, R. Supply chain social sustainability for developing nations: Evidence from India. *Resour. Conserv. Recycl.* **2016**, *111*, 42–52. [CrossRef]

22. Gegg, P.; Wells, V. The development of seaweed-derived fuels in the UK: An analysis of stakeholder issues and public perceptions. *Energy Policy* **2019**, *133*, 110924. [CrossRef]

23. Jean, R.J.B.; Sinkovics, R.R.; Hiebaum, T.P. The effects of supplier involvement and knowledge protection on product innovation in customer-supplier relationships: A study of global automotive suppliers in China. *J. Prod. Innov. Manag.* **2014**, *31*, 98–113. [CrossRef]

24. Parsa, S.; Roper, I.; Muller-Camen, M.; Szigetvari, E. Have labour practices and human rights disclosures enhanced corporate accountability? The case of the GRI framework. *Account. Forum* **2018**, *42*, 47–64. [CrossRef]

25. Sendlhofer, T.; Lernborg, C.M. Labour rights training 2.0: The digitalisation of knowledge for workers in global supply chains. *J. Clean. Prod.* **2018**, *179*, 616–630. [CrossRef]

26. Awan, U.; Kraslawski, A.; Huiskonen, J. Understanding influential factors on implementing social sustainability practices in Manufacturing Firms: An interpretive structural modelling (ISM) analysis. *Procedia Manuf.* **2018**, *17*, 1039–1048. [CrossRef]

27. Hoejmose, S.U.; Roehrich, J.K.; Grosvold, J. Is doing more doing better? The relationship between responsible supply chain management and corporate reputation. *Ind. Mark. Manag.* **2014**, *43*, 77–90. [CrossRef]

28. Merad, M.; Dechy, N.; Marcel, F. A pragmatic way of achieving Highly Sustainable Organisation: Governance and organisational learning in action in the public French sector. *Saf. Sci.* **2014**, *69*, 18–28. [CrossRef]

29. Fortunati, L.; O'Sullivan, J. Situating the social sustainability of print media in a world of digital alternatives. *Telemat. Inform.* **2019**, *37*, 137–145. [CrossRef]
30. Lamprinopoulou, C.; Tregear, A. Issues in supply chain management. *Ind. Mark. Manag.* **2000**. [CrossRef]
31. Maloni, M.J.; Brown, M.E. Corporate social responsibility in the supply chain: An application in the food industry. *J. Bus. Ethics,* **2006**, *68*, 35–52. [CrossRef]
32. Seuring, S.; Müller, M. From a literature review to a conceptual framework for sustainable supply chain management. *J. Clean. Prod.* **2008**, *16*, 1699–1710. [CrossRef]
33. Pagh, J.D.; Cooper, M. Supply Chain Management: Implementation Issues and Research Opportunities. *Int. J. Logist. Manag.* **1998**, *9*, 1. [CrossRef]
34. Bowersox, D.J.; Closs, D.J.; Cooper, M.B.; Bowersox, J.C. *Supply Chain Logistics Management*; McGraw-Hill: New York, NY, USA, 2013.
35. Zailani, S.; Jeyaraman, K.; Vengadasan, G.; Premkumar, R. Sustainable supply chain management (SSCM) in Malaysia: A survey. *Int. J. Prod. Econ.* **2012**, *140*, 330–340. [CrossRef]
36. Rajeev, A.; Pati, R.K.; Padhi, S.S.; Govindan, K. Evolution of sustainability in supply chain management: A literature review. *J. Clean. Prod.* **2017**, *162*, 299–314. [CrossRef]
37. García-Herrero, L.; De Menna, F.; Vittuari, M. Sustainability concerns and practices in the chocolate life cycle: Integrating consumers' perceptions and experts' knowledge. *Sustain. Prod. Consum.* **2019**, *20*, 117–127. [CrossRef]
38. Seuring, S.; Sarkis, J.; Müller, M.; Rao, P. Sustainability and supply chain management—An introduction to the special issue. *J. Clean. Prod.* **2008**, 1545–1551. [CrossRef]
39. Melnyk, S.A.; Narasimhan, R.; DeCampos, H.A. Supply chain design: Issues, challenges, frameworks and solutions. *Int. J. Prod. Res.* **2014**. [CrossRef]
40. Govindan, K.; Soleimani, H.; Kannan, D. Reverse logistics and closed-loop supply chain: A comprehensive review to explore the future. *Eur. J. Operational Res.* **2015**. [CrossRef]
41. Garmendia, E.; Prellezo, R.; Murillas, A.; Escapa, M. Gallastegui, Performance measurement for green supply chain managemM. Weak and strong sustainability assessment in fisheries. *Ecol. Econ.* **2010**, *70*, 96–106. [CrossRef]
42. Schaltegger, S.; Burritt, R.; Beske, P.; Seuring, S. Putting sustainability into supply chain management. *Supply Chain Manag.* **2014**. [CrossRef]
43. Ashby, A.; Leat, M.; Hudson-Smith, M. Making connections: A review of supply chain management and sustainability literature. *Supply Chain Manag.* **2012**. [CrossRef]
44. Wu, L.; Chuang, C.H.; Hsu, C.H. Information sharing and collaborative behaviors in enabling supply chain performance: A social exchange perspective. *Int. J. Prod. Econ.* **2014**, *148*, 122–132. [CrossRef]
45. Pandey, R. Essentials of Supply Chain Management. *OPSEARCH* **2001**, 238–239. [CrossRef]
46. Koçoğlu, İ.; İmamoğlu, S.Z.; İnce, H.; Keskin, H. The effect of supply chain integration on information sharing: Enhancing the supply chain performance. *Procedia Soc. Behav. Sci.* **2011**, *24*, 1630–1649. [CrossRef]
47. Pagell, M.; Shevchenko, A. Why research in sustainable supply chain management should have no future. *J. Supply Chain Manag.* **2014**, *501*, 44–55. [CrossRef]
48. Lindgreen, A.; Swaen, V.; Maon, F.; Andersen, M.; Skjoett-Larsen, T. Corporate social responsibility in global supply chains. *Supply Chain Manag. An Int. J.* 2009. [CrossRef]
49. Locke, E.A. Social Foundations of Thought and Action: A Social-Cognitive View. *Acad. Manag. Rev.* **1987**, 169–171. [CrossRef]
50. Bourdon, J. Real Punks and Pretenders: The Social Organization of a Counterculture. *J. Contemp. Ethnogr.* **1984**, *22*, 531–556. [CrossRef]
51. Fan, W.; Yang, S.; Pei, J.; Luo, H. Building trust into cloud. *Int. J. Cloud Comput. Serv. Sci. J.* **2012**. [CrossRef]
52. Ikram, M.; Zhou, P.; Shah, S.A.A.; Liu, G.Q. Do environmental management systems help improve corporate sustainable development? Evidence from manufacturing companies in Pakistan. *J. Clean. Prod.* **2019**, *226*, 628–641. [CrossRef]
53. Tolkamp, J.; Huijben, J.C.C.M.; Mourik, R.M.; Verbong, G.P.J.; Bouwknegt, R. User-centred sustainable business model design: The case of energy efficiency services in the Netherlands. *J. Clean. Prod.* **2018**, *182*, 755–764. [CrossRef]

54. Iqbal, W.; Altalbe, A.; Fatima, A.; Ali, A.; Hou, Y. A DEA approach for assessing the energy, environmental and economic performance of top 20 industrial countries. *Processes* **2019**, *7*, 902. [CrossRef]

55. Martin, R.A. Applications of Humor in Psychotherapy, Education, and the Workplace. *Psychol. Humor* **2007**, 335–371. [CrossRef]

56. Mani, V.; Gunasekaran, A.; Delgado, C. Supply chain social sustainability: Standard adoption practices in Portuguese manufacturing firms. *Int. J. Prod. Econ.* **2018**, *198*, 149–164. [CrossRef]

57. Ahi, P.; Searcy, C.; Jaber, M.Y. A Quantitative Approach for Assessing Sustainability Performance of Corporations. *Ecol. Econ.* **2018**, *152*, 336–346. [CrossRef]

58. Zaidi, S.A.H.; Mirza, F.M.; Hou, F.; Ashraf, R.U. Addressing the sustainable development through sustainable procurement: What factors resist the implementation of sustainable procurement in Pakistan? *Socio-Econ. Plan. Sci.* **2019**, 100671. [CrossRef]

59. Qazi, U.; Jahanzaib, M. An integrated sectoral framework for the development of sustainable power sector in Pakistan. *Energy Rep.* **2018**, *4*, 376–392. [CrossRef]

60. Atef, S.S.; Sadeqinazhad, F.; Farjaad, F.; Amatya, D.M. Water conflict management and cooperation between Afghanistan and Pakistan. *J. Hydrol.* **2019**, *570*, 875–892. [CrossRef]

61. Penny, A.J.; Ahmed Ali, M.; Farah, I.; Östberg, S.; Smith, R.L. A study of cross-national collaborative research: Reflecting on experience in Pakistan. *Int. J. Educ. Dev.* **2000**, *20*, 443–455. [CrossRef]

62. Rajak, S.; Vinodh, S. Application of fuzzy logic for social sustainability performance evaluation: A case study of an Indian automotive component manufacturing organization. *J. Clean. Prod.* **2015**, *108*, 1184–1192. [CrossRef]

63. Pagell, M.; Wu, Z. Building a more complete theory of sustainable supply chain management using case studies of 10 exemplars. *J. Supply Chain Manag.* **2009**, *452*, 37–56. [CrossRef]

64. Sabbaghi, N.; Sabbaghi, O. Sustainable supply chain management. In *Practical Sustainability: From Grounded Theory to Emerging Strategies*; Palgrave Macmillan: New York, NY, USA, 2011.

65. Brown, A.M.; Boogaerdt, M.H. Accounting for suburban tree information systems. *Corp. Soc. Responsib. Environ. Manag.* **2006**, *13*, 275–285. [CrossRef]

66. Crum, M.; Poist, R.; Carter, C.R.; Easton, P.L. Sustainable supply chain management: Evolution and future directions. *Int. J.Phys. Distrib. Logist. Manag.* **2011**. [CrossRef]

67. Ahi, P.; Searcy, C.; Jaber, M.Y. Energy-related performance measures employed in sustainable supply chains: A bibliometric analysis. *Sustain. Prod. Consum.* **2016**, *7*, 1–15. [CrossRef]

68. Carter, C.R.; Rogers, D.S.J.I. A framework of sustainable supply chain management: Moving toward new theory. *Int. J. Phys. Distrib. and Logist. Manag.* **2008**, *38*, 360–387. [CrossRef]

69. Wu, Z.; Jia, F. Toward a theory of supply chain fields—Understanding the institutional process of supply chain localization. *J. Oper. Manag.* **2018**, *58–59*, 7–41. [CrossRef]

70. Ahmad, N.; Zhu, Y.; Shafait, Z.; Sahibzada, U.F.; Waheed, A. Critical barriers to brownfield redevelopment in developing countries: The case of Pakistan. *J. Clean. Prod.* **2019**, *212*, 1193–1209. [CrossRef]

71. Sandler, S.; Sonderman, K.; Citron, I.; Bhutta, M.; Meara, J.G. Forced Labor in Surgical and Healthcare Supply Chains. *J. Am. Coll. Surg.* **2018**, *227*, 618–623. [CrossRef]

72. Cambero, C.; Sowlati, T. Assessment and optimization of forest biomass supply chains from economic, social and environmental perspectives—A review of literature. *Renew. Sustain. Energy Rev.* **2014**, *36*, 62–73. [CrossRef]

73. Ray, M.L.; Heeler, R.M. Analysis Techniques for Exploratory Use of the Multitrait-Multimethod Matrix. *Educ. Psychol. Meas.* **1975**, *35*, 255–265. [CrossRef]

74. Zhu, Q.; Sarkis, J.; Lai, K.-H. Confirmation of a measurement model for green supply chain management practices implementation. *Int. J. Prod. Econ.* **2008**, *111*, 261–273. [CrossRef]

75. Rose-Ackerman, S. (Ed.) *International Handbook on the Economics of Corruption*; Edward Elgar Publishing: Cheltenham, UK, 2006.

76. Asrar-Ul-Haq, M.; Kuchinke, K.P.; Iqbal, A. The relationship between corporate social responsibility, job satisfaction, and organizational commitment: Case of Pakistani higher education. *J. Clean. Prod.* **2017**, *142*, 2352–2363. [CrossRef]

77. Hassan, S.A.; Zaman, K. RETRACTED: Effect of oil prices on trade balance: New insights into the cointegration relationship from Pakistan. *Econ. Model.* **2012**, *29*, 2125–2143. [CrossRef]
78. Conaty, F.; Robbins, G. A stakeholder salience perspective on performance and management control systems in non-profit organisations. *Crit. Perspect. Account.* **2018**, 102052. [CrossRef]
79. Zulfiqar, F.; Thapa, G.B. Agricultural sustainability assessment at provincial level in Pakistan. *Land Use Policy* **2017**, *68*, 492–502. [CrossRef]

 *processes*

Article

L-Ascorbic Acid and Thymoquinone Dual-Loaded Palmitoyl-Chitosan Nanoparticles: Improved Preparation Method, Encapsulation and Release Efficiency

Nurhanisah Othman [1], Siti Nurul Ain Md. Jamil [1,2,*], Mas Jaffri Masarudin [3,4], Luqman Chuah Abdullah [5], Rusli Daik [6] and Nor Syazwani Sarman [1]

[1] Department of Chemistry, Faculty of Science, Universiti Putra Malaysia, Serdang 43400, Selangor, Malaysia; hanisahlab@gmail.com (N.O.); wanisarman96@gmail.com (N.S.S.)

[2] Center of Foundation Studies for Agricultural Science, Universiti Putra Malaysia, Serdang 43400, Selangor, Malaysia

[3] Department of Cell and Molecular Biology, Faculty of Biotechnology and Biomolecular Sciences, Universiti Putra Malaysia, Serdang 43400, Selangor, Malaysia; masjaffri@upm.edu.my

[4] Cancer Research Laboratory, Institute of Biosciences, Universiti Putra Malaysia, Serdang 43400, Selangor, Malaysia

[5] Department of Chemical and Environmental Engineering, Faculty of Engineering, Universiti Putra Malaysia, Serdang 43400, Selangor, Malaysia; chuah@upm.edu.my

[6] Department of Chemical Sciences, Faculty of Science and Technology, Universiti Kebangsaan Malaysia, Bangi 43600, Selangor, Malaysia; rusli.daik@ukm.edu.my

* Correspondence: ctnurulain@upm.edu.my; Tel.: +60-12-356-6076

Received: 18 June 2020; Accepted: 3 August 2020; Published: 26 August 2020

Abstract: Encapsulation of dual compounds of different characters (hydrophilic and hydrophobic) in single nanoparticles carrier could reach the site of action more accurately with the synergistic effect but it is less investigated. In our previous findings, combined-compounds encapsulation and delivery from chitosan nanoparticles were impaired by the hydrophilicity of chitosan. Therefore, hydrophobic modification on chitosan with palmitic acid was conducted in this study to provide an amphiphilic environment for better encapsulation of antioxidants; hydrophobic thymoquinone (TQ) and hydrophilic L-ascorbic acid (LAA). Palmitoyl chitosan nanoparticles (PCNPs) co-loaded with TQ and LAA (PCNP-TQ-LAA) were synthesized via the ionic gelation method. Few characterizations were conducted involving nanosizer, Fourier-transform infrared spectroscopy (FTIR), field-emission scanning electron microscopy (FESEM) and high-resolution transmission electron microscopy (HRTEM). UV–VIS spectrophotometry was used to analyze the encapsulation and release efficiency of the compounds in PCNPs. Successfully modified PCNP-TQ-LAA had an average particle size of 247.7 ± 24.0 nm, polydispersity index (PDI) of 0.348 ± 0.043 and zeta potential of 19.60 ± 1.27 mV. Encapsulation efficiency of TQ and LAA in PCNP-TQ-LAA increased to 64.9 ± 5.3% and 90.0 ± 0%, respectively. TQ and LAA in PCNP-TQ-LAA system showed zero-order release kinetics, with a release percentage of 97.5% and 36.1%, respectively. Improved preparation method, encapsulation and release efficiency in this study are anticipated to be beneficial for polymeric nanocarrier development.

Keywords: chitosan; co-loaded nanoparticles; hydrophobic modification; L-ascorbic acid; thymoquinone

---

1. Introduction

Tremendous illnesse treatments like those of cancers [1–3] and tuberculosis (TB) [4–6] require a combination of drugs to achieve positive progress, whether to cure, control or palliate the pain.

Although the standard therapies have been successful to some extent, multi-drug transportation could be eased by formulating them together rather than taking them individually, without reducing their effectiveness. Researchers have started to look at multi-drug delivery from single system formulation for various treatments [7–15]. Aside from knowing drug compatibility, a system definitely needs an outstanding carrier to content and deliver them excellently.

Nanoparticles (NPs) come to fulfil this request by acting as a promising carrier due to its large surface area that causes it to be more reactive [11,16,17]. Having a size of 10–1000 nm [18–21], NPs are classified as potent drug carrier as it can move more freely in the body. NPs uptake by cells was reported to be 15–250 times higher than 1–10 μm particles [22]. Furthermore, NPs have been broadly used in the biomedical sector to treat diseases like diabetes [23–25] and cancers [26–29]. As NPs carrier is now drawing promising outcomes for multi-drug therapy, the present study aims at developing smart chitosan-based nanoparticles (CNPs) for combined antioxidants, hydrophobic thymoquinone (TQ) and hydrophilic L-ascorbic acid (LAA).

The use of chitosan (CS) as a polymer for NPs production has been widely discovered [13,30–32]. Abundant sources of CS from chitin of crustacean shells, fungi and insects makes it accessible and cheap [33–35]. It is one of the most auspicious polymers for drug delivery, which is biodegradable and biocompatible [33,36,37]. CS can be degraded in vertebrates by enzymatic reactions, depending on the degree of deacetylation (DD) and molecular weight (MW); low DD and MW could assist in faster CS degradation [34,38]. Moreover, the cationic property of CNPs enriched its efficacy in internalizing negatively charged cell plasma membranes [18,39]. CNPs can be synthesized by several routes, but the ionic gelation method has been more meticulously discovered, involving crosslinking reactions of CS amine groups with various anionic crosslinkers [36,40]. However, ionic gelation implementation in producing CNPs was typically done by using sodium tripolyphosphate (TPP) crosslinker [30,36,41,42].

In addition, the presence of amino and hydroxyl groups in CS makes it highly modifiable for moieties addition [43]. Palmitic acid is one of the potential modifiers for hydrophobic sites addition. Studies by Sharma et al. stated that the transfection efficiency of CS for gene delivery was successfully enhanced once modified with palmitic acid [44]. In another study, palmitic acid was used in modifying CS for an improved controlled release of tamoxifen, a breast cancer drug [45]. Kuen et al. reported on the modification of CS with palmitic acid for more effective encapsulation of hydrophobic Silibinin into NPs. The encapsulation efficiency (EE) of Sibilibin in modified palmiltoyl-chitosan nanoparticles (PCNP) was increased twice as much, compared to the EE of Silibinin in unmodified chitosan nanoparticles (CNP), because the palmitoyl anchor provides a hydrophobic core for Silibinin to stay in [41]. The concept of having both hydrophilic and hydrophobic sites in one NP system could be implemented to contain a wider range of compatible drugs or compounds to elevate synergistic effects. Therefore, this study aims at developing modified-chitosan nanoparticles by using palmitic acid *N*-hydroxysuccinimide (NHS) ester for hydrophobic thymoquinone (TQ) and hydrophilic L-ascorbic acid (LAA) antioxidants, as illustrated in Figure 1 below.

TQ is an active hydrophobic component of *Nigella sativa*, known as black seed, that shows several pharmacological properties and potential therapeutic effects such as anti-inflammatory and antioxidant properties [46,47]. LAA, known as vitamin C, is a universal antioxidant that possesses various benefits in preventing and reducing the common cold [48]. Its potential for alleviating acute viral infections treatments [49,50] was also reported. The combination of LAA and TQ was proven to have an anticonvulsant property by showing synergistic effects in lessening pentylenetetrazole-induced seizures [51]. As TQ and LAA are both antioxidants, this powerful combination can scavenge reactive oxygen species (ROS). ROS are unstable radicals containing oxygen that can easily react with molecules in a cell. Its inhibition is important to maintain good health because an excessive amount of ROS in the body could lead to oxidative stress. Prolonged oxidative stress state could then prime the emergence of complications such as neurological disorders, hypertension and acute respiratory distress syndrome [52]. Hence, by taking sufficient antioxidants in a more effective formulation, along with practicing healthy life style, numerous illnesses could be prevented and a more productive life awaits.

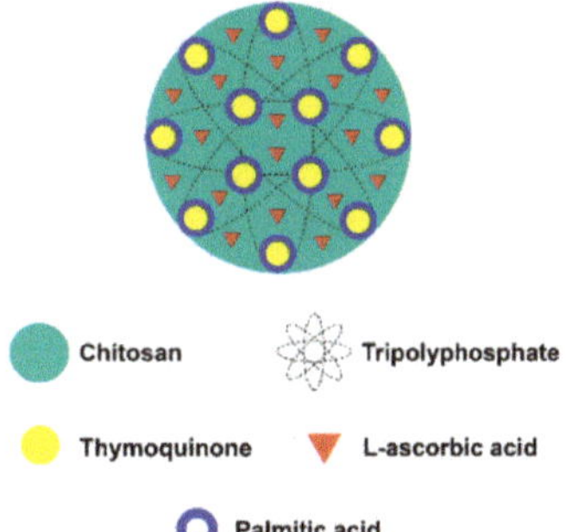

Figure 1. Proposed illustration of hydrophobically modified palmitoyl-chitosan nanoparticles encapsulated with hydrophobic thymoquinone and hydrophilic L-ascorbic acid, palmiltoyl-chitosan nanoparticles (PCNP)-thymoquinone (TQ)-L-ascorbic acid (LAA).

The present work demonstrates the synthesis and optimization of stable and homogenously dispersed palmitoyl-chitosan nanoparticles co-loaded with thymoquinone and L-ascorbic acid (PCNP-TQ-LAA). This study explores how the dual loaded compounds, hydrophobic TQ and hydrophilic LAA in a modified chitosan nanoparticles system is possible. Development and study of such systems are important, as a lot of treatment regimens involve a concoction of drugs/compounds of different physico–kinetic properties and dual administration of their combination are often plagued with many problems. This study attempts to be a proof-of-concept to pave way for finding a solution to this. As suggested in Figure 2, ammonium cations (from the chitosan chains) form ionic interaction with the phosphate anions (from the TPP crosslinker). Such interaction is possible because the positively charge chitosan can form an electrostatic attraction with negatively charge TPP. A similar interaction was suggested in previous research work [53]. Analysis on particle size, dispersion and zeta potential by the nanosizer, investigation on the presence of functional groups by Fourier-transform infrared spectroscopy (FTIR), surface morphologies by field-emission scanning electron microscopy (FESEM) and high-resolution transmission electron microscopy (HRTEM) together with encapsulation and release efficiency studies by UV-Vis spectrophotometry were conducted to characterize the NPs.

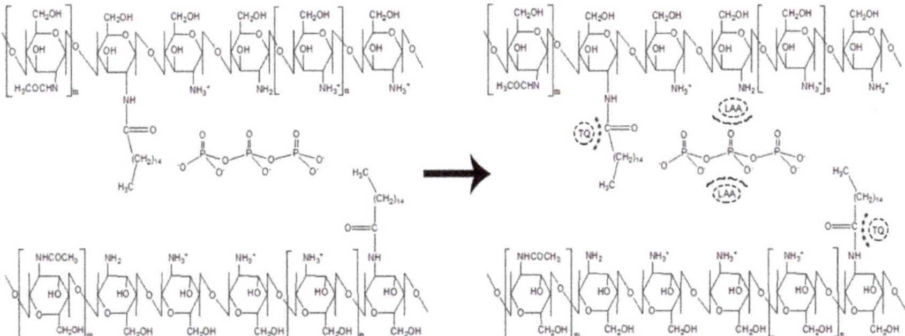

Figure 2. Proposed ionic interaction of pamiltoyl-chitosan nanoparticles, PCNP (**left**) and palmitoyl-chitosan nanoparticles encapsulated with thymoquinone and L-ascorbic acid, PCNP-TQ-LAA (**right**). "m" and "n" denote repetition of acetylated and deacetylated group of chitosan, respectively. (*) denotes hydrophobic–hydrophobic interaction between TQ and palmitic acid, while (~) denotes hydrophilic–hydrophilic interaction between L-ascorbic acid and tripolyphosphate.

2. Materials and Methods

2.1. Materials

CS with MW = 50,000–190,000 Da, ≥75% (deacetylated) was used as a carrier, and TPP with MW = 367.86 Da was used as the crosslinking agent. Both were purchased from Sigma-Aldrich (St. Louis, MO, USA). Glacial acetic acid, sodium hydroxide pallets, 5% w/v hydrochloric acid, L-ascorbic acid, dimethylsulfoxide (DMSO), sodium bicarbonate, sodium dodecyl sulfate and methanol were purchased from R&M Chemicals (Semenyih, Selangor, Malaysia). Thymoquinone, palmitic acid *N*-hydroxysuccinimide ester and picrylsulfonic acid solution were purchased from Sigma Aldrich (St. Louis, MO, USA). Absolute ethanol was purchased from Systerm. Phosphate Buffered Saline (PBS) pellet was purchased from MP Biomedicals (Solon, OH, USA). All chemicals are of analytical grade and were used without any further purification.

2.2. Synthesis of Palmitic Acid-Modified Chitosan, PCS

For the purpose of modifying chitosan to be more hydrophobic, palmitic acid NHS ester was added in the synthesis process. Chitosan (CS) powder was dissolved in 1.0% acetic acid and distilled water to form a concentration of 1.0 mg/mL CS master solution (CS MS). Separately, palmitic acid NHS ester was dissolved in absolute ethanol to a concentration of 0.36 mg/mL. Then, the CS solution was adjusted to pH 6. Palmitic acid NHS ester solution was added into the CS solution dropwise, with a volume ratio of 1:2 and the mixture was left for 20 h at 50 °C to react. After the incubation completed, the solution was adjusted to pH 9 by using 1 M sodium hydroxide. Then, it was centrifuged at 2200× *g* for 45 min to form a precipitate. Following precipitation, it was washed once with 50:50 acetone: absolute ethanol and twice with distilled water. The supernatant was removed from each centrifugation cycle. The precipitate was dried in an oven for 72 h at 50 °C. This process produced pellet, which is called palmitoyl-chitosan (PCS). The PCS pellet will be used to synthesize PCNP, PCNP-LAA, PCNP-TQ and PCNP-LAA-TQ.

2.3. Synthesis of Palmitic Acid-Modified Chitosan Nanoparticles, PCNP

Firstly, the PCS pellet was dissolved with 1.0% acetic acid and distilled water to a concentration of 1.0 mg/mL. It was diluted to two-fold to get 0.5 mg/mL PCS working solution. Then, the solution was adjusted to pH 5 by adding 1 M aqueous sodium hydroxide solution. The crosslinker, TPP was dissolved in distilled water to make a concentration of 0.7 mg/mL and altered to pH 2.0 by using 1.0 M hydrochloric acid. PCNP was formed by adding 600 µL of PCS to 250 µL of TPP. Then, the mixture was centrifuged at 13,000 rpm for 20 min to get purified PCNP (only for unencapsulated PCNP). Previously, it was found that 250 µL of TPP was an optimum volume to synthesize CNP and therefore, the same volume was used in this study to synthesize modified PCNP. That optimum TPP volume was mainly determined based on the smallest empty carrier produced [42]. This is because the expansion after encapsulation will be most minimal, which will enhance in vivo biological delivery.

2.4. Encapsulation of L-Ascorbic Acid and Thymoquinone into PCNP

For a single encapsulation process, LAA was prepared by first pouring approximately 10.0 mg of LAA into 10.0 mL of 0.7 mg/mL TPP solution, making an LAA stock concentration of 5.7 mM. Then, 250 µL of diluted LAA was added into 600 µL of 0.5 mg/mL PCS solution, to produce PCNP-LAA. Final LAA concentration that was used is 160 µM. Another single-loaded PCNP, PCNP-TQ, was prepared by dissolving approximately 2.0 mg of TQ in 2.0 mL of 99.0% DMSO, making a stock concentration of 6.1 mM. Only 100 µL of diluted TQ was added into 600 µL of 0.5 mg/mL of PCS. Lastly, 250 µL of 0.7 mg/mL of TPP was dropped into the PCS-TQ mixture to produce PCNP-TQ. Final TQ concentration that was used is 150 µM.

The encapsulation of both compounds, LAA and TQ, used to make PCNP-LAA-TQ is shown in Figure 1. Basically, the LAA and TQ solutions were prepared as mentioned previously. PCNP-TQ-LAA was prepared by mixing 100 µL of diluted TQ solution with 600 µL of 0.5 mg/mL of PCS for a while. Then, 250 µL of LAA-diluted solution was added into the mixture of TQ and PCS.

2.5. Quantification of Primary Amine Content in CNP and PCNP

Primary amine content determination was done by using a chemical called picrylsulfonic acid (TNBS reagent) according to modified methods [30,54]. The process started with preparation of the TNBS reagent, $NaHCO_3$, SDS and HCl at concentration of 0.05% (v/v), 0.1 M, 10.0% (w/v) and 1.0 M, respectively. First, standard solutions of CS were prepared by performing two-fold dilution of 0.5 mg/mL CS solution in 0.1 M $NaHCO_3$; each sample contained 50 µL of CS solution and 50 µL of $NaHCO_3$, serially diluted. This was followed by the addition of 50 µL of 0.05% (v/v) TNBS. Similar steps were applied to prepare standard solutions for PCS. Second, for sample solutions, 100 µL of CNP of different TPP volume (0 to 300 µL) were added into 100 µL of 0.05% (v/v) TNBS. All standards and sample solutions were incubated at 37 °C water bath for 3 h. Next, 100 µL of all standard and sample solutions were transferred into 96 well plates. This was followed by addition of 100 µL of 10.0% (w/v) sodium dodecyl sulfate and 75 µL of 1.0 M HCl into each occupied well and they were mixed well. Samples were then read with on a µQuant microplate reader (Bio-Tek Instruments, Winooski, VT, USA) at 335 nm. Absorbance values were analyzed by using Equation (2) to find the primary amine percentage available:

$$Percentage\ of\ available\ amines = \frac{\frac{Abs.\ of\ CNP}{Abs.\ of\ PCNP}}{\frac{Abs.\ of\ CS}{Abs.\ of\ PCS}} \times 100 \tag{1}$$

2.6. Physicochemical Characterizations of PCNP Samples

2.6.1. Detection of Functional Groups in PCNP Samples

Emergence and disappearance of functional groups in nanoparticle samples were detected by FTIR. Prior to analyzing samples by FTIR, they were freeze-dried using a Coolsafe 95-15 PRO freeze drier (ScanVac, Lynge, Denmark) for 48 h to pull out liquid content. FTIR was performed using a Perkin Elmer Spectrum 100 FTIR Spectrometer (Shelton, CT, USA) with a universal attenuated total reflectance (ATR) technique to identify the functional groups in the PCNPs. Samples were measured in the range of 280–4000 cm^{-1} at 25 °C.

2.6.2. Particle Size Distribution, PSD Study

PSD study was performed by using dynamic light scattering (DLS) technique to analyze particle size, dispersity in the sample and zeta potential. Prepared NPs were diluted with distilled water to produce 40% v/v solution. Then, the sample was analyzed by using Zetasizer 3000HSA (Malvern Instruments, Worcestershire, UK). pH of the system was around 6–6.5, near to pKa of chitosan. Three different synthesis batches (N = 3) were used in this study to obtain the average particle size, polydispersity index (PDI) and zeta potential.

2.6.3. Surface Morphology of PCNP Samples by Field-Emission Scanning Electron Microscopy (FESEM)

The surface morphologies of PCNP, PCNP-TQ, PCNP-LAA and PCNP-TQ-LAA were observed using FESEM analysis. Samples were prepared as for DLS study, which were then sonicated for five minutes. Small volumes of samples were dropped on cleaned stubs and left to dry for three days in a 50 °C oven. The dried samples were coated with platinum by using JEOL JEC-3000FC auto fine coater (Tokyo, Japan). Finally, the samples were analyzed under an electron microscope, JEOL JSM 7600F (Tokyo, Japan). The NP diameters and particle size distributions were calculated using Image J software by National Institutes of Health (version 1.52a) from the FESEM image analysis of 50 individual particles [55].

2.6.4. Surface Morphology of PCNP Samples by High-Resolution Transmission Electron Microscopy (HRTEM)

HRTEM was used to determine morphology and clarify functions of detected structures; also, to measure particle size and check its uniformity. Firstly, samples were prepared as for DLS study, followed by sonication for fifteen minutes. Then, a tiny droplet of each sample was applied onto a formvar/carbon film, 400 mesh copper TEM grids. After that, the samples were letft to dry for 7 h under a lamp, followed by the analysis process by JEOL JEM 2100F Field Emission TEM (Tokyo, Japan).

2.6.5. Encapsulation Efficiency Study of Thymoquinone and L-Ascorbic Acid into PCNP

The encapsulation efficiency (EE) was calculated by comparing the difference in absorbance of a total compound and free compound. Total compound refers to compound solution only, while free compound refers to the compound that is unencapsulated in PCNP-TQ, PCNP-LAA or PCNP-TQ-LAA. Both solutions contained the same concentration of the compound. The % EE indicates the percentage of compound successfully encapsulated in the carrier; it was calculated using the following formula [56]:

$$Encapsulation\ efficiency,\ \% = \frac{Abs.\ of\ total\ compound - Abs.\ of\ free\ compound}{Abs.\ of\ total\ compound} \times 100 \qquad (2)$$

The absorbance was measured using Lambda 35 UV–Vis spectrophotometer (PerkinElmer, Waltham, MA, USA) at wavelengths of 257 nm and 267 nm for LAA and TQ, respectively. The wavelengths were determined based on the highest peak detected. Triplicate test (N = 3) analysis of single and dual compounds loaded in PCNP were tested.

2.6.6. Preliminary Study of Thymoquinone and L-Ascorbic Acid In Vitro Release from PCNP

The in vitro release study of thymoquinone (TQ) and L-ascorbic acid (LAA) loaded PCNP formulations were conducted first by freeze-drying all samples using a Coolsafe 95-15 PRO freeze drier (ScanVac, Lynge, Denmark) for 48 h. Then, the pellet was loaded into a quartz cuvette containing phosphate-buffered saline (PBS) pH 7.4. The cuvette was later inserted into a UV–Vis spectrophotometer (PerkinElmer, Waltham, MA, USA) and set to wavelengths of 257 nm and 267 nm for LAA and TQ, respectively. The sample was left in the UV–Vis spectrophotometer for 48 h, with auto data recording set for every 1 min (time drive option).

A calibration curve for each compound is needed in order to convert the absorbance reading to the concentration of the compound released. After considering the dilution factor, the percentage of compound released (%) could then be measured using the following formula:

$$Compound\ released,\ \% = \frac{Released\ compound\ concentration}{Encapsulated\ compound\ concentration} \times 100 \qquad (3)$$

Obtained release profiles were analyzed with zero order, first order, Higuchi, Hixson–Crowell and Korsmeyer–Peppas models.

Zero-order kinetic model is as follows [57]:

$$m_t = m_b + k_0 t, \qquad (4)$$

where m_t is the amount of compound released at time t, m_b is the amount of compound in solution before release (usually 0) and k_0 is the zero-order rate constant.

First-order kinetic model is as follows [57]:

$$\ln(m_0 - m_t) = \ln(m_0) - k_1 t, \qquad (5)$$

where m_0 is the amount of compound in the formulation before the dissolution and k_1 is the first-order release rate constant.

Higuchi model is as follows [58]:

$$m_t = k_H \sqrt{t}, \tag{6}$$

where k_H is Higuchi rate constant.

Hixson–Crowell model is as follows [59]:

$$\sqrt[3]{m_0} - \sqrt[3]{m_{left}} = k_{H-C}t, \tag{7}$$

where m_{left} is the amount of compound left in the formulation over time t and k_{H-C} is Hixson–Crowell rate constant.

Korsmeyer–Peppas model is as follows [60]:

$$\log\left(\frac{m_t}{m_\infty}\right) = \log k_{K-P} + n \log t \tag{8}$$

where m_∞ is the amount of compound released after an infinitive time, k_{K-P} is Korsmeyer–Peppas rate constant and n is the parameter indicative of the release mechanism.

3. Results and Discussion

3.1. Quantification of Hydrophobic Palmitic Acid Functionalization on Chitosan by (TNBS) Assay

The amount of primary amine content in chitosan (CS) before and after modified with palmitic acid are shown in Table 1 and Figure 3. At optimum TPP volume of 250 μL, 25% of free amine content decreased from 88.35 ± 5.78% (CNP) to 62.99 ± 2.18% (PCNP). Decreased in the free primary amine groups of CS with TPP indicates increased in the crosslinking reactions between cationic amine groups of CS and anionic TPP. In other words, functionalization of the amine group of CS with hydrophobic palmitic acid NHS was a success and this outcome is similarly seen in previous studies [30,41]. This is essential to ensure adequate sites for hydrophobic thymoquinone encapsulation later. As shown in Figure 4, the palmitic acid only conjugated the amine sites of CS partially and that is crucial because some of the amine sites will be utilized by TPP crosslinker to interact during the PCNPs synthesis later [61].

Table 1. Percentage of free amine and significancy test (*t*-test) between chitosan nanoparticles (CNP) and PCNP at different sodium tripolyphosphate (TPP) volume. (a) significant **** with a *p* value of <0.0001, (b) significant *** with a *p* value of 0.0004, (c) significant **** with a *p* value of <0.0001, (d) significant **** with a *p* value of <0.0001, (e) significant **** with a *p* value of <0.0001, (f) significant *** with a *p* value of 0.0004, and (g) significant *** with a *p* value of 0.0004.

Label	Volume of TPP (μL)	Percentage of Free Amine (%)					*t*-Test
		CNP	Standard Deviation	PCNP	Standard Deviation	*p*-Value	Significance between CNP and PCNP Reading at Different TPP Volume
a	0	100.00	0	84.76	0.77	<0.0001	****
b	50	97.41	1.15	77.11	2.93	0.0004	***
c	100	94.05	0.94	74.04	0.79	<0.0001	****
d	150	93.47	1.59	71.61	1.78	<0.0001	****
e	200	92.54	2.16	64.72	2.14	<0.0001	****
f	250	84.87	2.81	62.99	2.18	0.0004	***
g	300	92.38	4.70	58.17	2.43	0.0004	***

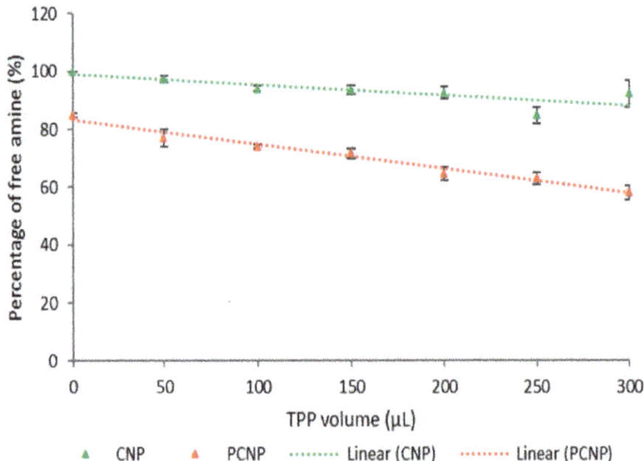

Figure 3. Percentage of free amine content in chitosan before (CNP) and after (PCNP) modification across a range of TPP volume. The free amine content in PCNP was lesser than in CNP. The loss of primary amine in CNP was due to the conjugation of palmitic acid during modification.

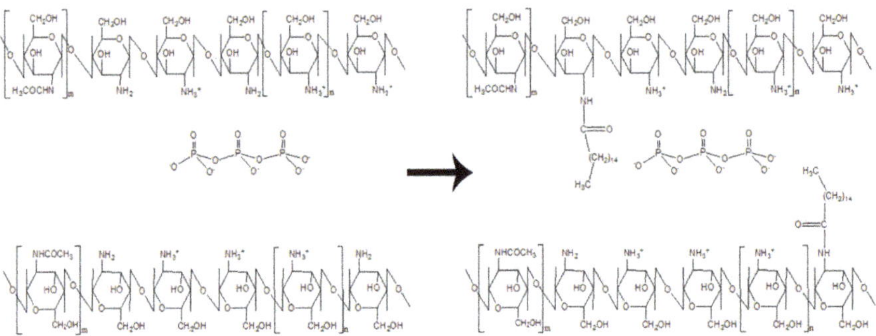

Figure 4. Conjugation of palmitic acid to amine group of chitosan. (**left**) Structure refers to unmodified CNP, and (**right**) structure refers to modified PCNP. "m" and "n" denote repetition of acetylated and deacetylated group of chitosan, respectively.

3.2. Functional Groups Determination by Fourier-Transform Infrared Spectroscopy (FTIR)

ATR-FTIR spectra of the NPs samples; PCNP, PCNP-TQ, PCNP-LAA and PCNP-TQ-LAA are shown in Figure 5 in the range of 4000–650 cm^{-1}. A number of moieties were recognized based on the peaks and the numerical values of transmitted beams are reported in Table 2. Based on Table 2, PCNP, PCNP-LAA and PCNP-TQ-LAA have the same set of peaks. The FTIR interpretation was done by referring to the Virtual Textbook of Organic Chemistry, Infrared Spectroscopy topic by William Reusch [62].

Table 2. Percent transmittance of functional groups existed in PCNP, PCNP-TQ, PCNP-LAA and PCNP-TQ-LAA.

Peak	Functional Group	Wavenumber (cm^{-1})	Percent Transmittance (%)			
			PCNP	PCNP-TQ	PCNP-LAA	PCNP-TQ-LAA
a	Alcohol OH (H bonded) and NH stretch (2° amine)	3399	7.47	4.15	3.17	6.18
b	Amine NH$_2$ scissor (1° amine), carboxylic acid C=O (amide) and alkene C=C stretch	1647	24.20	20.24	12.46	21.96
c	Amine NH$_2$ scissor (1° amine)	1563	26.91	-	10.89	18.99
d	Carboxylic acid NH (amide) bend	1542	-	62.19	-	-
e	Alkane CH$_2$ and CH$_3$ deformation and carboxylic acid C-O-H bend	1414	42.41	-	24.14	36.45
f	Alkane CH$_2$ and CH$_3$ deformation	1379	-	56.40	-	-
g	Amine C-N and carboxylic acid C-O stretch, P=O stretch	1106	3.19	9.01	8.52	7.02
h	Alkene =C-H and =CH$_2$ bend	903	12.77	28.42	26.88	18.32
i	Amine NH$_2$ and N-H wagging (shifts on H-bonding)	718	64.79	63.52	77.46	86.57

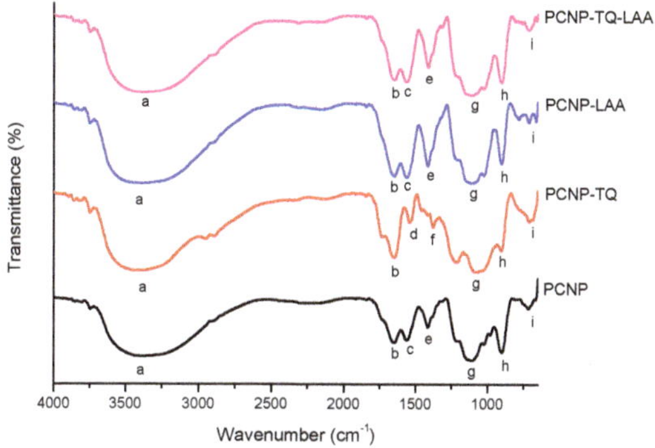

Figure 5. Fourier-transform infrared spectroscopy (FTIR) spectra of PCNP, PCNP-TQ, PCNP-LAA and PCNP-TQ-LAA. The labelled peaks that represent certain functional groups are stated in Table 2.

PCNP-TQ, on the other hand, has extra peak d (1542 cm^{-1}) and peak f (1379 cm^{-1}) that represent carboxylic acid NH (amide) bend and alkane CH$_2$ and CH$_3$ deformation, respectively. The emergence of these peaks could be supported by the chemical structure of TQ itself (refer to Supplementary Materials, Figure S1). TQ has many C=O components that might bind with the amine of palmitic acid-NHS and CS. This answers the emergence of peak d and indirectly proves that hydrophobic TQ had successfully interacted with hydrophobic palmitic acid-NHS, which enhanced its encapsulation. Peak f in PCNP-TQ, on the other hand, emerged to resemble CH$_2$ and CH$_3$ deformation in conjunction with TQ association with palmitic acid NHS.

Peak a at 3399 cm^{-1} appeared in all samples and it suggests an overlay of alcohol OH (H bonded) with secondary amine NH stretch. The strong and broad peak a was majorly contributed by OH (H bonded) stretch, while NH stretch was weak. For peak a, encapsulated PCNPs had stronger bands as compared to PCNP because PCNP has the least source of OH. PCNP-LAA showed the strongest band because LAA has the highest OH sites that could have bonded with the OH from TPP (refer to Supplementary Materials, Figure S1 for LAA chemical structure). On the other hand, NH contributed to the emergence of the peak a and CS played an important role in supplying NH for all PCNPs conditions.

Peak b at 1647 cm^{-1} appeared in all samples and they are primary amine, C=O amide and alkene C=C stretch. It also represents N–C=O group resulted in the addition of palmitic acid to chitosan during the hydrophobic modification. This N–C=O group was less significant in unmodified chitosan samples as reported by Othman et al., 2018 in Table 1 [42]. Therefore, this justifies the successful insertion of palmitic acid to chitosan. Besides, similar palmitoyl-chitosan nanoparticles have been reported to exhibit palmitic acid in this range [41,45].

3.3. Particle Size Distribution by Zetasizer

The particle size distribution of all samples (PCNP, PCNP-TQ, PCNP-LAA and PCNP-TQ-LAA) were analyzed by zetasizer. The size, polydispersity index (PDI) and zeta potential of particles were successfully determined as shown in Table 3 and Figure 6. Particle size resembles the size of nanoparticles, while PDI resembles homogeneity of particles distribution; lower PDI samples are made up of more uniform particles size and therefore, they are more monodisperse [30]. Zeta potential, on the other hand, indicates the surface charge of nanoparticles that develops at the particle–liquid interface (slipping plane).

Overall, PCNP had the smallest particle size average, 92.6 ± 8.6 nm because it has not been encapsulated with any antioxidant yet. Therefore, the crosslinked PCS with TPP resulted in the smallest particles, with a size less than 100 nm. PNCP also had the lowest PDI value, 0.277 ± 0.103; this indicates that it had the most even dispersion of particles as compared to other samples. Studies reported that incorporation of compounds or drugs into nanoparticles resulted in increased of particle size [45,63]. Likewise, in this study, the encapsulated nanoparticle samples, PCNP-TQ, PCNP-LAA and PCNP-TQ-LAA showed an increment in size. PCNP-LAA had a slightly larger particles, 165.8 ± 12.9 nm than PCNP-TQ, 158.3 ± 13.9 nm because LAA has a higher molecular weight, 176.12 g/mol, while TQ is 164.20 g/mol. Furthermore, the final concentration of LAA used in the PCNP-LAA and PCNP-TQ-LAA formulation was 160 μM; which is 10 μM more than TQ encapsulated in PCNP-TQ and PCNP-TQ-LAA. Additionally, TQ also has more double bonds as compared to LAA; hence, it may have a smaller radius as it is more compact (refer to Supplementary Materials, Figure S1). For PCNP-TQ-LAA, the average particle size was the biggest, 247.7 ± 24.0 nm, and it signifies that LAA and TQ had been efficaciously encapsulated. Besides that, the PDI value of PCNP-TQ-LAA was 0.348 ± 0.043 and it is lower than PCNP-TQ, 0.374 ± 0.052 and PCNP-LAA, 0.392 ± 0.087. This remarks that dual loaded PCNP had more unvarying particles size against single loaded PCNP-TQ and PCNP-LAA due to more complex crosslinking formation.

For zeta potential, the values among samples did not differ much. It can be concluded that the encapsulation of thymoquinone, TQ and L-ascorbic acid, LAA did not contribute to a significant change of the nanoparticle surface charge. Therefore, it proves that they did not conjugate to the nanoparticle, instead they were encapsulated. In addition, the zeta potentials of PCNPs were almost similar with reported studies, which were in the range of +20 mV to +30 mV [46,64,65]. Moreover, the positively charged PCNPs could provide better interaction with negatively charged mucosal membrane; this could later facilitate the PCNPs delivery and cellular uptake [46,66].

Table 3. Particle size, polydispersity index and zeta potential of all samples; PCNP, PCNP-TQ, PCNP-LAA and PCNP-TQ-LAA.

Sample	Particle Size (nm)	PDI	Zeta Potential (mV)
PCNP	92.6 ± 8.6	0.277 ± 0.103	22.35 ± 1.48
PCNP-TQ	158.3 ± 13.9	0.374 ± 0.052	19.45 ± 1.20
PCNP-LAA	165.8 ± 12.9	0.392 ± 0.087	20.60 ± 1.13
PCNP-TQ-LAA	247.7 ± 24.0	0.348 ± 0.043	19.60 ± 1.27

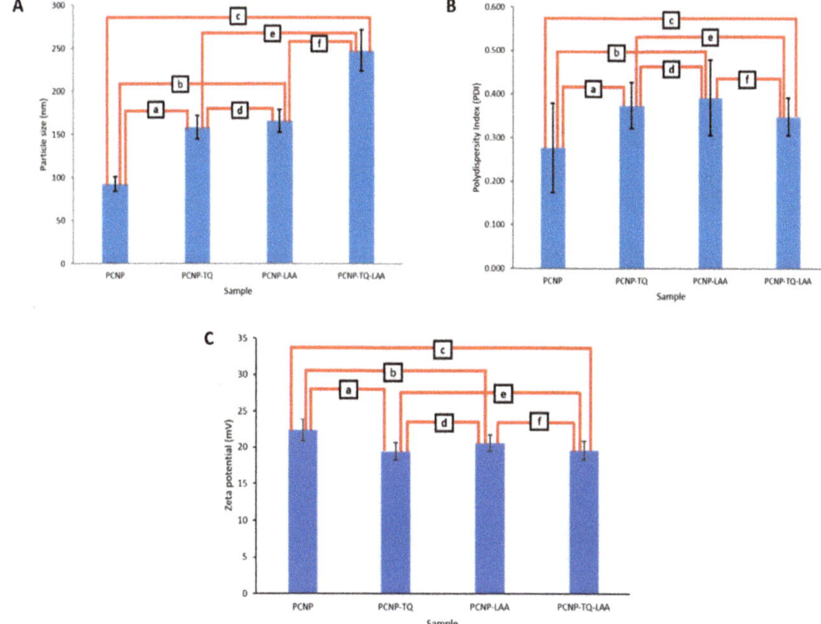

Figure 6. (**A**) Particle size of PCNP, PCNP-TQ, PCNP-LAA and PCNP-TQ-LAA, with *t*-test conducted to see the significance of particle size changes between samples. The confidence level was set to 95% and the results are labelled as a to f. (a) Significant ** with a *p*-value of 0.0022, (b) significant ** with a *p*-value of 0.0012, (c) significant *** with a *p*-value of 0.0005, (d) not significant with a *p*-value of 0.5320, (e) significant ** with a *p*-value of 0.0051, and (f) significant ** with a *p*-value of 0.0065. (**B**) Polydispersity index and significance test (*t*-test) of PCNP, PCNP-TQ, PCNP-LAA and PCNP-TQ-LAA. (a) Not significant with a *p*-value of 0.2168, (b) not significant with a *p*-value of 0.2129, (c) not significant with a *p*-value of 0.3279, (d) not significant with a *p*-value of 0.7820, (e) not significant with a *p*-value of 0.5437, and (f) not significant with a *p*-value of 0.4818. (**C**) Zeta potential and significance test (*t*-test) of PCNP, PCNP-TQ, PCNP-LAA and PCNP-TQ-LAA. The differences between all samples were insignificant with (a–f) *p*-values of 0.1649, 0.3161, 0.1851, 0.4284, 0.9146 and 0.4936, respectively.

3.4. Surface Morphologies of Nanoparticles

Morphologies of empty PCNP, single loaded PCNP-TQ and PCNP-LAA and dual loaded PCNP-TQ-LAA were performed by FESEM and HRTEM. FESEM images were taken at 50,000 magnification as shown in Figure 7A, while HRTEM images were taken at 10,000 magnification as shown in Figure 7C. All samples were performed at least thrice.

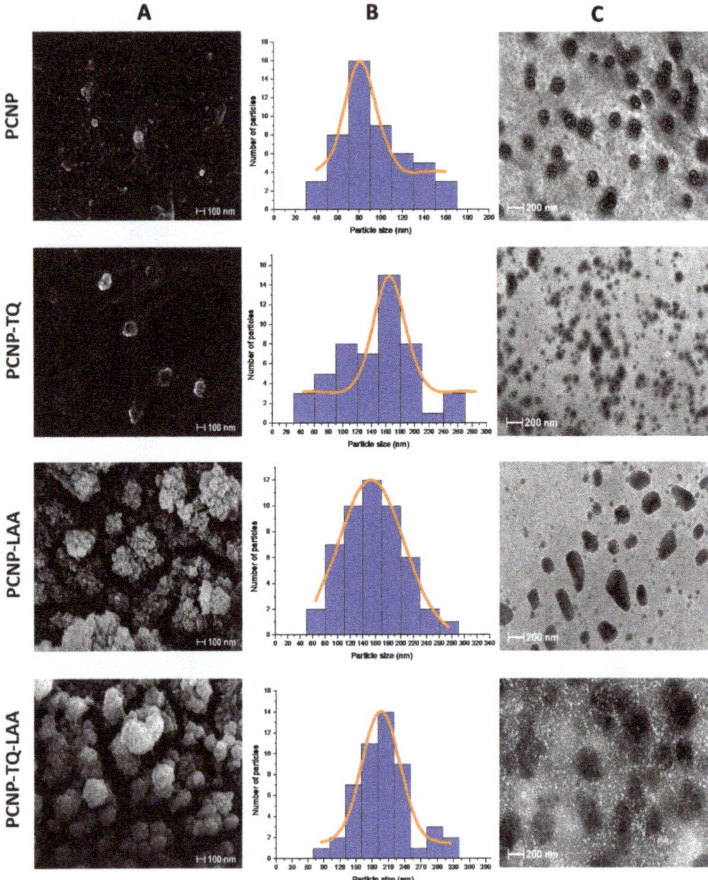

Figure 7. Field-emission scanning electron microscopy (FESEM) images (column **A**), normal distribution of particles based on FESEM images (column **B**) and high-resolution transmission electron microscopy (HRTEM) images (column **C**) of PCNP, PCNP-TQ, PCNP-LAA and PCNP-TQ-LAA samples (from top to bottom row). Particle sizes increased as the encapsulated number of antioxidant increased (from empty PCNP to PCNP-TQ-LAA).

Based on FESEM images Figure 7A, the shape of nanoparticles is spherical with uneven surface texture. As reported by a few researchers, the spherical shape of nanoparticles establishes more efficacious therapeutics delivery process by having large surface areas [11,16,17]. The formation of particles was clearly seen and the particle sizes were in range if compared with PSD study. Figure 7B shows a histogram and normal distribution curves of particle size based on the FESEM images. The bin range for PCNP is 20 while the other bin range is 30. The average particle size of PCNP was 81.49 ± 14.33 nm. PCNP had the smallest particle size average in comparison with PCNP-TQ, 165.32 ± 21.80 nm, PCNP-LAA, 153.50 ± 51.89 nm and PCNP-TQ-LAA, 195.90 ± 34.58 nm. For PCNP-TQ particle size distribution (refer to Figure 7B, PCNP-TQ), the second most measured particles have a size range of 90–110 nm. This clarifies that there are few particles unencapsulated and it matches with encapsulation efficiency percentage of TQ as stated in Table 4, which is only 41.3 ± 0.6%. On the other hand, PCNP-LAA has better particle dispersion as compared to PCNP-TQ because of the selectivity of chitosan that readily interacted with LAA which is of the same nature, hydrophilic [67].

HRTEM images, on the other hand, show layers of compounds by looking at the different color tones. Black particles represent the nanoparticles that have an average size that is almost the same as FESEM and PSD results. Besides, there are lots of tiny white pods in PCNP and PCNP-TQ which are palmitic acid micelles, the chemical used to modify chitosan to hydrophobic. The existence of palmitic acid in the system could be proven by looking at the FTIR interpretation. However, once PCNP was loaded with LAA, the white pods were less obvious and solid lines outlining those particles appeared to separate them from the surrounding. Meanwhile in PCNP-TQ-LAA, the image of particles became opaque, which indicated that they were fully occupied.

In addition, small white pods were also seen outside of the black particles. They are globular unreacted palmitic acid micelles. The amount of palmitic acid loaded might be excessive; therefore, they leached out from the NPs. It was proposed that the palmitic acid will attach to chitosan as an anchor on the surface of it; instead, it formed circular shapes. The palmitic acid anchor might be too long, which caused the formation of a circle as a result of joint ends.

3.5. Encapsulation Efficiency (EE)

This study was conducted to quantify how much of the loaded compound was successfully encapsulated. The concentration of TQ and LAA loaded in each sample PCNP-TQ, PCNP-LAA and PCNP-TQ-LAA were 150 µM and 160 µM, respectively. These optimized concentrations were used according to the previous study on unmodified chitosan nanoparticles done by Othman et al. (2018) in Section 3.5 [42]. According to Table 4 and Figure 8, LAA showed better EE as compared to TQ in both single and dual system PCNP with EE values of 73.0 ± 2.6% (116.8 µM) and 90.0 ± 0.0% (143.9 µM), respectively. These EE of LAA were much higher as compared to its EE before the modification by palmitic acid. By looking at Section 3.5, Table 4 by Othman et al., 2018, the EE of LAA in unmodified CNP-LAA and CNP-TQ-LAA were 69.3 ± 1.8% (110.9 µM) and 22.8 ± 3.2% (36.5 µM), respectively [42].

Table 4. The concentration of TQ and LAA loaded and encapsulated in PCNP samples.

Compound	Loaded Compound Concentration (µM)	EE (%)	Standard Deviation	Encapsulated Compound Concentration (µM)
TQ in PCNP-TQ	150	41.3	0.6	62.0
TQ in PCNP-TQ-LAA		64.9	5.3	97.3
LAA in PCNP-LAA	160	73.0	2.6	116.8
LAA in PCNP-TQ-LAA		90.0	0.0	143.9

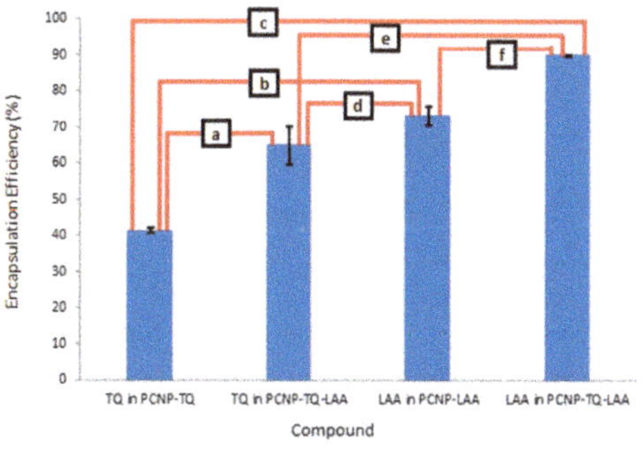

Figure 8. The percentage encapsulation of TQ in PCNP-TQ, TQ in PCNP-TQ-LAA, LAA in PCNP-LAA, and LAA in PCNP-TQ-LAA. A *t*-test was conducted to see the significance of the percentage encapsulation changes between samples. The confidence level was set to 95% and the results are labelled as a to f. (a) Significant ** with a *p*-value of 0.0097, (b) significant *** with a *p*-value of 0.0005, (c) significant **** with *p*-value of <0.0001, (d) not significant with a *p*-value of 0.0762, (e) significant ** with a *p*-value of 0.0080, and (f) significant ** with *p*-value of 0.0031.

On the other hand, only 41.3 ± 0.6% (62.0 µM) and 64.9 ± 5.3% (97.3 µM) of TQ were encapsulated in PCNP-TQ and PCNP-TQ-LAA, respectively. Lower EE of TQ compared with LAA in PCNPs was contributed by the hydrophobic nature of the compound itself, which requires stronger energy to interact with the main polymer, hydrophilic chitosan [33,46,68]. However, by having the palmitic acid modifier that holds TQ in place, the EE of TQ in the modified PCNPs were found to be higher compared to EE of TQ in the unmodified CNPs. As stated by Othman et al., 2018 in Section 3.5, Table 4, the EE of TQ in unmodified CNP-TQ and CNP-TQ-LAA were 68.7 ± 4.8% (103.1 µM) and 35.6 ± 3.6% (53.4 µM), respectively [42].

Following the modification, both TQ and LAA showed an increase in the EE for the dual loaded system. The EE of TQ in modified PCNP-TQ-LAA increased to about 29% compared to unmodified CNP-TQ-LAA. Meanwhile, the EE of LAA in modified PCNP-TQ-LAA increased to about 67% compared to unmodified CNP-TQ-LAA. This proves that the modification of chitosan by palmitic acid NHS was able to augment the EE of antioxidants. Same observations were seen in the increment of drugs EE when palmitic acid was used as a modifier in chitosan [30,41]. This clarifies that the palmitic acid is a highly potential modifier in polymeric nanocarrier for encapsulation enhancement of pharmaceuticals.

3.6. Preliminary Study of Thymoquinone and L-Ascorbic Acid In Vitro Release from PCNP

Release study was conducted to quantify how much of the encapsulated compound was successfully released from the PCNP carrier. The percentage and concentration of compounds released were included in Table 5. Release of TQ from the dual loaded system, PCNP-TQ-LAA was 37% more than the release of TQ from single loaded PCNP-TQ. LAA also showed higher release percentage from the dual loaded system as compared to a single loaded system, PCNP-LAA by around 19%. This trend indicates that when TQ and LAA were encapsulated together in a carrier, they could be released more efficiently.

Table 5. Release percentage and concentration of thymoquinone (TQ) and L-ascorbic acid (LAA) from PCNP samples.

Label	Release of	Released		
		Percentage (%)	**Concentration (µM)**	**Total Time (h)**
A	TQ from PCNP-TQ	60.6	37.6	43.9
B	TQ from PCNP-TQ-LAA	97.5	94.9	43.9
C	LAA from PCNP-LAA	17.0	19.8	33.7
D	LAA from PCNP-TQ-LAA	36.1	52.0	33.7

The zero-order, first-order, Higuchi, Hixson–Crowell and Korsmeyer–Peppas plots were made for each release sample (A, B, C and D as labelled in Table 5). Release kinetic of a sample was determined by selecting the highest correlation value, R^2 from the best fit line of all models. Figure 9 summarizes the best release kinetic model for all samples. Only sample C follows the Higuchi model, while the rest follow the zero-order model. The zero-order explains the constant release rate of a pharmaceutical product. It means TQ and LAA were able to be released controllably from the dual loaded system, PCNP-TQ-LAA. This indirectly tells us that the system could help people predict how long to expect a pharmaceutical product to work.

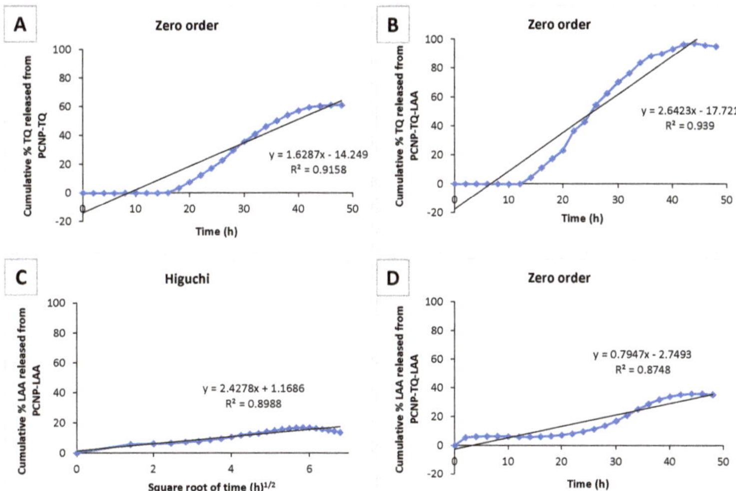

Figure 9. Release kinetics summary of TQ and LAA from all samples. (**A**) Release of TQ from PCNP-TQ, (**B**) release of TQ from PCNP-TQ-LAA, (**C**) release of LAA from PCNP-TQ-LAA and (**D**) release of LAA from PCNP-TQ-LAA. (**A,B,D**) follow zero-order release kinetics, while (**C**) follows the Higuchi release kinetic.

4. Conclusions

In this study, modification of biodegradable and biocompatible chitosan with palmitic acid NHS prior to nanoparticles formation had resulted in average particle size of around 250 nm for dual loaded thymoquinone and L-ascorbic acid palmitoyl-chitosan nanoparticles, PCNP-TQ-LAA. The robust and easy modification had also successfully increased the encapsulation efficiency of thymoquinone and L-ascorbic acid in the dual loaded system by about 29% and 67%, respectively; therefore, less loaded antioxidants were unencapsulated. This is supported by FTIR spectra of particular peaks to resemble related functional groups and also by morphology images using FESEM and HRTEM. It could also be proven that a 2:1 volume ratio of chitosan to palmitic acid NHS is remarkably appropriate to obtain partial hydrophobic sites in chitosan for more efficient encapsulation of antioxidants. The achievement of modifying this carrier with better capability of encapsulating dual classes hydrophilic and hydrophobic antioxidants has contributed to improvization in multi-drug therapy. As the dual loaded antioxidants TQ and LAA showed zero- order release kinetics, the polymeric nanoparticles carrier could potentially be reused with different sets of drugs for more effective delivery.

Supplementary Materials: The following are available online at http://www.mdpi.com/2227-9717/8/9/1040/s1, Figure S1: Chemical structure of thymoquinone, TQ (left) and L-ascorbic acid, LAA (right).

Author Contributions: Conceptualization and methodology: N.O., S.N.A.M.J. and M.J.M.; software and data curation: N.O. and N.S.S.; experimental work: N.O.; reagents/materials/analysis tools and validation: S.N.A.M.J., M.J.M., L.C.A. and R.D.; writing: N.O., S.N.A.M.J. and M.J.M. All authors have read and agreed to the published version of the manuscript.

Funding: This research was funded by Universiti Putra Malaysia, UPM-GP-IPS, project code UPM/700-2/1/GP-IPS/2017/9546900, with vote number 9546900. The APC was funded by Research Management Center of Universiti Putra Malaysia.

Acknowledgments: Thanks are due to Chemistry Department, Faculty of Science, Universiti Putra Malaysia, Cell and Molecular Biology Department, Faculty of Biotechnology and Biomolecular Sciences, Universiti Putra Malaysia, Chemical and Environmental Engineering Department, Faculty of Engineering, Universiti Putra Malaysia and Chemical Sciences Department, Faculty of Science and Technology, Universiti Kebangsaan Malaysia for providing the research facilities.

Conflicts of Interest: The authors declare no conflict of interest.

References

1. Chen, S.; Wang, Z.; Huang, Y.; Barr, S.A.O.; Wong, R.A.; Yeung, S.; Chow, M.S.S. Ginseng and Anticancer Drug Combination to Improve Cancer Chemotherapy: A Critical Review. *Hindawi* **2014**, *2014*. [CrossRef]
2. Sun, J.; Wei, Q.; Zhou, Y.; Wang, J.; Liu, Q.; Xu, H. A systematic analysis of FDA-approved anticancer drugs. *BMC Syst. Biol.* **2017**, *11*. [CrossRef]
3. Wong, C.C.; Cheng, K.; Rigas, B. Preclinical Predictors of Anticancer Drug Efficacy: Critical Assessment with Emphasis on Whether Nanomolar Potency Should Be Required of Candidate Agents. *J. Pharm. Exp. Ther.* **2012**, *341*, 572–578. [CrossRef]
4. Xu, Y.; Wu, J.; Liao, S.; Sun, Z. Treating tuberculosis with high doses of anti-TB drugs: Mechanisms and outcomes. *Ann. Clin. Microbiol. Antimicrob.* **2017**, *16*, 1–13. [CrossRef] [PubMed]
5. Zhu, C.; Liu, Y.; Hu, L.; Yang, X.M.; He, Z. Molecular mechanism of the synergistic activity of ethambutol and isoniazid against Mycobacterium tuberculosis. *J. Biol. Chem.* **2018**, *293*, 16741–16750. [CrossRef]
6. Rey-jurado, E.; Tudó, G.; Antonio, J.; González-martín, J. Synergistic effect of two combinations of antituberculous drugs against Mycobacterium tuberculosis. *Tuberculosis* **2012**, *92*, 260–263. [CrossRef] [PubMed]
7. Liu, Q.; Zhang, J.; Sun, W.; Xie, Q.R.; Xia, W.; Gu, H. Delivering hydrophilic and hydrophobic chemotherapeutics simultaneously by magnetic mesoporous silica nanoparticles to inhibit cancer cells. *Int. J. Nanomed.* **2012**, *7*, 999–1013.
8. Ma, Y.; Liu, D.; Wang, D.; Wang, Y.; Fu, Q.; Fallon, J.K.; Yang, X.; He, Z.; Liu, F. Combinational delivery of hydrophobic and hydrophilic anticancer drugs in single nanoemulsions to treat MDR in cancer. *Mol. Pharm.* **2014**, *11*, 2623–2630. [CrossRef] [PubMed]
9. Jingou, J.; Shilei, H.; Weiqi, L.; Danjun, W.; Tengfei, W.; Yi, X. Preparation, characterization of hydrophilic and hydrophobic drug in combine loaded chitosan/cyclodextrin nanoparticles and in vitro release study. *Colloids Surf. B Biointerfaces* **2011**, *83*, 103–107. [CrossRef] [PubMed]
10. Naderinezhad, S.; Amoabediny, G.; Haghiralsadat, F. Co-delivery of hydrophilic and hydrophobic anticancer drugs using biocompatible pH-sensitive lipid-based nano-carriers for multidrug-resistant cancers. *RSC Adv.* **2017**, *7*, 30008–30019. [CrossRef]
11. Hao, S.; Wang, B.; Wang, Y. Porous hydrophilic core/hydrophobic shell nanoparticles for particle size and drug release control. *Mater. Sci. Eng. C* **2015**, *49*, 51–57. [CrossRef] [PubMed]
12. Kaur, G.; Mehta, S.K.; Kumar, S.; Bhanjana, G.; Dilbaghi, N. Coencapsulation of Hydrophobic and Hydrophilic Antituberculosis Drugs in Synergistic Brij 96 Microemulsions: A Biophysical Characterization. *J. Pharm. Sci.* **2015**, *104*, 2203–2212. [CrossRef] [PubMed]
13. Calvo, P.; Remunan-Lopez, C.; Vila-Jato, J.L.; Alonso, M.J. Novel hydrophilic chitosan-polyethylene oxide nanoparticles as protein carriers. *J. Appl. Polym. Sci.* **1997**, *63*, 125–132. [CrossRef]
14. Dini, E.; Alexandridou, S.; Kiparissides, C. Synthesis and characterization of cross-linked chitosan microspheres for drug delivery applications. *J. Microencapsul.* **2003**, *20*, 375–385. [CrossRef]
15. Liu, H.; He, J. Simultaneous release of hydrophilic and hydrophobic drugs from modified chitosan nanoparticles. *Mater. Lett.* **2015**, *161*, 415–418. [CrossRef]
16. Jamil, A.; Lim, H.N.; Yusof, N.A.; Ahmad Tajudin, A.; Huang, N.M.; Pandikumar, A.; Moradi Golsheikh, A.; Lee, Y.H.; Andou, Y. Preparation and characterization of silver nanoparticles-reduced graphene oxide on ITO for immunosensing platform. *Sens. Actuators B Chem.* **2015**, *221*, 1423–1432. [CrossRef]
17. Vijayan, V.; Reddy, K.R.; Sakthivel, S.; Swetha, C. Optimization and charaterization of repaglinide biodegradable polymeric nanoparticle loaded transdermal patchs: In vitro and in vivo studies. *Colloids Surf. B Biointerfaces* **2013**, *111*, 150–155. [CrossRef] [PubMed]
18. Biswas, A.K.; Islam, R.; Choudhury, Z.S.; Mostafa, A.; Kadir, M.F. Nanotechnology based approaches in cancer therapeutics. *Adv. Nat. Sci. Nanosci. Nanotechnol.* **2014**, *5*. [CrossRef]
19. Rizvi, S.A.A.; Saleh, A.M. Applications of nanoparticle systems in drug delivery technology. *Saudi Pharm. J.* **2018**, *26*, 64–70. [CrossRef]
20. Jeevanandam, J.; Barhoum, A.; Chan, Y.S.; Dufresne, A.; Danquah, M.K. Review on nanoparticles and nanostructured materials: History, sources, toxicity and regulations. *Beilstein J. Nanotechnol.* **2018**, *9*, 1050–1074. [CrossRef]

21. Mudshinge, S.R.; Deore, A.B.; Patil, S.; Bhalgat, C.M. Nanoparticles: Emerging carriers for drug delivery. *Saudi Pharm. J.* **2011**, *19*, 129–141. [CrossRef] [PubMed]

22. Emeje, M.O.; Obidike, I.C.; Akpabio, E.I.; Ofoefule, S.I. *Nanotechnology in Drug Delivery*; IntechOpen: London, UK, 2012.

23. Ahad, A.; Al-Saleh, A.A.; Akhtar, N.; Al-Mohizea, A.M.; Al-Jenoobi, F.I. Transdermal delivery of antidiabetic drugs: Formulation and delivery strategies. *Drug Discov. Today* **2015**, *20*, 1217–1227. [CrossRef] [PubMed]

24. Sharma, G.; Sharma, A.R.; Nam, J.S.; Doss, G.P.C.; Lee, S.S.; Chakraborty, C. Nanoparticle based insulin delivery system: The next generation efficient therapy for Type 1 diabetes. *J. Nanobiotechnol.* **2015**, *13*, 1–13. [CrossRef] [PubMed]

25. Wong, C.Y.; Al-Salami, H.; Dass, C.R. Potential of insulin nanoparticle formulations for oral delivery and diabetes treatment. *J. Control. Release* **2017**, *264*, 247–275. [CrossRef] [PubMed]

26. Nayak, D.; Minz, A.P.; Ashe, S.; Rauta, P.R.; Kumari, M.; Chopra, P.; Nayak, B. Synergistic combination of antioxidants, silver nanoparticles and chitosan in a nanoparticle based formulation: Characterization and cytotoxic effect on MCF-7 breast cancer cell lines. *J. Colloid Interface Sci.* **2016**, *470*, 142–152. [CrossRef] [PubMed]

27. Bahrami, B.; Hojjat-farsangi, M.; Mohammadi, H.; Anvari, E. Nanoparticles and targeted drug delivery in cancer therapy. *Immunol. Lett.* **2017**, *190*, 64–83. [CrossRef]

28. Grobmyer, S.R.; Zhou, G.; Gutwein, L.G.; Iwakuma, N.; Sharma, P.; Hochwald, S.N. Nanoparticle delivery for metastatic breast cancer. *Nanomed. Nanotechnol. Biol. Med.* **2012**, *8*, S21–S30. [CrossRef]

29. Li, C.; Kuo, T.; Su, H.; Lai, W.; Yang, P.-C.; Chen, J.-S.; Wang, D.-Y.; Wu, Y.C.; Chen, C.-C. Fluorescence-Guided Probes of Aptamer-Targeted Gold Nanoparticles with Computed Tomography Imaging Accesses for in Vivo Tumor Resection. *Sci. Rep.* **2015**, *5*. [CrossRef] [PubMed]

30. Masarudin, M.J.; Cutts, S.M.; Evison, B.J.; Phillips, D.R.; Pigram, P.J. Factors determining the stability, size distribution, and cellular accumulation of small, monodisperse chitosan nanoparticles as candidate vectors for anticancer drug delivery: Application to the passive encapsulation of [14C]-doxorubicin. *Nanotechnol. Sci. Appl.* **2015**, *8*, 67–80. [CrossRef]

31. Zhao, L.M.; Shi, L.E.; Zhang, Z.L.; Chen, J.M.; Shi, D.D.; Yang, J.; Tang, Z.X. Preparation and application of chitosan nanoparticles and nanofibers. *Braz. J. Chem. Eng.* **2011**, *28*, 353–362. [CrossRef]

32. Elgadir, M.A.; Uddin, S.; Ferdosh, S.; Adam, A.; Jalal, A.; Chowdhury, K.; Islam, Z. Impact of chitosan composites and chitosan nanoparticle composites on various drug delivery systems: A review. *J. Food Drug Anal.* **2014**, *23*, 619–629. [CrossRef]

33. Szymanska, E.; Winnicka, K. Stability of Chitosan—A Challange for Pharmaceutical and Biomedical Applications. *Mar. Drugs* **2015**, *13*, 1819–1846. [CrossRef] [PubMed]

34. Dash, M.; Chiellini, F.; Ottenbrite, R.M.; Chiellini, E. Chitosan—A versatile semi-synthetic polymer in biomedical applications. *Prog. Polym. Sci.* **2011**, *36*, 981–1014. [CrossRef]

35. Ifuku, S. Chitin and Chitosan Nanofibers: Preparation and Chemical Modifications. *Molecules* **2014**, *19*, 18367–18380. [CrossRef] [PubMed]

36. Sreekumar, S.; Goycoolea, F.M.; Moerschbacher, B.M.; Rivera-Rodriguez, G.R. Parameters influencing the size of chitosan-TPP nano- and microparticles. *Sci. Rep.* **2018**, *8*, 1–11. [CrossRef]

37. Desai, P.; Patlolla, R.R.; Singh, M. Interaction of nanoparticles and cell-penetrating peptides with skin for transdermal drug delivery. *Mol. Membr. Biol.* **2010**, *27*, 247–259. [CrossRef] [PubMed]

38. Kofuji, K.; Qian, C.; Nishimura, M.; Sugiyama, I.; Murata, Y.; Kawashima, S. Relationship between physicochemical characteristics and functional properties of chitosan. *Eur. Polym. J.* **2005**, *41*, 2784–2791. [CrossRef]

39. Banquy, X.; Suarez, F.; Argaw, A.; Rabanel, J.; Grutter, P.; Giasson, S. Effect of mechanical properties of hydrogel nanoparticles on macrophage cell uptake. *Soft Matter* **2009**, *5*, 3984–3991. [CrossRef]

40. Mahapatro, A.; Singh, D.K. Biodegradable nanoparticles are excellent vehicle for site directed in-vivo delivery of drugs and vaccines. *J. Nanobiotechnol.* **2011**, *9*, 55. [CrossRef]

41. Kuen, C.; Fakurazi, S.; Othman, S.; Masarudin, M. Increased Loading, Efficacy and Sustained Release of Silibinin, a Poorly Soluble Drug Using Hydrophobically-Modified Chitosan Nanoparticles for Enhanced Delivery of Anticancer Drug Delivery Systems. *Nanomaterials* **2017**, *7*, 379. [CrossRef]

42. Othman, N.; Masarudin, M.J.; Kuen, C.Y.; Dasuan, N.A.; Abdullah, L.C.; Md. Jamil, S.N.A. Synthesis and Optimization of Chitosan Nanoparticles Loaded with L -Ascorbic Acid and Thymoquinone. *Nanomaterials* **2018**, *8*, 920. [CrossRef] [PubMed]

43. Kyzas, G.Z.; Bikiaris, D.N. Recent modifications of chitosan for adsorption applications: A critical and systematic review. *Mar. Drugs* **2015**, *13*, 312–337. [CrossRef] [PubMed]

44. Sharma, D.; Singh, J. Synthesis and Characterization of Fatty Acid Grafted Chitosan Polymer and Their Nanomicelles for Nonviral Gene Delivery Applications. *Bioconjug. Chem.* **2017**, *28*, 2772–2783. [CrossRef] [PubMed]

45. Thotakura, N.; Dadarwal, M.; Kumar, R.; Singh, B.; Sharma, G.; Kumar, P.; Prakash, O.; Raza, K. Chitosan-palmitic acid based polymeric micelles as promising carrier for circumventing pharmacokinetic and drug delivery concerns of tamoxifen. *Int. J. Biol. Macromol.* **2017**, *102*, 1220–1225. [CrossRef]

46. Alam, S.; Khan, Z.I.; Mustafa, G.; Kumar, M.; Islam, F.; Bhatnagar, A.; Ahmad, F.J. Development and evaluation of thymoquinone—Encapsulated chitosan nanoparticles for nose-to-brain targeting: A pharmacoscintigraphic study. *Int. J. Nanomed.* **2012**, *7*, 5705–5718. [CrossRef] [PubMed]

47. Goyal, S.N.; Prajapati, C.P.; Gore, P.R.; Patil, C.R.; Mahajan, U.B.; Sharma, C.; Talla, S.P.; Ojha, S.K. Therapeutic potential and pharmaceutical development of thymoquinone: A multitargeted molecule of natural origin. *Front. Pharmacol.* **2017**, *8*. [CrossRef]

48. Hemilä, H.; Chalker, E. Vitamin C for preventing and treating the common cold. *Cochrane Libr. Cochrane Rev.* **2017**. [CrossRef]

49. Gonzalez, M.J.; Miranda-Massari, J.R.; Berdiel, M.J.; Duconge, J.; Rodriguez-Lopez, J.L.; Hunninghake, R.; Cobas-Rosario, V.J. High Dose Intraveneous Vitamin C and Chikungunya Fever: A Case Report. *J. Orthomol. Med.* **2015**, *29*, 154–156.

50. Hemilä, H. Vitamin C and Infections. *Nutrients* **2017**, *9*, 339. [CrossRef]

51. Ullah, I.; Badshah, H.; Naseer, M.I.; Lee, H.Y.; Kim, M.O. Thymoquinone and Vitamin C Attenuates Pentylenetetrazole- Induced Seizures Via Activation of GABA B1 Receptor in Adult Rats Cortex and Hippocampus. *Neuromol. Med.* **2014**. [CrossRef]

52. Birben, E.; Sahiner, U.M.; Sackesen, C.; Erzurum, S.; Kalayci, O. Oxidative Stress and Antioxidant Defense. *World Allergy Organ. J.* **2012**, *5*, 9–19. [CrossRef] [PubMed]

53. Loutfy, S.A.; El-din, H.M.A.; Elberry, M.H.; Allam, N.G.; Hasanin, M.T.M.; Abdellah, A.M. Synthesis, characterization and cytotoxic evaluation of chitosan nanoparticles: In vitro liver cancer model. *Adv. Nat. Sci. Nanosci. Nanotechnol.* **2016**, *7*. [CrossRef]

54. Hanahan, D.; Weinberg, R.A. Review Hallmarks of Cancer: The Next Generation. *Cell* **2011**, *144*, 646–674. [CrossRef] [PubMed]

55. Subri, N.N.S.; Cormack, P.A.G.; Md. Jamil, S.N.A.; Abdullah, L.C.; Daik, R. Synthesis of poly(acrylonitrile-co-divinylbenzene-co-vinylbenzyl chloride)-derived hypercrosslinked polymer microspheres and a preliminary evaluation of their potential for the solid-phase capture of pharmaceuticals. *J. Appl. Polym. Sci.* **2017**, *135*, 1–9. [CrossRef]

56. Shi, Y.; Wan, A.; Shi, Y.; Zhang, Y.; Chen, Y. Experimental and mathematical studies on the drug release properties of aspirin loaded chitosan nanoparticles. *BioMed Res. Int.* **2014**. [CrossRef] [PubMed]

57. Bohrey, S.; Chourasiya, V.; Pandey, A. Polymeric nanoparticles containing diazepam: Preparation, optimization, characterization, in-vitro drug release and release kinetic study. *Nano Converg.* **2016**, *3*. [CrossRef]

58. Higuchi, T. Mechanism of Sustained—Action Medication. *J. Pharm. Sci.* **1963**.

59. Hixson, A.W.; Crowell, J.H. Dependence of Reaction Velocity upon Surface and Agitation. *Ind. Eng. Chem.* **1931**, *23*, 1160–1168. [CrossRef]

60. Korsmeyer, R.W.; Gurny, R.; Doelker, E.; Buri, P.; Peppas, N.A. Mechanisms of solute release from porous hydrophilic polymers. *Int. J. Pharm.* **1983**, *15*, 25–35. [CrossRef]

61. Balan, V.; Redinciuc, V.; Tudorachi, N.; Verestiuc, L. Biotinylated N-palmitoyl chitosan for design of drug loaded self-assembled nanocarriers. *Eur. Polym. J.* **2016**, *81*, 284–294. [CrossRef]

62. Reusch, W. Virtual Textbook of Organic Chemistry. Available online: https://www2.chemistry.msu.edu/faculty/reusch/VirtTxtJml/Spectrpy/InfraRed/infrared.htm#ir1 (accessed on 11 February 2020).

63. Ramalingam, P.; Ko, Y.T. Improved oral delivery of resveratrol from N-trimethyl chitosan-g-palmitic acid surface-modified solid lipid nanoparticles. *Colloids Surf. B Biointerfaces* **2015**, *139*. [CrossRef] [PubMed]

64. Zhao, Y.; Du, W.; Wu, H.; Wu, M.; Liu, Z.; Dong, S. Chitosan/sodium tripolyphosphate nanoparticles as efficient vehicles for enhancing the cellular uptake of fish—Derived peptide. *J. Food Biochem.* **2018**. [CrossRef] [PubMed]

65. Jang, K.-I.; Lee, H.G. Stability of Chitosan Nanoparticles for L-Ascorbic Acid during Heat Treatment in Aqueous Solution. *J. Agric. Food Chem.* **2008**, *56*, 1936–1941. [CrossRef] [PubMed]

66. Du, X.J.; Wang, J.L.; Iqbal, S.; Li, H.J.; Cao, Z.T.; Wang, Y.C.; Wang, J. The Effect of Surface Charge on Oral Absorption of Polymeric Nanoparticles. *Biomater. Sci.* **2018**. [CrossRef] [PubMed]

67. Thaipong, K.; Boonprakob, U.; Crosby, K.; Cisneros-Zevallos, L.; Hawkins Byrne, D. Comparison of ABTS, DPPH, FRAP, and ORAC assays for estimating antioxidant activity from guava fruit extracts. *J. Food Compos. Anal.* **2006**, *19*, 669–675. [CrossRef]

68. Ong, Y.S.; Yazan, L.S.; Ng, W.K.; Noordin, M.M.; Sapuan, S.; Foo, J.B.; Tor, Y.S. Acute and subacute toxicity profiles of thymoquinone-loaded nanostructured lipid carrier in BALB/c mice. *Int. J. Nanomed.* **2016**, 5905–5915. [CrossRef] [PubMed]

 *processes*

Article

Potential Cultivation of *Lactobacillus pentosus* from Human Breastmilk with Rapid Monitoring through the Spectrophotometer Method

Toan Nguyen-Sy [1], Guo Yong Yew [2], Kit Wayne Chew [3], Thi Dong Phuong Nguyen [1,*],
Thi Ngoc Thu Tran [1], Thi Dieu Huong Le [1], Chau Tuan Vo [4], Hoang Kim Pham Tran [5],
Muhammad Mubashir [6,*] and Pau Loke Show [2,*]

[1] Department of Chemical Engineering and Environment, University of Technology and Education-The University of Danang, 48 Cao Thang St., Danang 550000, Vietnam; thutoantamly@gmail.com (T.N.-S.); ttnthu@ute.udn.vn (T.N.T.T.); ltdhuong@ute.udn.vn (T.D.H.L.)

[2] Department of Chemical and Environmental Engineering, Faculty of Science and Engineering, University of Nottingham Malaysia, Jalan Broga, Semenyih 43500, Selangor Darul Ehsan, Malaysia; keby5ygy@nottingham.edu.my

[3] School of Energy and Chemical Engineering, Xiamen University Malaysia, Sepang 43900, Selangor Darul Ehsan, Malaysia; kitwayne.chew@xmu.edu.my

[4] Department of Biology and Environmental Science, The Danang University of Science and Education, 459 Ton Duc Thang St., Danang 550000, Vietnam; vctuan@ued.udn.vn

[5] Le Quy Don High School for the Gifted Danang, Danang 550000, Vietnam; kim76nk2003@gmail.com

[6] Department of Chemical Engineering and Chemistry, Eindhoven University of Technology, 5600 MB Eindhoven, The Netherlands

* Correspondence: ntdphuong@ute.udn.vn (T.D.P.N.); m.mubashir@tue.nl (M.M.); pauloke.show@nottingham.edu.my (P.L.S.)

Received: 8 June 2020; Accepted: 15 July 2020; Published: 29 July 2020

Abstract: The present study focused on the development of a new method to determine the lag phase of Lactobacillus in breast milk which was attained during the 1st, 3rd, and 6th month (M1, M3, and M6). The colonies' phylogenetic analysis, derived from the 16S rRNA gene sequences, was evaluated with genus *Lactobacillus pentosus* and achieved a similarity value of 99%. Raman spectroscopy in optical densities of 600 nm (OD600) were used for six consecutive days to observe the changes of the cell growth rate. The values of OD600 were well fitted with the regression model. From this work, M1 was found to be the longest lag phase in 18 h, and it was 17% to 27% longer compared to M3 and M6, respectively. However, the samples of M3 and M6 showed the shortest duration in reaching 0.5 of OD600 nm (16 h) which was enhanced by 80% and 96% compared to M1, respectively. These studies will be of significance when applied in determining the bacteria growth curve and in assessing the growth behavior for the strain in human breast milk.

Keywords: Lactobacillus isolation; lag phase; bacteria sequencing; breast milk

---

1. Introduction

Bacterial growth can be presented in four phases: lag phase, exponential phase, stationary phase, and death phase. The comprehension of lag phase plays an important role in many aspects of the biotechnological field. As known, once bacteria are transferred to a new medium, the metabolite energy will be depleted in the cells and cause termination of the cell growth [1,2]. This phenomenon is explained by the differences between original media and the new culture. On the other hand, there is a lag phase where bacteria might adapt to the new media's conditions. During this phase, organisms increase officially in size but have no change in number.

The determination of the microbial growth through the optical density (OD) is an effective and conventional parameter to be observed from the growth phases of microbials. According to Ren et al. [3], the work adapted Raman spectroscopy of OD600 to enable the continuous analysis of the growth phases which is found to be marker-free, and subsequently the technique obtained different metabolic states from the strain cells through spectra during the different phases. In general, OD600 is specified to determine the relationship between the colony forming unit (CFU) and growing phases. Modelling the bacterial growth curve has been reported by several research works [4–7]. The growth curve enables other incubation production derived by microorganisms to observe the cell concentrations and make comparison [8,9]. The prediction of the cell growth rate can be applied on the growth model in which it can be derived from the beginning of the transferring time [1,7,8]. Meanwhile, the lag phase can be ignored from the transferring time and by obtaining a regression model to trace the lag phase. These studies will be applied in determining the bacteria growth curve and to determine the growth behavior for the strain in human breast milk.

Breast milk has been known to contain a significant amount of beneficial microorganism for cell culture [10,11]. Besides, parents with no prior drug prescriptions during pregnancy, the mother's breastmilk will contain serotonin which is the natural compound for assisting newborns to sleep [12,13]. While the concurrent microbiota from breastmilk contains probiotic that may have a positive influence on the immunity and digestive system depending on the number of Ig-secreting cells [14]. The drugs which may affect the newborns are mostly by the psychoactive drugs, such as antidepressants, lithium, carbamazepine, and valproate, which further decrease the beneficial strains that emerge from breastmilk [15]. Lactobacillus strains are one of the most beneficial bacteria which are highly considered as antimicrobial and probiotic potential organisms [10,11,16–18]. Breast milk contains strains of beneficial bacteria, such as Lactobacillus, to sustain a high-quality digestive system for the newborn [19–21].

For determining bacterial growing phases, a small volume of microorganism will be cultured in a selective medium, and their OD600 will be observed over the time. The determination of lag phases of microbes is therefore difficult to determine with the high accuracy of time. The observation from the growing phases of a microbial in a medium need to be done for several days, while to observe the changes in OD600 through Raman spectrophotometer will be performed in a few hours. The need to understand the growth kinetics of Lactobacillus strain in human breastmilk isolates from Man, Rogosa, and Sharpe (MRS) medium is primarily useful for cultivation of lactobacilli and for the study on the influence of time and prediction on the cell growth concentration. Moreover, the data may be used to simulate mathematical models for problems involving the rate of growth and relationship of the specific growth rate with the substrate during the strains' growth period [22]. The bacteria cell growth kinetics are an autocatalytic reaction according to each cell metabolite cycle. Growth kinetics would directly influence the dry cell weight, optical density, cell on plate per counts, at the same time, indirectly influencing the nutrient content such as protein, adenosine triphosphate (ATP) and deoxyribonucleic acids (DNA) molecules [23]. In the culture growth, there are several uncertainties that exist in the growth kinetic data as the microorganism will adapt to the changing environmental. The growth kinetics model may resolve the analytical difficulty when the experiment is not in the conventional growth pattern and allow for comparison with the developed mathematical model toward the experiment data. Therefore, the bacteria growth species in the same medium (breastmilk) would have a control data based on the mathematical equation from this experiment.

In this study, the lag phases from the microorganism was determined through the mathematical regression model which was able to verify the accuracy deviation. Besides, lag phase is related to the time when the bacteria have not started to divide, and this period is known as the preparation stage for bacteria to harvest nutrients while adapting to the new environmental condition [24]. Therefore, this research was conducted with two main objectives: (i) to propose a simple method for determining the lag phase and (ii) to compare the growth phases of bacteria from breastmilk to develop the proposed method.

2. Materials and Methods

2.1. Lactobacillus Isolation

Breast milk from the 1st, 3rd, and 6th month (hereby denoted as M1, M3, and M6) were collected each to fill 3 bottles of 125 mL of milk samples (not during the postpartum period) from selected healthy female patients from Danang Hospital, Vietnam. Prior to inoculating the microorganism in MRS selective broth (Merck, HoChiMinh City, Vietnam) as an enrichment media, these samples of 15 mL were centrifuged at 5000× g for 5 min to withdraw 1 mL of supernatant into 15 mL of MRS broth. This broth with microbial was incubated for 24 h at 37 °C. After incubation, 100 µL of enriched microbial was spread on MRS agar with 0.5% of calcium carbonate by a glass spreader, and incubated again at 37 °C under anaerobic condition for 48 h. The 30 milky white colonies in total from M1, M3, and M6 sprouted on an agar plate were chosen to characterize their biochemical properties such as Gram staining, catalase, oxidase test, mobility, and indole production. The colonies, which were isolated and distinguished as Lactobacillus bacteria, were kept in glycerol 20% (Merck, Vietnam) as a stock culture at −20 °C for further phylogenetic analysis. All samples' data presented were performed in triplicate.

2.2. Biochemical Screening, PCR Reaction

The Gram staining, catalase, oxidase test, motility, and indole test was performed as a beginning of Lactobacillus screening. The colonies consisted of Gram-positive, catalase-negative, oxidase test-negative, and indole-negative were selected for DNA extraction. The microorganism enriched overnight in sub-cultured media at 30 °C after it was extracted by Kit AquaPure Genomic Isolation (Bio Rad, Hercules, CA, USA). Foremost, centrifugation at 13,000 rpm for 10 min was performed on the enriched cells for collecting the pellet and further treated by 300 µL of Genomic DNA Lysis Solution. Then 1.5 µL of RNase was added in this suspension to be homogenized by vortex, and then the homogenous suspension was incubated at 37 °C for 5 min. The solution was further adjusted using 100 µL of Protein Precipitation Buffer and vortexed for 20 s for incubated suspension. The solution obtained was centrifuged at 13,000 rpm for 5 min to harvest the supernatant and transferred to a new tube. Three hundred microliters of isopropanol (2-propanol) was added to this tube to precipitate the DNA which was harvested by centrifugation at 13,000 rpm for 15 min. The DNA precipitation was rinsed by 1 mL of 70% ethanol, then was centrifuged 13,000 rpm for 5 min. The DNA precipitation was dried at room temperature and was adjusted to 50 µL of DNA hydration solution. The DNA extraction product was examined by gel electrophoresis on 0.8 *w/v* of agarose gel (Bio Rad, America) and was stored at −40 °C for further PCR reaction.

For PCR reaction, the 16S rRNA gene segment amplification reaction for sequencing using primers 27F (5′-AGAGTTTGATCCTGGCTCAG-3′) and 1492R (5′-GGTTACCTTGTTACGACTT-3) was performed. The PCR component consisted of 6 µL of master mix, 10 µmol of per primer, 50 ng of genomic DNA, and 12 µL of distilled water. After being denatured for 95 min and treated with the PCR thermal cycle for 95 °C for 1 min, subsequently, at a temperature of 55 °C for 1 min, and, lastly, at 72 °C for 1.5 min, this sequence was repeated for 30 cycles. The PCR reaction was then incubated at 72 °C for 10 min. Finally, PCR products were tested by electrophoresis with 0.8% *w/v* of agarose gel and were sequenced by the First BASE Laboratories Sdn Bhd, Malaysia. As a result, sequencing identification were checked with the GeneBank to manipulate the advanced BLAST which was available at the US National Library of Medicine [25].

2.3. The Regression Model Analysis

One colony of M1, M3, and M6 chosen from the sub-culture, which well indicated the Lactobacillus sequencing, was dropped in an Erlenmeyer flask containing 100 mL of MRS broth. A single cell was inoculated from a Petri dish into 5 mL of the MRS broth in test tube and incubated at 37 ± 2 °C for the observation on the cell growth. The cells reached ~0.3 and was determined using UV-Vis spectrophotometer (Shimadzu 1800) at OD600 for a period of 7 days. The Raman spectrometry was performed through cell aliquots in triplicate according to Ren et al. [3].

The software Sigmaplot Verion 14 was used to determine the regression equation. The growth of Lactobacillus was presented as the exponential growth curve. Generally, the growth equation was described as follows:

$$F = a[1 - \exp(-bt)] \tag{1}$$

where a—the potential growth in OD600; b—constant; t—time in hour.

Most microorganisms require a duration to adapt to the new conditions and environmental surroundings. The lag phase is represented by the time shown in Equation (1). The growth curve was modified to Equation (2) as follows:

$$F = y_0 + a[1 - \exp(-bt)] \tag{2}$$

where y_0—OD600, a—growth in OD600; b—constant; t—time in hour.

2.4. Lag Phase Determination

Since the lag phase duration consumed hours to days of duration, depending on the growing conditions or different microbial strains, the exponential growth phase of microbials was supposed to start at a time point out of zero. The lag phase duration would be presented as Equation (3) below:

$$t_{lag} = -b^{-1} \ln \sqrt{\frac{y_0 + a}{a}} \tag{3}$$

where y_0, a, and b are similar as in Equations (1) and (2).

2.5. Determination of 0.5 OD600

Microbial growth after the lag phase are varying; hence, the growing time to reach 0.5 OD600 was determined after the lag phase. Equation (4) is described as follows:

$$t_{0.5\ OD} = -b^{-1} \ln \frac{y_0 + a - 0.5}{a} \tag{4}$$

where y_0, a, and b are similar as in Equations (1) and (2).

3. Result and Discussion

3.1. Lactobacillus Identification

The 30 colonies isolated from M1, M3, and M6 resulted in the biochemical tests to identify its strains, such as lactobacilli, and displayed a similarity of more than 99% to a type of Lactobacillus pentosus and 99.51% to a type of Lactobacillus plantarium (Table 1). The samples collected from different time points was for studying how time affects the lag phase activity such as metabolic and cell size. The bacteria population growth may affect the reduction in nutrient content in the breastmilk. All bacterial 16S rDNA gene sequences were entrusted in GenBank using the accession numbers MT026914, MT026915, MT026916, and MT026917. This bacteria strain is a class of Bacilli belonging to the family of Lactobacillaceae which is commonly found in human body intestines. This strain assists to break down lactose molecules from milk. The chemical composition from the breast milk supported the Lactobacillus sp. growth and enzyme activity which could be further transferred and used to produce daily goods such as yogurt drinks [26,27]. Apart from obtaining this strain of bacteria in the fermentation of milk goods, cereal- and soy-based foods in a reported experiment have shown that breastmilk associated with glucose and lipids have the ability to culture healthy and beneficial bacteria strains [28]. Lactic acids are commonly found for in production of yogurt drink in the form of

Lactobacillus sp. strain which could survive in high acidity environments such as gastric and intestinal juices [29].

Table 1. The phylotypes of colonies isolated from 1st month (M1), 3rd month (M3), and 6th month (M6) breast milk.

Colonies Isolated	Number of Identical	Sequence Length, Similarity	Most Similar Type Strain	Accession Number
M1	8	1231, 99.51%	*Lactobacillus plantatirum*	MT026917
M3	12	1000, 99.9%	*Lactobacillus plantatirum*	MT026916
M6	10	1233, 99%	*Lactobacillus pentosus*	MT026915

3.2. Lactobacilli Growth Curve

The growth curves of M1, M3, and M6 modeled by the regression model as described in Equation (2) is shown in Table 2. All results were well fitted to the model with a high correlation ($R^2 >$ 0.92). It is clear to see that M1 had the highest growth rate, equivalent to 2.08, followed by M3. This result showed that the Lactobacillus growth rate with the highest rate was in the first stage of the mother's milk during month 1 and slowed down in the 3rd and 6th months. However, the potential OD600 of M1 and M3 were in high similarity with 1.348 and 1.329, respectively. The value was slightly higher than M6 which was 0.948. The potential value in the optical density also showed a declining trend. According to the literature, this current experiment is the first to report a growth rate of *L. pentosus* in breast milk decreasing over time, creating an impact on the relevant field of studies.

Table 2. Regression model for growth of *L. pentosus* (exponential rise to maximum, single, 3 parameter), run by Sigmaplot Version 14. The equation described is Equation (2) where y_0, a, and b are constants and t is the cultured time.

	y_0	a	b	R^2	Potential Absorbance at OD600
M1	−0.732	2.08	0.26	0.984	1.348
M3	−0.26	1.588	0.1015	0.926	1.329
M6	−0.328	1.276	0.021	0.971	0.984

3.3. Lag Phase Determination

By obtaining the model for determination of the lag phases of M1, M3, and M6, which is seen in Table 3, M1 was shown as having the longest time for the lag phase with 16.8 h; this is longer than M3 and M6 by 4.6 and 2.9 h, respectively. In contrast, to reach 0.5 OD600 after lag phases, M1 had the shortest duration of time which was 18 h, while the others were 51–55% higher compared to M1 for M3 and M6 (Table 3).

Table 3. The duration for growth phases of *L. pentosus* from breast milk determined by regression model (hours).

(Colonies)	Lag Phase	Reaching 0.5 OD600	Reaching 0.5 OD600 after Lag Phase
M1	16.8 ± 0.21	34.8 ± 0.18	18.0 ± 1.81
M3	12.2 ± 0.34	44.6 ± 0.65	32.3 ± 0.38
M6	13.9 ± 0.10	49.1 ± 1.27	35.2 ± 0.97

Indeed, the model could be referenced to understand the ability to adapt a new environment for other bacteria. It is noted that M1 had the longest lag phase compared to the rest. This result mentioned that the bacteria in the early stages of a mother's milk takes a longer duration to be activated, but then shortens the period overtime. The reaching of 0.5 OD600 was supposed to see the potential of the early growth of bacteria, which is also an important factor for culturing most strains of the bacteria.

In this research, M1 had the shortest time to reach 0.5 OD600, confirming that Lactobacillus in the early stage of mother's milk provided the highest growth ability. The Lactobacillus in M1 reached 0.5 OD600 as the fastest among the three samples but was also considered as the highest potential in OD600. This suggests that M1 had an optimum condition which enhanced the Lactobacillus growth rate. The mechanism of Lactobacillus in breast milk made it less active over time.

3.4. Applied of Regression Model in Determining Lag Phase of Lactobacillus

Equation (1), as shown in Figure 1, has been applied in many models regarding growth curves and other fields which increases gradually compared to the potential of growth by observing OD600 [4–8]. However, these models could not deplete the lag phase. This study has concluded that the bacteria noticeably consumed several hours for the lag phase that ranged from 12.2–16.8 h. By fitting the regression model, the onset of the exponential phase time was determined as described by the lag phase in Equation (3). Therefore, the mathematical model value was supported by the practical experiment value. It is noteworthy that the lag phase requires a period of more than half a day before going into the exponential phase (log phase). The equation developed can be fitted into a general concept for biological growth and alternative phenomenological approach for other studies based on similar experimental practices to assist in those studies.

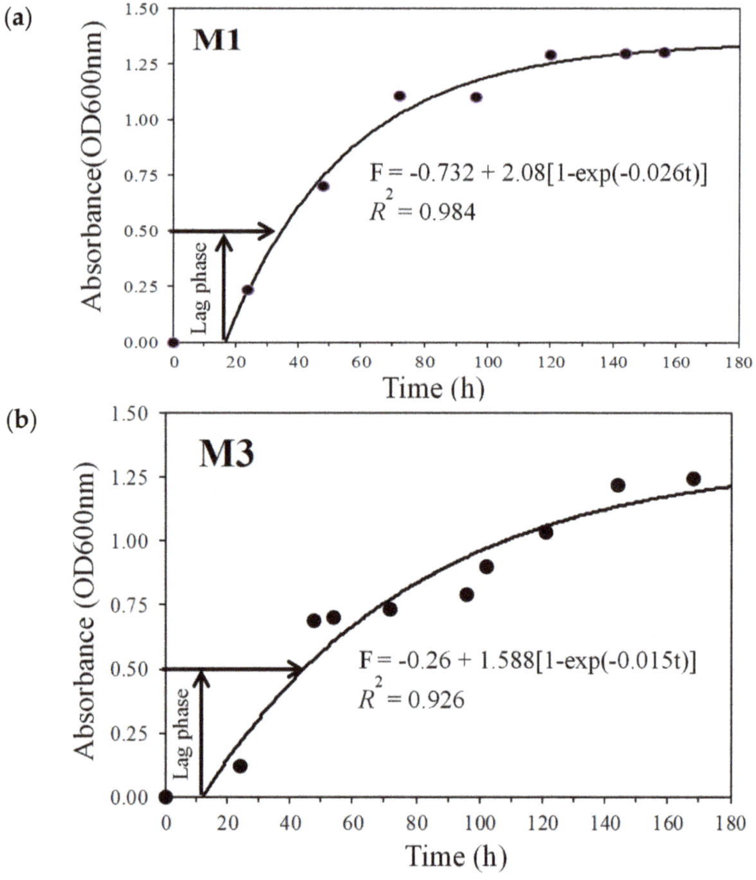

Figure 1. *Cont.*

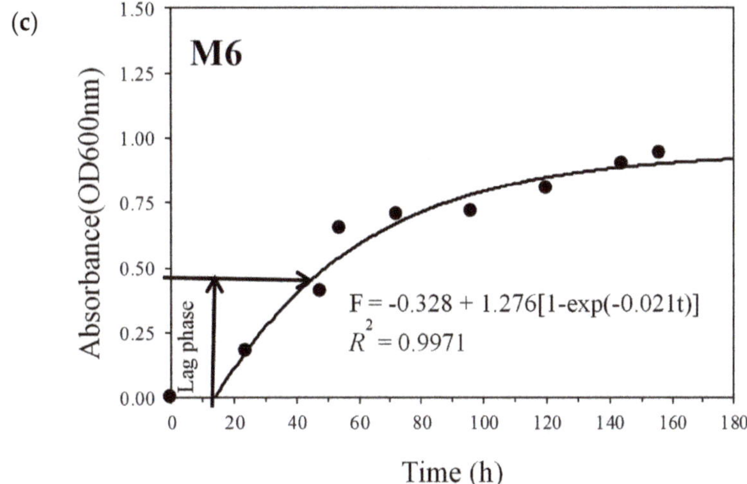

Figure 1. The growth curve of the 1st (**a**), 3rd (**b**), and 6th (**c**) month of breast milk fitted to the regression model in Equation (2). The black points were observed OD600, and the continuous curve line was modelled by Sigmaplot Version 14.

4. Conclusions

The growth rate of Lactobacillus in breast milk was suited well according to the regression model with high correlation ($R^2 > 0.92$). It is proposed that the model could be applied to estimate the lag phase as well as early growth phases of bacteria. In addition, Lactobacillus pentosus isolated from the first month of the breast milk has the most potential as a medium for growth. The early stages of the exponential phase were found to decrease in trend after the third month of the breast milk which may conclude the nutrient degradation from the breast milk. Future research could be performed on different strains of bacteria using later or intermediate stages of breast milk while applying other mathematical orders and equations to predict the growth pattern.

Author Contributions: Conceptualization, T.D.P.N. and P.L.S.; methodology, T.N.-S. and G.Y.Y.; software, T.N.T.T. and T.D.H.L.; validation, G.Y.Y. and K.W.C.; formal analysis, M.M.; investigation, C.T.V. and H.K.P.T.; data curation, T.N.-S. and T.D.P.N.; writing—original draft preparation, T.N.-S.; writing—review and editing, G.Y.Y. and K.W.C.; supervision, T.D.P.N., M.M. and P.L.S.; project administration, K.W.C. and P.L.S.; funding acquisition, T.D.P.N. and P.L.S. All authors have read and agreed to the published version of the manuscript.

Funding: This research was funded by the Vietnam Vingroup Innovation Foundation under grant number VINIF.2019.TS.63. The genome sequencing was provided and validated by First BASE Laboratories Sdn Bhd, Malaysia. This work was supported by the Fundamental Research Grant Scheme, Malaysia [FRGS/1/2019/STG05/UNIM/02/2] and MyPAIR-PHC-Hibiscus Grant [MyPAIR/1/2020/STG05/UNIM/1].

Conflicts of Interest: The authors declare no financial or commercial conflict of interest.

References

1. Yew, G.Y.; Tham, T.C.; Show, P.-L.; Ho, Y.-C.; Ong, S.K.; Law, C.L.; Song, C.; Chang, J.-S. Unlocking the Secret of Bio-additive Components in Rubber Compounding in Processing Quality Nitrile Glove. *Appl. Biochem. Biotechnol.* **2020**, *191*, 1–28. [CrossRef] [PubMed]
2. Yeang, C.-H. Integration of Metabolic Reactions and Gene Regulation. *Mol. Biotechnol.* **2011**, *47*, 70–82. [CrossRef] [PubMed]
3. Ren, Y.; Ji, Y.; Teng, L.; Zhang, H. Using Raman spectroscopy and chemometrics to identify the growth phase of Lactobacillus casei Zhang during batch culture at the single-cell level. *Microb. Cell Factories* **2017**, *16*, 233. [CrossRef] [PubMed]
4. Baranyi, J. Stochastic modelling of bacterial lag phase. *Int. J. Food Microbiol.* **2002**, *73*, 203–206. [CrossRef]

5. Baranyi, J.; Roberts, T.A. A dynamic approach to predicting bacterial growth in food. *Int. J. Food Microbiol.* **1994**, *23*, 277–294. [CrossRef]
6. Zwietering, M.; De Koos, J.; Hasenack, B.; De Witt, J.; Van't Riet, K. Modeling of bacterial growth as a function of temperature. *Appl. Environ. Microbiol.* **1991**, *57*, 1094–1101. [CrossRef]
7. Zwietering, M.; Jongenburger, I.; Rombouts, F.; Van't Riet, K. Modeling of the bacterial growth curve. *Appl. Environ. Microbiol.* **1990**, *56*, 1875–1881. [CrossRef]
8. Cheng, W.; Padre, A.T.; Sato, C.; Shiono, H.; Hattori, S.; Kajihara, A.; Aoyama, M.; Tawaraya, K.; Kumagai, K. Changes in the soil C and N contents, C decomposition and N mineralization potentials in a rice paddy after long-term application of inorganic fertilizers and organic matter. *Soil Sci. Plant Nutr.* **2016**, *62*, 212–219. [CrossRef]
9. García, C.; Rendueles, M.; Díaz, M. Microbial amensalism in Lactobacillus casei and Pseudomonas taetrolens mixed culture. *Bioprocess Biosyst. Eng.* **2017**, *40*, 1111–1122. [CrossRef]
10. Martín, R.; Jiménez, E.; Olivares, M.; Marín, M.; Fernández, L.; Xaus, J.; Rodríguez, J. Lactobacillus salivarius CECT 5713, a potential probiotic strain isolated from infant feces and breast milk of a mother–child pair. *Int. J. Food Microbiol.* **2006**, *112*, 35–43. [CrossRef]
11. Martín, R.; Olivares, M.; Marín, M.L.; Fernández, L.; Xaus, J.; Rodríguez, J.M. Probiotic potential of 3 lactobacilli strains isolated from breast milk. *J. Hum. Lact.* **2005**, *21*, 8–17. [CrossRef]
12. Vitale, S.G.; Laganà, A.S.; Muscatello, M.R.A.; La Rosa, V.L.; Currò, V.; Pandolfo, G.; Zoccali, R.A.; Bruno, A. Psychopharmacotherapy in pregnancy and breastfeeding. *Obstet. Gynecol. Surv.* **2016**, *71*, 721–733. [CrossRef] [PubMed]
13. Lagana, A.S.; Triolo, O.; D'Amico, V.; Cartella, S.M.; Sofo, V.; Salmeri, F.M.; Bokal, E.V.; Spina, E. Management of women with epilepsy: From preconception to post-partum. *Arch. Gynecol. Obstet.* **2016**, *293*, 493–503. [CrossRef]
14. Rinne, M.; Kalliomaki, M.; Arvilommi, H.; Salminen, S.; Isolauri, E. Effect of probiotics and breastfeeding on the bifidobacterium and lactobacillus/enterococcus microbiota and humoral immune responses. *J. Pediatr.* **2005**, *147*, 186–191. [CrossRef]
15. Ram, D.; Gowdappa, B.; Ashoka, H.; Eiman, N. Psychopharmacoteratophobia: Excessive fear of malformation associated with prescribing psychotropic drugs during pregnancy: An Indian perspective. *Indian J. Pharmacol.* **2015**, *47*, 484. [CrossRef]
16. Anandharaj, M.; Sivasankari, B. Isolation of potential probiotic Lactobacillus oris HMI68 from mother's milk with cholesterol-reducing property. *J. Biosci. Bioeng.* **2014**, *118*, 153–159. [CrossRef]
17. Huang, H.; Song, X.; Yang, S. Development of a RecE/T-Assisted CRISPR–Cas9 Toolbox for Lactobacillus. *Biotechnol. J.* **2019**, *14*, 1800690. [CrossRef]
18. Li, F.; Zhou, H.; Zhou, X.; Yi, R.; Mu, J.; Zhao, X.; Liu, W. Lactobacillus plantarum CQPC05 Isolated from Pickled Vegetables Inhibits Constipation in Mice. *Appl. Sci.* **2019**, *9*, 159. [CrossRef]
19. Todorov, S.D.; Dicks, L.M.T. Parameters affecting the adsorption of plantaricin 423, a bacteriocin produced by Lactobacillus plantarum 423 isolated from sorghum beer. *Biotechnol. J.* **2006**, *1*, 405–409. [CrossRef]
20. Splechtna, B.; Nguyen, T.-H.; Zehetner, R.; Lettner, H.P.; Lorenz, W.; Haltrich, D. Process development for the production of prebiotic galacto-oligosaccharides from lactose using β-galactosidase from Lactobacillus sp. *Biotechnol. J.* **2007**, *2*, 480–485. [CrossRef]
21. Tajabadi, N.; Ebrahimpour, A.; Baradaran, A.; Rahim, R.A.; Mahyudin, N.A.; Manap, M.Y.A.; Bakar, F.A.; Saari, N. Optimization of γ-aminobutyric acid production by Lactobacillus plantarum Taj-Apis362 from honeybees. *Molecules* **2015**, *20*, 6654–6669. [CrossRef] [PubMed]
22. Nor, Z.O.; El-Enshasy, H.A.; Roslinda, A.M.; Sarmidi, M.R.; Ramlan, A.A. Kinetics of cell growth and functional characterization of probiotic strains Lactobacillus delbrueckii and Lactobacillus paracasei isolated from breast milk. *Dtsc. Lebensm.-Rundsch.* **2009**, *105*, 444–450.
23. Sakthiselvan, P.; Meenambiga, S.S.; Madhumathi, R. Kinetic Studies on Cell Growth. In *Cell Growth*; IntechOpen: London, UK, 2019.
24. Bertrand, R.L. Lag phase is a dynamic, organized, adaptive, and evolvable period that prepares bacteria for cell division. *J. Bacteriol.* **2019**, *201*, 1–21. [CrossRef]
25. NCBI. *Basic Local Alignment Search Tool (BLAST)*; National Library of Medicine: Bethesda, MA, USA, 2020.

26. Delgado, S.; Guadamuro, L.; Flórez, A.B.; Vázquez, L.; Mayo, B. Fermentation of commercial soy beverages with lactobacilli and bifidobacteria strains featuring high β-glucosidase activity. *Innov. Food Sci. Emerg. Technol.* **2019**, *51*, 148–155. [CrossRef]
27. Jha, A.K.; Prasad, K. Biosynthesis of metal and oxide nanoparticles using Lactobacilli from yoghurt and probiotic spore tablets. *Biotechnol. J* **2010**, *5*, 285–291. [CrossRef] [PubMed]
28. Lee, E.; Jung, S.-R.; Lee, S.-Y.; Lee, N.-K.; Paik, H.-D.; Lim, S.-I. Lactobacillus plantarum strain ln4 attenuates diet-induced obesity, insulin resistance, and changes in hepatic mrna levels associated with glucose and lipid metabolism. *Nutrients* **2018**, *10*, 643. [CrossRef] [PubMed]
29. Nishinari, K.; Fang, Y.; Nagano, T.; Guo, S.; Wang, R. 6-Soy as a food ingredient. In *Proteins in Food Processing*, 2nd ed.; Yada, R.Y., Ed.; Woodhead Publishing: Philadelphia, PA, USA, 2018; pp. 149–186. [CrossRef]

Article

Conversion of Lignocellulosic Corn Agro-Waste into Cellulose Derivative and Its Potential Application as Pharmaceutical Excipient

Md. Saifur Rahman [1,2], Md. Ibrahim H. Mondal [1,*], Mst. Sarmina Yeasmin [3], M. Abu Sayeed [1], Md Ashraf Hossain [4] and Mohammad Boshir Ahmed [1,2,5,*]

[1] Department of Applied Chemistry and Chemical Engineering, University of Rajshahi, Rajshahi 6205, Bangladesh; saifurrahman@gist.ac.kr (M.S.R.); drmdabusayeed@gmail.com (M.A.S.)
[2] Gwangju Institute of Science and Technology, School of Materials Science and Engineering, Gwangju 61005, Korea
[3] Bangladesh Council of Scientific and Industrial Research Laboratories, Rajshahi 6206, Bangladesh; lisabcsir@yahoo.com
[4] Department of Materials Science and Engineering, Korea University, Seoul 02841, Korea; ashraf3521@gmail.com
[5] Center for Green Technologies, School of Civil and Environmental Engineering, University of Technology Sydney, 15, Broadway, Sydney, NSW 2007, Australia
[*] Correspondence: mihmondal@gmail.com (M.I.H.M.); mohammad.ahmed@gist.ac.kr (M.B.A.)

Received: 26 May 2020; Accepted: 15 June 2020; Published: 19 June 2020

Abstract: Lignocellulosic biomass is widely grown in many agricultural-based countries. These are typically incinerated or discarded in open spaces, which further may cause severe health and environmental problems. Hence, the proper utilization and conversion of different parts of lignocellulosic biomasses (e.g., corn wastes derived leave, cob, stalk, and husk) into value-added materials could be a promising way of protecting both health and environments. In addition, they have high-potential for myriads applications (e.g., pharmaceuticals, cosmetics, textiles, and so on). In this context, herein, we isolated holocellulose (a mixture of alpha α, beta β, and gamma γ cellulose) from corn waste, and then it was converted into carboxymethyl cellulose (CMC). Subsequently, the prepared CMC was evaluated successfully to be used as a pharmaceutical excipient. Different characterization tools were employed for structural, morphological, and thermal properties of the extracted holocellulose and synthesized CMC. Results showed that the highest yield of CMC was obtained 187.5% along with the highest degree of substitution (DS i.e., 1.83) in a single stage (i.e., size reduction technique) with the lowest particle size of holocellulose (100 μm). This happened due to the use of a single stage instead of multiple stages. Finally, extracted CMC was successfully used as a pharmaceutical excipient with promising results compared to commercially available pharmaceutical-grade CMC.

Keywords: lignocellulosic biomass; holocellulose; CMC; degree of substitution; excipient

1. Introduction

Lignocellulosic biomass is the most abundant resource in nature with immense potential for numerous applications [1]. Among different lignocellulosic biomasses, agricultural waste-based biomass consists of cellulose (35–50%), hemicellulose (20–35%), and lignin (10–25%) [2]. Corn (*Zea mays*) is one of the abundant cereal crops that is cultivated extensively across the world, which produces huge lignocellulosic biomass. The world's total corn production was 985,889.6 (1000 MT) from 2012 to 2017. USA was the top country for producing corn in the period of 2012–2017 with the production capacity of 343,167.8 (1000 MT), followed by the China, Brazil, Argentina, and Ukraine which produced 216,787.0, 82,400.0, 30,550.0, 26,321.0 (1000 MT) of corn, respectively [3]. Therefore, the processing of matured corn can produce large amount of waste. It is estimated that 1 kg of dry corn may yield up to 150 g of cobs, 220 g of leaves, and 50 g of stalks [4]. However, corn wastes namely leave, cob, stalk, and husk are the major biomass matters that often remain unutilized in the harvested fields. Not only are these corn wastes being used as a cooking fuel in some rural areas of many countries, but also often they are causing environmental problems due to inapt waste control such as on-site burning as well as landfilling. On the other hand, wastes from corn contain good quality cellulosic matters in their cell walls. Generally, cellulose is a linear and high molecular weight polymer that neither melts nor dissolves readily in water and many organic solvents. This characteristic makes cellulose ineffective in most of the industrial uses. Notwithstanding, cellulose can be transformed into valuable chemical feedstock (e.g., ethanol, lactic acid, furfural, and fermentable sugars including glucose and xylose) as it is susceptible to chemical and enzymatic derivatization reactions [5,6]. Cellulose contains three -OH groups in each of its anhydroglucose units. Among them, primary -OH at C-6 and two secondary ones at C-2 and C-3 can take part in typical reactions such as esterification, etherification, and oxidation. Cellulose derivatives have been obtained by reacting to some (or all) -OH groups of anhydroglucose units [6,7]. Carboxymethylation of cellulose is a common conversion process which provides versatile water-swellable or water-soluble polymers and intermediates with variable characteristics [8,9].

Carboxymethyl cellulose (CMC) is produced through the reaction between alkali cellulose swollen in aqueous NaOH and monochloroacetic acid in the surplus of alcohol (Figure 1) [10]. Hydroxyl groups of the anhydrous glucose unit (AGU) are substituted by the sodium carboxymethyl groups in C-2, C-3, and C-6, of which substitution slightly dominates at C-2 position [11]. CMC has been synthesized by many researchers from different cellulosic sources such as paper sludge, hyacinth, wood residue, cotton linters, and bagasse [12–14]. Due to the polyelectrolyte character of CMC, it has many applications, such as being widely used in the food industry, detergents, cosmetics, pharmaceuticals, textiles, paper, adhesives, and ceramic industries [15]. The degree of substitution (DS) is considered a significant property of CMC particularly for its solubility in water, and the highest theoretical DS of CMC is considered as 3. It is reported that commercially available CMC has a DS value of 0.4 to 1.5 [16]. However, to achieve higher DS of CMC, several parameters such as the solvent system, the concentration of NaOH, monochloroacetic acid (MCA), temperature, reaction time, and the different steps of carboxymethylation need to tune properly [17]. Generally, in every case, several steps have been performed to gain the higher DS. Therefore, an alternative approach such as minimization of steps to get higher DS is preferable.

CMC is an anionic derivative that is being largely used in oral, ophthalmic, injectable, and topical pharmaceutical formulations as an excipient. For solid dosage forms, CMC is used primarily as a binder or matrix former. When CMC is used as a binder, then it yields softer granules with good compressibility which form tough tablets with moderate strength [18]. On the other hand, commercially available microcrystalline cellulose-based excipients are extracted from hardwoods and also from purified cotton. Therefore, it is considered an expensive process that can further trigger investigations of finding cheaper resources for similar excipients preparation [19]. To the best of our knowledge, no report has been found in literature where a pharmaceutical excipient was prepared from the corn wastes based on lignocellulosic biomass. Henceforth, extracting CMC as well as excipient from the lignocellulosic agriculture waste (i.e., corn waste) could be considered an effective way for the reduction of production costs as well as process barriers as these biomasses are widely available and free of cost or at a very negligible price. In addition, such kind of low-cost material i.e., CMC can be further used to produce tablets at an industrial scale.

Hence, the main objective of this study is to extract CMC from low-cost corn wastes with higher DS by the utilization of a single-stage size-reduction method for lowering the extraction cost. The subsequent objective of this study is to use the synthesized CMC as a pharmaceutical excipient by testing its feasibility by performing different characterizations.

2. Material and Methods

2.1. Materials

Corn wastes (*Zea mays*) including leaves, cob, stalk, and husk were collected in the harvesting season (i.e., June–July) from the Wheat Research Center Rajshahi, Rajshahi Division, Bangladesh. All the chemicals such as pharmaceutical excipient grade CMC (commercial), sodium hydroxide, monochloroacetic acid (MCA), acetic acid, ammonium oxalate, sulfuric acid, and hydrochloric acid were purchased from Sigma Aldrich, Bangladesh, and they were in the highest purity.

2.2. Preparation of Sample

Defective parts and foreign materials from the corn wastes were removed, followed by cutting into small pieces and dried in sunlight for several days to minimize the intrinsic moisture. The corn leave, cob, stalk, and husk were dried in the oven (FC-610, Toyo Seisakusho Co., Ltd., Chiba, Japan) at 105 °C for several hours and grounded into powder using a disk mill (FFC-15). Later, the powder was screened into three different particle sizes (100, 400, and 700 μm) by using a GFL Orbital Shaker (Model: 3017, Germany) and stored in a silica-containing desiccator for further use.

2.3. Estimation of Fatty and Waxy Matters, Pectic Substances, and Lignin

The dried powder sample was immersed in n-hexane- ethanol mixture in the solid to liquor ratio of 2: 200 for 10 h. The suspension was then filtered and washed with the residue with the fresh n-hexane-ethanol mixture. After drying the residue, fatty and waxy matters were calculated using the following formula.

$$\% \text{ of fatty and waxy matters} = \frac{y \times 100}{x}, \tag{1}$$

where y is the loss in weight and x is the initial weight of the sample. The dewaxed and defatted powder was then heated with an ammonium oxalate solution (0.5% *w/v*)) in a liquor ratio of 0.1:10 at 80 °C for 3 consecutive days in a heating mantle. The level of the solution kept constant by adding hot DI water simultaneously. Finally, the suspension was filtered and the residue was washed with DI water and dried at 105 °C for getting pectic matters percentages:

$$\% \text{ of pectic matters} = \frac{y \times 100}{x}. \tag{2}$$

The dewaxed and depectinized dried powder was then treated with 72% sulfuric acid with solid to acid ratio of 1:15 at ambient condition. The mixture was kept for 1.0 h and diluted by DI water up to 3% acid solution. Subsequently, the mixture was refluxed for 4.0 h and kept overnight. The mixture was filtered and washed thoroughly with hot DI water and the residue dried until reaching the constant weight at 105 °C. The residue was considered as lignin, and the powdered sample from the filtrate was the delignified sample [20].

2.4. Isolation of Holocellulose

Holocellulose was isolated with a slight modification of the previously reported method [20]. Briefly, a suitable amount of dewaxed and depectinized powder corn wastes were treated with a 0.7% $NaClO_2$ solution at pH 4, and at 90–95 °C for 90 min with a liquor ratio of 1:80 (*w/v*). After being washed with DI water, chlorite holocellulose was treated with a 0.2% $Na_2S_2O_5$ solution for 15 min. Subsequently, the holocellulose containing solution was again filtered and washed thoroughly with distilled water, and finally dried at 60 °C to get the holocellulose.

2.5. Estimation of α-Cellulose and Hemicellulose (β and γ-Celluloses)

The α-cellulose and hemicellulose amount were determined to measure the total quantity in corn wastes [20]. Briefly, 1.0 g of the dried chlorite holocellulose was treated with an 18% NaOH (*w/v*) solution for 2 h in the ratio of 1:100 (*w/v*). The mixture was then filtered and washed thoroughly with

2% acetic acid solution, followed by hot water. The residue was dried at 105 °C until reaching constant weight. The dried residue (i.e., α-cellulose) was deducted from the weight of the holocellulose to estimate the amount of hemicellulose.

An equal volume of filtrate and 3N H_2SO_4 was mixed and placed in a water bath for several minutes at 90 °C to coagulate the β-cellulose. The mixture was kept for 12 h to settle the precipitate. Afterward, the precipitate was separated and dried at 105 °C to obtain β-cellulose. Finally, the γ–cellulose was estimated by deducting the amount of α-cellulose and β-cellulose from the initial weight of the holocellulose. Here, diluted acids were used to extract as well as estimate the components of holocellulose from the lignocellulosic wastes due to its economic and environmental feasibility [21].

2.6. Conversion of Holocellulose into Carboxymethyl Cellulose

The general schematic diagram and synthesis route of CMC from the holocellulose of corn wastes is depicted in Figure 1. The synthesis of CMC includes two consecutive steps namely, alkalization and etherification. During alkalization, holocellulose was suspended in ethanol, and 30% (*w/v*) NaOH was added slowly for half an hour with vigorous stirring at room temperature, and the stirring was continued for an hour. In the etherification step, MCA (120%) was added gently to the slurry before placing into the water bath at 50 °C and heated for 3.6 h with intermittent stirring. The synthesized CMC was filtered and washed with 70% (*v/v*) alcohol to minimize the unwanted leftover and dried at 65 °C.

The outline of the mechanism for preparing CMC from holocellulose of corn wastes followed by two consecutive steps including basification as well as etherification are documented in Figure 1.

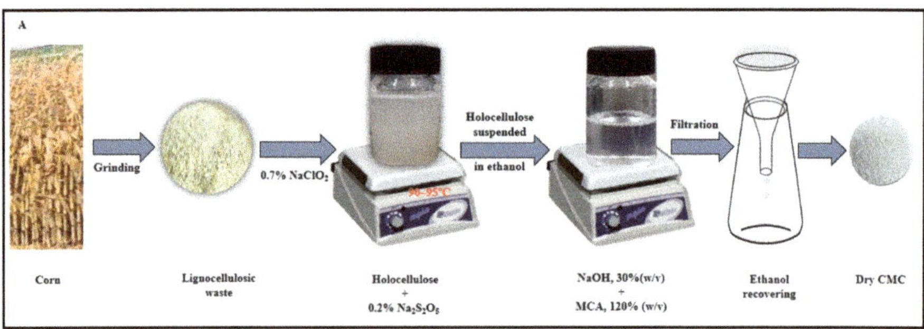

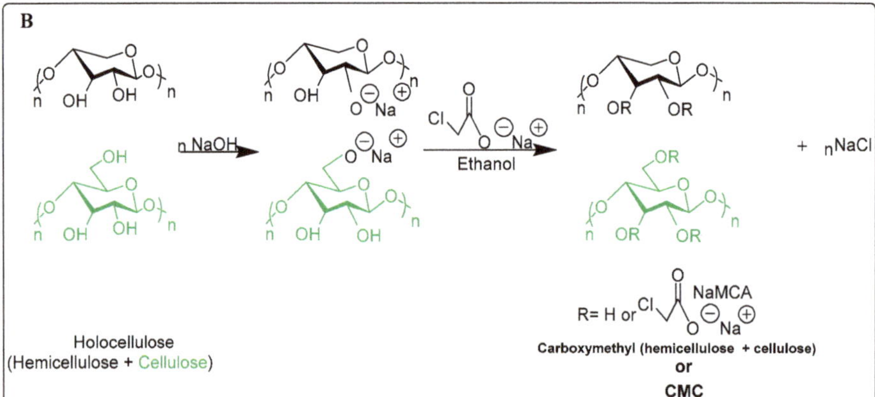

Figure 1. Schematic diagram of synthesizing of carboxymethyl cellulose (CMC) from holocellulose (**A**). Reaction mechanism of carboxymethylation of holocellulose (hemicellulose and cellulose) (**B**).

2.7. Estimation of Yield (%) of CMC

The yield of CMC was calculated based on a dry weight basis, where the weight of dried CMC was divided by the weight of holocellulose:

$$\text{CMC yield, } \% = \frac{y}{x} \times 100, \tag{3}$$

where y and x is the weight of moisture-free CMC (g) and holocellulose (g), respectively.

2.8. Determination of Degree of Substitution (DS)

Before the determination of DS, the prepared CMC was acidified by using the modified protocols as described elsewhere [22,23]. Briefly, 2 g of CMC powder was placed in 100 mL of a beaker, and 40 mL of 95% ethanol was added, followed by 10 min agitation. Then, 2.5 mL of nitric acid was added and the solution was boiled in a hotplate. Subsequently, after removing the solution from the hotplate, it was stirred for 15 min. The liquid solution was decanted by using a vacuum pump and washed with 80% ethanol for several times. Then, the residue was washed with a small quantity of methanol and filtered. To analyze the DS, 1.0 g of dried CMC was added to 100 mL of distilled water, and 12.50 mL of 1 N NaOH was added with agitation. After completely dissolving the mixture, it was then titrated by 1 N HCl in the presence indicator phenolphthalein. The DS of CMC was calculated by the utilization of the following equations [23]:

$$O = \frac{PQ - RS}{T}, \tag{4}$$

$$DS = \frac{0.162 \times O}{1 - 0.058 \times O}, \tag{5}$$

where

O = milli-equivalents of used HCl per gram of specimen;
P = volume of NaOH;
Q = concentration in the normality of NaOH;
R = volume of consumed HCl;
S = concentration in normality of HCl;
T = specimen grams;
162 is the molecular weight of the anhydrous glucose unit and 58 is the net increment in the anhydrous glucose unit for every substituted carboxymethyl group.

2.9. Determination of Molecular Weight

CMC powder was dissolved in 0.8 M NaOH aqueous solution to measure the molecular weight by using an Ostwald viscometer. From the value of intrinsic viscosity, the molecular weight of the CMC was calculated by using the Mark–Houwink–Sakurada equation i.e., $[\eta] = K \times M \times a$ [19]. Where K, a, $[\eta]$, and M are the constant for solvent, polymer shape factor, intrinsic viscosity, and molecular weight of CMC, respectively.

2.10. Structural Morphological and Thermal Study

FTIR spectrum analysis of the extracted holocellulose and synthesized CMC were performed by Fourier transform infrared (FTIR) spectroscopy (Perkin-Elmer 240C, Waltham, MA 02451, USA) in between 400 and 4000 cm^{-1}. For surface morphology of the dried samples, they were sputter-coated with gold for 10 min and then analyzed by using a scanning electron microscope (SEM) (Model-S 3400 N, VP SEM, Hitachi, Japan) using 20 kV accelerating voltage. The thermogravimetric analyses (TGA) of the samples were carried out using a Shimadzu TGA-50 system (Kyoto 604-8511, Japan) under a nitrogen atmosphere. The heating rate was 20 °C/min, and the temperature range was 25 to 600 °C.

2.11. Physicochemical Characteristics of Prepared CMC (Excipient) (with DS 1.83)

2.11.1. Moisture Content

The prepared CMC was dried in an oven at 105° for 120 min. Following this, the dried CMC was cooled in a desiccator until the ambient temperature was reached, and the final weight was taken. The moisture content was determined by this following equation [23]:

$$\text{Moisture content } (\%) = \frac{y}{w_0} \times 100, \tag{6}$$

where y and x is the final and initial weight of CMC (g), respectively.

2.11.2. Flow Properties of CMC

Flow properties of CMC were measured by performing several tests such as bulk and tap densities, true density, porosity, angle of repose, Carr's index, and Hausner's ratio. Detailed descriptions of these tests are given below.

2.11.3. Bulk and Tap Densities

A suitable amount of CMC powder (g) was poured in a 100 mL calibrated graduate cylinder and placed in a bulk density apparatus (LOGAN TAP-2S, New Jersey 08873 USA). After lightly tapping the cylinder, the occupied volume V_0 was estimated. After, that the cylinder was tapped 500 times for measuring the tap density and calculated by using the following relationship, respectively [24–26]:

$$\text{Bulk density (BD)} = \frac{w}{V_0}, \tag{7}$$

$$\text{Tap density (TD)} = \frac{w}{V_{500}}, \tag{8}$$

where w is the weight of CMC powder, V_0 is volume before tapping, and V_{500} is the volume of CMC powder after 500 times tapping.

2.11.4. True Density

The true density of the CMC powder was calculated by using a calibrated Quantachrome pycnometer (Quantachrome Corporation, FL, USA). CMC powder was dried at ambient temperature overnight under reduced pressure before analysis. True density was estimated by using the following formula [25,26]:

$$\text{True density (TD)} = \frac{w}{v}, \tag{9}$$

where w is the weight of the CMC powder and v is true volume of the CMC powder.

2.11.5. Porosity

The porosity of the CMC powder was determined according to the following equation [25,26]:

$$\text{True density (TD)} = 1 - \frac{\text{tap density}}{\text{true density}} \times 100. \tag{10}$$

2.11.6. Angle of Repose

The angle of repose of the CMC powder was measured by using a funnel and a Petri dish. At first, the funnel was fixed with a funnel holder. For making a cone, the CMC powder was allowed to emanate freely through the funnel. The height and diameter of the cone were recorded by a measuring scale and determined by using the following formula [25,26]:

$$\tan \theta \ = \frac{2h}{D}, \tag{11}$$

where D is diameter of the cone and h is the height of the cone.

2.11.7. Carr's Index and Hausner's Ratio

The Carr's index (CI) [25] and the Hausner ratio (HR) [26] were estimated by using the value tap and bulk density:

$$CI = \frac{\text{tap density} - \text{bulk density}}{\text{tap density}} \times 100, \tag{12}$$

$$HR = \frac{\text{tap density}}{\text{bulk density}}. \tag{13}$$

3. Results and Discussion

3.1. Chemical Composition, DS, Yield, and Molecular Weight of CMC

The chemical composition of corn waste residue including holocellulose (i.e., α, β, and γ- cellulose), lignin, pectic matter, fatty and waxy matter were estimated, and their results are shown in Table 1. From Table, it can be observed that the α-cellulose contents (i.e., 41.2%) were dominant among all other constituents, followed by β-cellulose and γ-cellulose contents. The amount of lignin, fatty and waxy matter, and pectic matters were 19.4%, 2.6%, and 3.6%, respectively. However, all types of holocellulose (i.e., α, β, and γ-cellulose) were converted into CMC through carboxymethylation. The synthetic route and reaction mechanism of synthesizing CMC from holocellulose is depicted in Figure 1.

Table 1. Composition of corn waste residue (Each test was performed at least three times and average and standard deviation were considered).

Holocellulose			Lignin, wt%	Fatty and Waxy Matters, wt%	Pectic Matters, wt%	Others, wt%
α-Cellulose, wt%	β-Cellulose, wt%	γ-Cellulose, wt%				
41.2 ± 1.1	15.2 ± 0.9	14.7 ± 1.0	19.4 ± 1.4	2.6 ± 0.2	3.6 ± 0.3	3.3 ± 0.5

Corn wastes were converted into CMC depending on different particle sizes. The values of DS, yield (wt%), and molecular weight of the obtained CMC are illustrated in Figures 2–4 respectively. From Figure 2, it can be seen that the DS of the prepared CMC was greatly dependent on the particle size of the starting material (holocellulose). Therefore, it shows a trade-off relationship with the particle size of the holocellulose i.e., the values of DS gradually increased with the decreasing size of the holocellulose. The highest DS value of 1.83 was obtained from the lowest particle size of 100 μm.

The highest yield (i.e., 182.55%) of CMC was found with the lowest particle size of holocellulose i.e., 100 μm, whereas the yield declined with increasing particle size of holocellulose (Figure 3). One might postulate that we have reported a higher yield of CMC in this study. This was highly desirable as the anhydrous glucose unit (molecular weight 162 g/moL) substituted into the sodium carboxymethyl groups (molecular weight 80 g/moL) based on the DS value. Therefore, the higher yield was highly desirable. However, we did a theoretical calculation and compared the data with the experimental findings. We found that theoretical yield was slightly higher (i.e., 7.0–8.6%) for theoretical mass yield, which indicated that successful substitution of the hydroxyl groups occurred by the sodium carboxymethyl groups. This result shows the resemblance with the principle that the reduced particle generates a larger surface area, which increases the chance of collisions between reactants and holocellulose. Therefore, the yield of CMC was increased by decreasing the holocellulose particle sizes [27].

From Figure 4, it can be noticed that the molecular weight of the prepared CMCs was significantly increased with the increasing DS value. From Figure 2, it was mentioned that the DS value was increased with the decreasing of holocellulose particle size. Since the smaller holocellulose particle size has a greater surface area, the excessive amount of reactants can infiltrate into the holocellulose at a time. In addition, Wang et al. mentioned that the etherification process significantly relies on the availability of the activated hydroxyl groups of AGU as well as the approachability of reactants [28]. Thus, higher DS, as well as molecular weight, were obtained due to the increasing number of -OH groups substituted by the sodium carboxymethyl group. The highest molecular weight of 457,910 Da was obtained with DS value of 1.83, whereas the lowest molecular weight of 100,388 Da was yielded with DS value of 0.34, as depicted in Figure 4. The molecular weight of the AGU was 162 g/mol, and the net gain in the AGU for every substituted sodium carboxymethyl group was 80 g/mol. Likewise, the carboxymethyl group is weightier than the -OH group, hence, the molecular weight of the CMC increased [23].

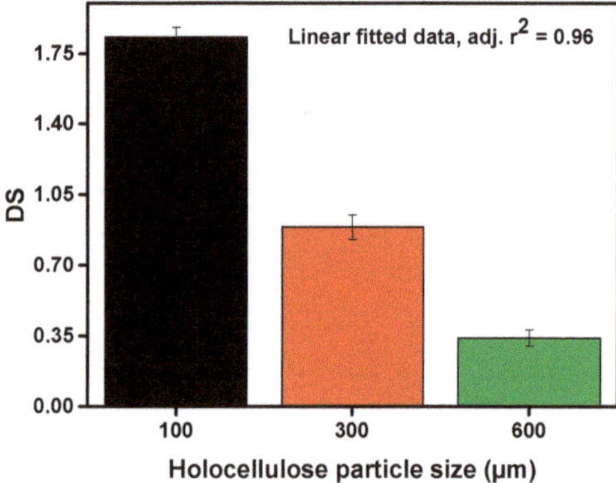

Figure 2. Degree of substitution of the synthesized CMC depending on varying particle size of holocellulose.

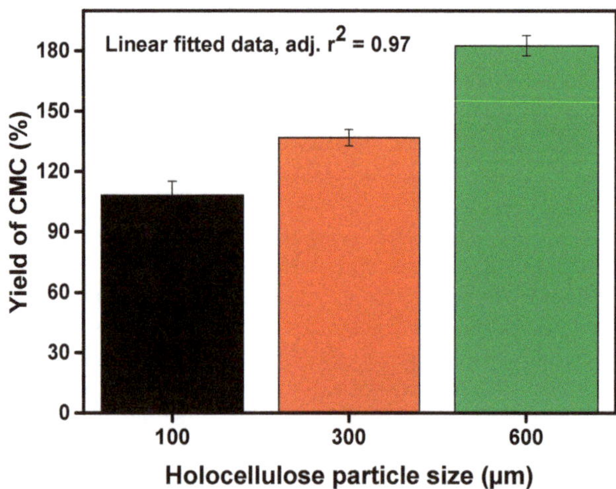

Figure 3. Converted amount of CMC as yield (wt%) via varying particle size of holocellulose.

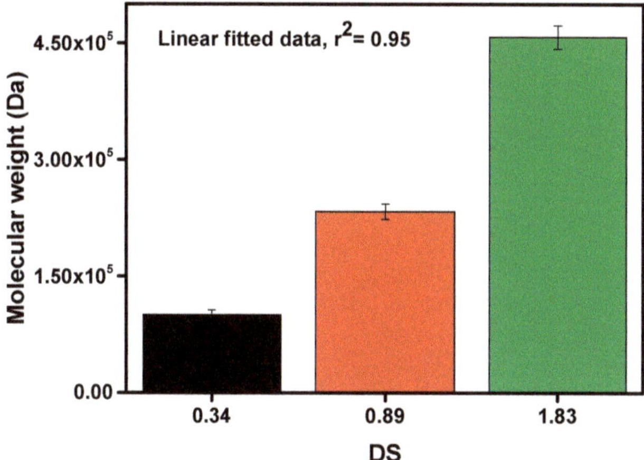

Figure 4. Estimation of molecular weight of prepared carboxymethyl cellulose (CMC) with different degrees of substitution value.

3.2. Structural Characterization

Different characterization techniques were used to characterize the synthesized CMC and holocellulose extracted from corn waste. For example, FTIR spectroscopy was used to analyze the surface functional groups present in the extracted holocellulose and synthesized CMC (Figure 5). From the FTIR spectra of the corn wastes powder and holocellulose, a distinct difference was noticed. Significant differences were observed in the wavenumber in between 1800 and 1000 cm^{-1}. Several peaks such as at wavenumber 1741 cm^{-1} (due to C=O stretching), 1637 cm^{-1} (due to carbonyl stretching conjugate with aromatic rings), and 1513 cm^{-1} (due aromatic C=C stretch) were not found in the extracted holocellulose [29]. Most specifically, peak at wavenumber 1250 cm^{-1} (due to C-O stretching vibration) was significantly reduced, and peak at wavenumber 1513 cm^{-1} (due to aromatic C=C stretch) was absent in the holocellulose. This phenomenon indicated that the lignin was removed during the extract process [30]. On the other hand, the wavenumbers at around 1423 and 1640 cm^{-1} were due to -CH$_2$ bending and O-H bending vibration of the absorbed water, respectively [31].

From the spectra of CMC, a broad absorption band at 3436 cm^{-1} was found, which indicated the presence of -OH group, and a band at 2928 cm^{-1} was attributed to the C-H stretching vibration [10,32]. In addition, a new and strong wavenumber at 1620 cm^{-1} was found, which confirmed the stretching vibration of carboxyl groups (COO$^-$), and a peak at 1424 cm^{-1} assigned to the salts of carboxyl groups [22]. The peaks at around 1327 and 1116 cm^{-1} can be assigned to -OH bending vibration and -C-O-C stretching, respectively. A wavelength of 898 cm^{-1} was found, which was due to 1 and 4-β glycosides of cellulose [11].

For the characterization of cellulose-based materials, SEM is one of the general techniques for imaging the microstructure and morphology of the materials. The morphologies of isolated holocellulose and synthesized CMC were also observed using an optical microscope, depicted in Figure 6. Ribbon shaped or rod-like morphology was found for the synthesized CMC, which is similar with the reported literature [33]. Furthermore, from Figure 6, it can also be observed that the surface morphology of extracted holocellulose is smoother with very low damage. In contrast, the morphology of the prepared CMC was more extended, rough, and collapsed [34]. In addition, the isolated holocellulose was further treated with sodium hydroxide during carboxymethylation, thus the ruptured surface was obtained from the synthesized CMC [35]. From Figure 6, it can be found that the size of the particles of holocellulose and CMC were in the range of 1.0–3.5 and 1.5–3.5 μm (approximately), respectively.

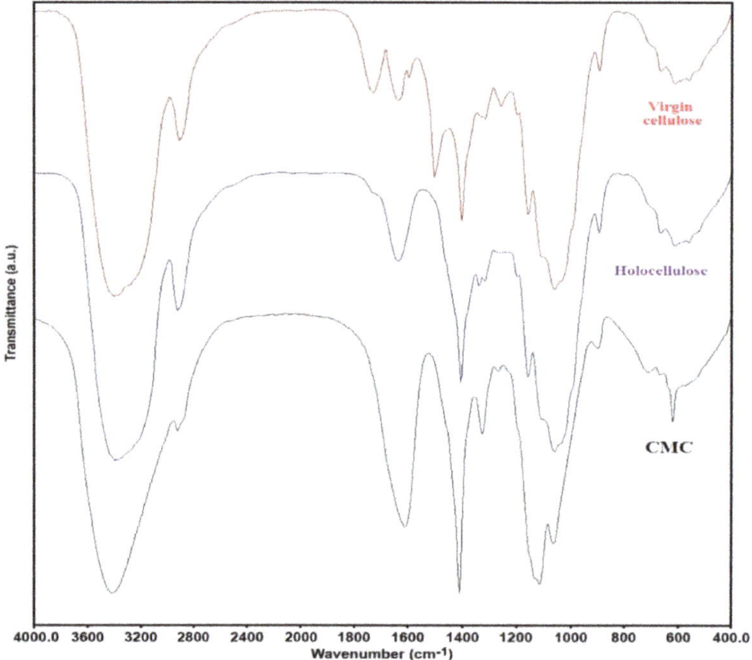

Figure 5. Structural properties of holocellulose and CMC (with degree of substitution 1.83). FTIR spectra of virgin cellulose, isolated holocellulose, and synthesized CMC.

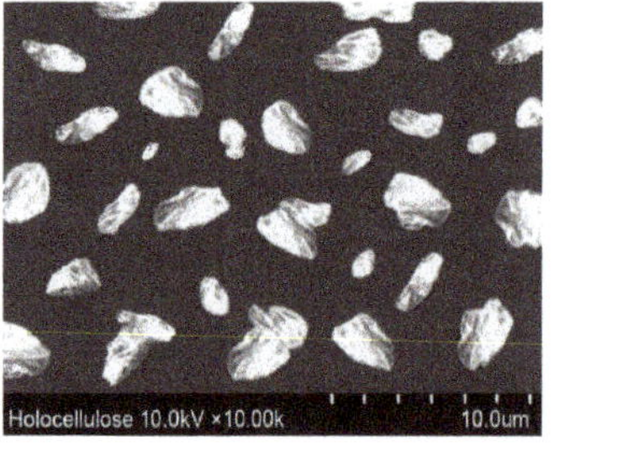

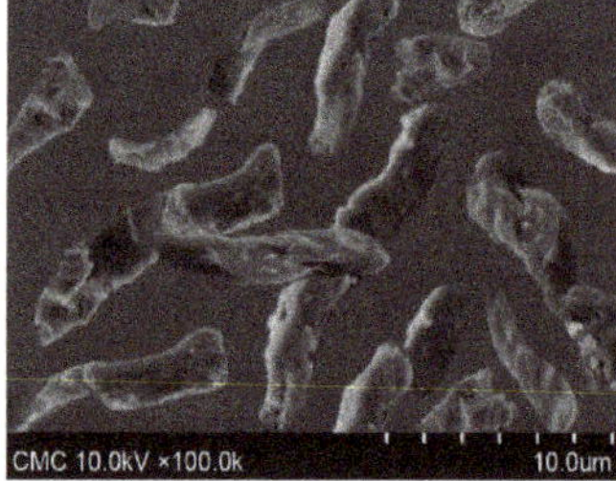

Figure 6. Surface morphology of isolated holocellulose and synthesized CMC (with DS 1.83).

XRD analysis is one of the prime methods of physical research. This technique was employed to investigate the degree of crystallinity before and after the carboxymethylation of cellulose [36]. XRD analysis of the samples is presented in the Figure 7. The obtained peaks of the samples were analogous to both crystalline and amorphous phases [37]. From the figure, it can be inferred that that holocellulose is more crystalline than CMC. More clearly, holocellulose gave five peaks at $2\theta = 14.2°$, $22.1°$, $27.4°$, $31.6°$, and $45.4°$ and the peaks were sharp, which indicated the presence of more crystalline phases on its core structure. On the other hand, in the CMC diffractogram, less number of peaks were found after carboxymethylation in comparison with the diffractogram of holocellulose in Figure 7. Therefore, the presence of more amorphous structures in the CMC compared to holocellulose can be seen since the characteristic peaks at $2\theta = 14.2°$, $22.1°$, and $27.4°$ became broader and intensity was reduced significantly. In addition, it can be observed that the typical peak at $2\theta = 31.6°$ for extracted holocellulose disappeared in CMC [9]. On the contrary, the peak at $2\theta = 45.4°$ reduced dramatically for CMC. This was due to the decrease in crystallinity in CMC as holocellulose was transformed from the crystalline to highly amorphous phase after carboxymethylation [38]. Furthermore, the peak at $2\theta = 14.2°$ still appeared in CMC, which indicated the presence of some sort of crystallinity, although the peak intensity was not sharp enough as holocellulose. Even though the percentage of crystallinity of synthesized CMC was not quantitatively determined, it can be supposed that the CMC adopted a disordered arrangement as compared to isolated holocellulose. This characteristic can be attributed to the presence of the carboxymethyl moieties which substituted the hydrogen atoms of the hydroxyl groups of cellulose [39]. Finally, holocellulose was treated with the alkaline solution during the carboxymethylation process, and as a result they swelled and showed tension with neighboring crystallites of cellulose molecules [40].

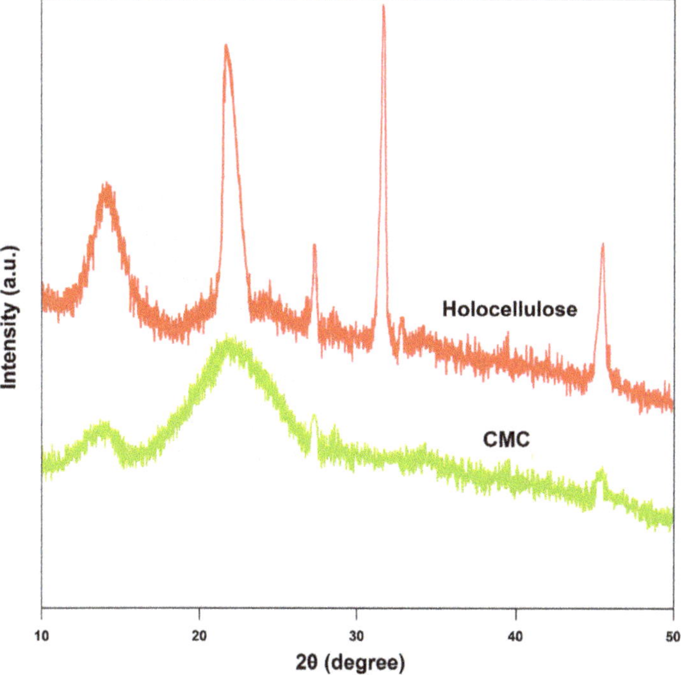

Figure 7. X-ray diffractogram of extracted holocellulose and synthesized CMC (with degree of substitution of 1.83).

TGA is a technique by which the thermal stability of a material can be analyzed, where the material is decomposed by the heat, and bonds are broken within the molecule [41]. During the test, when the maximum degradation occurs at a certain temperate, it is considered as an indicator of the stability of the material. The TGA graph of holocellulose and CMC is shown in Figure 8. From the figure, the weight loss of moisture, volatile compounds, and carbohydrate polymers during the carbonization phase can be observed.

According to the figure, 2.3% weight loss was observed in the case of holocellulose. This was due to evaporation of absorbed water. Lin et al. [42] reported that the decomposition temperatures of hemicellulose and cellulose were in the ranges of 200–315 °C and 360–400 °C, respectively. In our case, we observed the degradation stage of hemicellulose and cellulose in the ranges of ~300 °C and ~380–400 °C, respectively. The weight loss of holocellulose was around 79% in the temperature range of 340 to 387 °C, but holocellulose started to degrade at around 300 °C, which was due to the thermal decomposition of glycoside linkages of cellulose (i.e., hemicellulose). In addition, the final change was observed after 380 °C, which was due to the decomposition of α-cellulose [43,44].

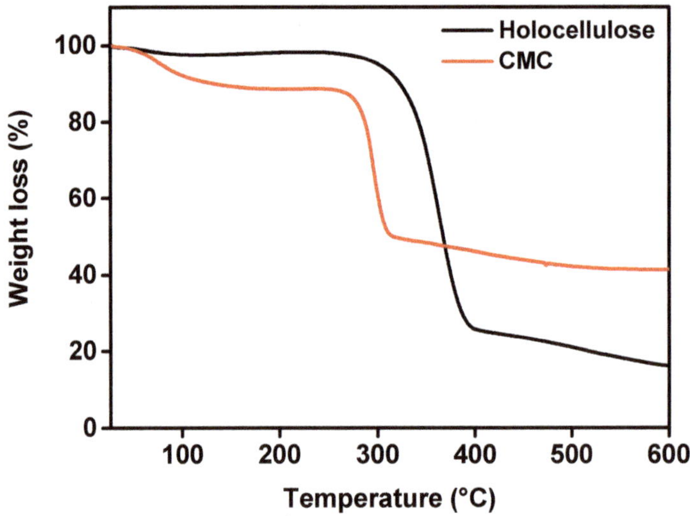

Figure 8. TGA curves of holocellulose and synthesized CMC (with degree of substitute 1.83).

On the other hand, CMC contained about 11.0% moisture. This primarily indicates that the CMC was more hygroscopic, and this is basically due to the presence of the carboxyl groups. The main decomposition of synthesized CMC was initiated at above 100 °C. This was due to the release of moisture from the CMC by the breaking of the hydrogen bond. Subsequently, the second and third stages decomposition was held between 283 and 318 °C. Finally, the weight loss was about 40%, which was probably due to the depolymerization of CMC by forming H_2O, CO, CO_2, and CH_4 [45]. In the temperature of 283 and 318 °C, CMC was decarboxylated as it contained COO^- groups in their structure. In addition, as shown in Figure 8, the thermal decomposition temperature further occurred from about 318 to 377 °C, which indicated a further loss of the mass up to 42.2%. The rate of weight loss gradually enhanced with the increase in temperature. This indicated the presence of more non-volatile fraction in CMC. More clearly, from the previous discussion, we confirmed that CMC had more irregular fractions of the amorphous structure, together with the sodium carboxymethyl unit. This unit was quite tough to break down at the referred temperature. Therefore, we assume that CMC should have higher stability than holocellulose (mass basis). This was mainly due to the presence of sodium ions in the residual mass, the degradation is quite tough. Henceforth, the mass residual mass (in the form of char) was expected to increase significantly, together with higher degradation temperature [46,47]. Therefore, in our case, we found that around 42.2% residual mass was even at a higher temperature compared to the holocellulose residual mass (i.e., ~21%).

3.3. Feasibility as Pharmaceuticals Excipient

Generally, microcrystalline cellulose is being used as excipient, which is extracted from hardwood and also from purified cotton. This process is considered an expensive process. Therefore, investigations are going on to find out the cheaper resources like CMC [19]. In general, the excipient has no medicinal functions. Excipients are being utilized in many ways in the pharmaceutical industries. They can be used as a binder, disintegrator, coating material, diluent, lubricant, and so on [19]. Especially, during the manufacturing of tablets, physical parameters of the excipient plays a crucial role. Herein, we compared the physicochemical characteristics of synthesized CMC with commercial CMC (as excipient). The physicochemical characteristics of synthesized CMC powder and commercial CMC are shown in Table 2. Moisture content plays a crucial role in the case of flow properties of the powder. The effect of moisture on powder flowability depends on the amount of water and its distribution. It is quite common that the flowability of the powder often decreases with the increase of moisture content. Because a higher amount of water will increase the thickness of the adsorbed liquid layer, it increases the strength of liquid bridges. As a result, the powder becomes more cohesive and tends to form agglomerates [48]. It is clear from the table that the physicochemical properties of synthesized CMC power remained within acceptable limits with not more than 10% variations [49]. However, the bulk density rendered an estimate of the ability of a material to flow, which is related to compressibility.

Table 2. Physicochemical characteristics of synthesized CMC as an excipient.

Characteristics	CMC Powder (with DS 1.83)	CMC Powder (Excipient Grade with DS 0.9)
Moisture content	1.36 ± 0.00	0.94 ± 0.01
Bulk density (g/mL)	0.50 ± 0.004	0.52 ± 0.003
Tap density (g/mL)	0.51 ± 0.003	0.55 ± 0.001
True density (g/mL)	1.75 ± 0.007	1.84 ± 0.005
Porosity	70.82	80.03
Angle of repose	38.55 ± 1.70	41.27±1.20
Carr's index	8.14 ± 1.20	7.24 ± 1.10
Hausner's ratio	1.37 ± 0.010	1.18 ± 0.004

Each test was performed at least three times and average and standard deviation were considered.

On the other hand, the tap density is a measure of the property of any sample of how well a powder can be packed in a confined space on recurring tapping. Bulk and tap density values of synthesized CMC showed almost similar behavior with the commercial CMC, as depicted in Table 2. It is generally considered that the better the potential for a material to flow and to re-arrange under compression, the higher the bulk and tapped densities [50]. In general, the higher true density of a powder reflects the better compressibility. The true density of synthesized CMC powder was approximately the same compared with commercial CMC. The total porosity or void fraction of a powder is the measurement of the void or empty space between the particles as well as pores within the particles of a material. In addition, it is a fraction of the volume of the void over the total volume, between 0–1 or as a percentage between 0–100%. The porosity value of the synthesized CMC was similar to the commercial CMC, which reveals the presence of poly-sized particles. Hence, synthesized CMC can easily be compressed during tablet making. The angle of repose of powder is another important criterion that gives a qualitative assessment of its internal and cohesive frictions. Angles of up to 40° show reasonable flow potential of the solid powders. On the other hand, samples with angles greater than 50° exhibit poor or absent flow [51]. In this study, the angle of repose of the synthesized CMC was found to be 38.5, which indicates there was no significant difference when compared with commercial CMC. To evaluate the flow properties of the powder, the Carr index and Hausner ratio have been commonly used. Carr's compressibility index exerts a clue of how much powder can be compressed, whereas Hausner index measures the cohesion between particles and the particle flows inversely with the values of the Carr index and Hausner ratio [50,52]. In the case of Carr's index, values ranged between 5 to 10, 12 to 16, 18 to 21, and 23 to 28, indicating excellent, good, fair and poor flow properties of the material, respectively [50]. The Carr's index of the synthesized CMC powder lies in the range of 5 to 10, which indicates its excellent flow properties. By contrast, the Hausner ratio (<1.20) often indicates good flowability of material, whereas a value of 1.5 or higher suggests a poor flow display by the material [52]. The Hausner ratio of the synthesized CMC showed a good flow property with value at around 1.3 in comparison with commercial CMC. Finally, it can be inferred from the discussion that the newly synthesized CMC from the agro waste can be a great source of commercial excipient.

4. Conclusions

CMC was successfully synthesized from extracted holocellulose. The highest yield of 182.55% was obtained with lower particle size (100 µm) of the holocellulose. The higher yield was reported due to the smaller particle sizes with DS of 1.83, which significantly contributed to incorporate more carboxylic groups in the CMC. Hence, it was used as a starting material for CMC synthesis. Different characterization analyses showed that CMC was successfully synthesized with ideal structural, morphological, as well as thermal properties. It was also found that the synthesized CMC had a good pharmaceutical excipient compatibility with a commercial excipient. The synthesized CMC with DS value of 1.83 can be a promising pharmaceutical excipient than other CMC bases excipients. Hence, this study has shown an efficient way of converting corn waste (lignocellulosic biomass) into value-added material, which can be a great aspect to be used as a pharmaceutical excipient, together with further applicability in real industrial application such as tablet making and other fields.

Author Contributions: Data curation, M.S.R. and M.S.Y.; Formal analysis, M.S.R., M.A.H. and M.B.A.; Investigation, M.S.R.; Methodology, M.S.R.; Resources, M.I.H.M. and M.S.Y.; Software, M.S.R., M.S.Y., M.A.H. and M.A.B.; Supervision, M.I.H.M. and M.A.S.; Writing–original draft, M.S.R.; Writing–review and editing, M.I.H.M. and M.B.A. All authors have read and agree to the published version of the manuscript.

Funding: This research received no external funding.

Conflicts of Interest: The authors declare no conflict of interest.

References

1. Dong, H.; Zheng, L.; Yu, P.; Jiang, Q.; Wu, Y.; Huang, C.; Yin, B. Characterization and application of lignin–carbohydrate complexes from lignocellulosic materials as antioxidants for scavenging in vitro and in vivo reactive oxygen species. *ACS Sustain. Chem. Eng.* **2020**, *8*, 256–266. [CrossRef]

2. Arora, A.; Nandal, P.; Singh, J.; Verma, M.L. Nanobiotechnological advancements in lignocellulosic biomass pretreatment. *Mater. Sci. Energy Technol.* **2020**, *3*, 308–318. [CrossRef]

3. Tigchelaar, M.; Battisti, D.S.; Naylor, R.L.; Ray, D.K. Future warming increases probability of globally synchronized maize production shocks. *Proc. Natl. Acad. Sci. USA* **2018**, *115*, 6644–6649. [CrossRef] [PubMed]

4. Zhang, Y.; Ghaly, A.E.; Li, B. Physical Properties of Corn Residues. *Am. J. Biochem. Biotechnol.* **2012**, *8*, 44–53.

5. Lin, W.; Xing, S.; Jin, Y.; Lu, X.; Huang, C.; Yong, Q. Insight into understanding the performance of deep eutectic solvent pretreatment on improving enzymatic digestibility of bamboo residues. *Bioresour. Technol.* **2020**, *306*, 123163. [CrossRef]

6. Giuliano, A.; Barletta, D.; De Bari, I.; Poletto, M. Techno-economic assessment of a lignocellulosic biorefinery co-producing ethanol and xylitol or furfural. *Comput. Aided Chem. Eng.* **2018**, *43*, 585–590.

7. Rop, K.; Mbui, D.; Njomo, N.; Karuku, G.N.; Michira, I.; Ajayi, R.F. Biodegradable water hyacinth cellulose-graft-poly (ammonium acrylate-co-acrylic acid) polymer hydrogel for potential agricultural application. *Heliyon* **2019**, *5*, e01416. [CrossRef] [PubMed]

8. Lin, W.; Chen, D.; Yong, Q.; Huang, C.; Huang, S. Improving enzymatic hydrolysis of acid-pretreated bamboo residues using amphiphilic surfactant derived from dehydroabietic acid. *Bioresour. Technol.* **2019**, *293*, 122055. [CrossRef]

9. Klemm, D.; Heublein, B.; Fink, H.P.; Bohn, A. Cellulose: Fascinating biopolymer and sustainable raw material. *Angew. Chem.* **2005**, *44*, 3358–3393. [CrossRef]

10. Santos, D.M.D.; Bukzem, A.D.L.; Ascheri, D.P.R.; Signini, R.; Aquino, G.L.B.D. Microwave-assisted carboxymethylation of cellulose extracted from brewer's spent grain. *Carbohydr. Polym.* **2015**, *131*, 125–133. [CrossRef] [PubMed]

11. Rose, G.P.; Viera, G.R.F.; De Rosana, A.M.N.; Da Carla, S.M.; Julia, G.V. Synthesis and characterization of methyl cellulose from sugar cane bagasse cellulose. *Carbohydr. Polym.* **2007**, *67*, 182–189.

12. He, X.; Wu, S.; Fua, D.; Nia, J. Preparation of sodium CMC from paper sludge. *J. Chem. Technol. Biotechnol.* **2009**, *84*, 427–434. [CrossRef]

13. Gulati, I.; Park, J.; Maken, S.; Lee, M.G. Production of carboxymethylcellulose fibers from waste lignocellulosic sawdust using NaOH/NaClO$_2$ pretreatment. *Fiber. Polym.* **2014**, *15*, 680–686. [CrossRef]

14. Nagieb, Z.; Sakhawy, M.E.; Samir, K. Carboxymethylation of cotton linters in alc. Medium. *Int. J. Polym. Mater.* **2001**, *50*, 163–173. [CrossRef]

15. Yang, X.H.; Zhu, W.L. Viscosity properties of sodium carboxymethylcellulose solutions. *Cellulose* **2007**, *14*, 409–417. [CrossRef]

16. Almlöf, H.; Schenzel, K.; Germgård, U. Carboxymethyl Cellulose Produced at Different Mercerization Conditions and Characterized by NIR FT Raman Spectroscopy in Combination with Multivariate Analytical Methods. *Bioresources* **2013**, *8*, 1918–1932.

17. Khullar, R.; Varshney, V.K.; Naithani, S.; Heinze, T.; Soni, P.L. Carboxymethylation of Cellulosic Material (Average Degree of Polymerization 2600) Isolated from Cotton(Gossypium) Linters with Respect to Degree of Substitution and Rheological Behavior. *J. Appl. Polym. Sci.* **2005**, *96*, 1477–1482. [CrossRef]

18. Dürig, T.; Karan, K. *Binders in Wet Granulation, Handbook of Pharmaceutical Wet Granulation, Theory and Practice in a Quality by Design Paradigm*; Academic Press: Cambridge, MA, USA, 2019; Chapter 9; pp. 317–349.

19. Jacques, E.R.; Alexandridis, P. Tablet Scoring: Current Practice, Fundamentals, and Knowledge Gaps. *Appl. Sci.* **2019**, *9*, 3066. [CrossRef]

20. Mondal, M.I.H.; Haque, M.M.U. Effect of grafting methacrylate monomers onto jute constituents with a potassium per sulfate initiator catalyzed by Fe (II). *J. Appl. Polym. Sci.* **2007**, *103*, 2369–2375. [CrossRef]

21. Da Silva, A.R.G.; Giuliano, A.; Errico, M.; Rong, B.G.; Barletta, D. Economic value and environmental impact analysis of lignocellulosic ethanol production: Assessment of different pretreatment processes. *Clean Technol. Environ. Policy* **2019**, *21*, 637–654. [CrossRef]

22. Bono, A.; Ying, P.H.; Yan, F.Y.; Muei, C.L.; Sarbatly, R.; Krishnaiah, D. Synthesis and characterization of carboxymethyl cellulose from palm kernel cake. *Adv. Natl. Appl. Sci.* **2009**, *3*, 5–11.

23. Mondal, M.I.H.; Alam, A.B.M.F. Utilization of cellulosic wastes in textile and garment industries: 2. Synthesis and characterization of cellulose acetate from knitted rag. *J. Polym. Environ.* **2013**, *21*, 280–285.

24. Das, S.; Mondal, S.; Ghosh, S. Interaction of cationic gemini surfactant tetramethylene-1,4-bis(dimethyltetradecylammonium bromide) with anionic polyelectrolyte sodium carboxymethyl cellulose, with two different molar masses, in aqueous and aquo-organic (isopropanol) media. *RSC Adv.* **2016**, *6*, 30795–30803. [CrossRef]

25. Carr, R.L. Classifying flow properties of solids. *Chem. Eng.* **1965**, *72*, 69–72.

26. Hausner, H.H. Friction conditions in a mass of metal powder. *Int. J. Powder Metall.* **1967**, *3*, 7–13.

27. Fu, Q.S.; Xue, Y.Q.; Cui, Z.X.; Wang, M.F. Study on the Size-Dependent Oxidation Reaction Kinetics of Nanosized Zinc Sulfide. *J. Nanomater.* **2014**, *2014*, 1–8. [CrossRef]

28. Wang, Y.; Wang, X.; Xie, Y.; Zhang, K. Functional nanomaterials through esterification of cellulose: A review of chemistry and application. *Cellulose* **2018**, *25*, 3703–3731. [CrossRef]

29. Shi, J.; Xing, D.; Li, J. FTIR Studies of the Changes in Wood Chemistry from Wood Forming Tissue under Inclined Treatment. *Energy Procedia* **2012**, *16*, 758–762. [CrossRef]

30. Sun, X.F.; Xu, F.; Sun, R.C.; Fowler, P.; Baird, M.S. Characteristics of Degraded Cellulose Obtained from Steam-Exploded Wheat Straw. *Carbohydr. Res.* **2005**, *340*, 97–106. [CrossRef]

31. Chowdhury, Z.Z.; Chandran, R.; Jahan, A.; Khalid, K.; Rahman, M.M.; Al-Amin, M.; Akbarzadeh, O.; Badruddin, I.A.; Khan, T.; Kamangar, S. Extraction of Cellulose Nano-Whiskers Using Ionic Liquid-Assisted Ultra-Sonication: Optimization and Mathematical Modelling Using Box–Behnken Design. *Symmetry* **2019**, *11*, 1148. [CrossRef]

32. Basuny, M.; Ali, I.O.; El-Gawad, A.A.; Bakr, M.F.; Salama, T.M. A fast green synthesis of Ag nanoparticles in carboxymethyl cellulose (CMC) through UV irradiation technique for antibacterial applications. *J. Sol-Gel Sci. Technol.* **2015**, *75*, 530–540. [CrossRef]

33. Klinpituksa, P.; Kosaiyakanon, P. Superabsorbent polymer based on sodium carboxymethyl cellulose grafted polyacrylic acid by inverse suspension polymerization. *Int. J. Polym. Sci.* **2017**, *2017*, 1–6. [CrossRef]

34. Rachtanapun, P.; Simasatitkul, P.; Chaiwan, W.; Watthanaworasakun, Y. Effect of sodium hydroxide concentration on properties of carboxymethyl rice starch. *Int. Food Res. J.* **2012**, *19*, 923–931.

35. Donald, A.M.; Kato, K.L.; Perry, P.A.; Waigh, T.A. Scattering Studies of the Internal Structure of Starch Granules. *Starch—Stärke* **2001**, *53*, 504–512. [CrossRef]

36. Beyene, D.; Chae, M.; Dai, J.; Danumah, C.; Tosto, F.; Demesa, A.G.; Bressler, D.C. Characterization of cellulase-treated fibers and resulting cellulose nanocrystals generated through acid hydrolysis. *Materials* **2018**, *11*, 1272. [CrossRef]

37. Park, S.; Baker, J.O.; Himmel, M.E.; Parilla, P.A.; Johnson, D.K. Cellulose crystallinity index: Measurement techniques and their impact on interpreting cellulase performance. *Biotechnol. Biofuels* **2010**, *3*, 10. [CrossRef]

38. Parid, M.D.; Rahman, N.A.A.; Baharuddin, A.S.; Mohammed, M.A.P.; Johari, A.M.; Razak, S.Z.A. Synthesis and characterization of carboxymethyl cellulose from oil palm empty fruit bunch stalk fibres. *BioRes* **2018**, *13*, 535–554.

39. Johar, N.; Ahmad, I.; Dufresne, A. Extraction, preparation and characterization of cellulose fibres and nanocrystals from rice husk. *Ind. Crop. Prod.* **2012**, *37*, 93–99. [CrossRef]

40. Fang, J.M.; Fowler, P.; Tomkinson, J.; Hill, S. The Preparation and characterisation of a series of chemically modified potato starches. *Carbohydr. Polym.* **2002**, *47*, 245–252.

41. Loof, D.; Hiller, M.; Oschkinat, H.; Koschek, K. Quantitative and qualitative analysis of surface modified cellulose utilizing TGA-MS. *Materials* **2016**, *9*, 415. [CrossRef]

42. Lin, B.J.; Chen, W.H. Sugarcane bagasse pyrolysis in a carbon dioxide atmosphere with conventional and microwave-assisted heating. *Front. Energy Res.* **2015**, *3*, 1–9.

43. Seo, J.M.; Cho, D.; Park, W.H.; Han, S.O.; Hwang, T.W.; Choi, C.H.; Jun, S.J. Fiber surface treatments for improvement of the interfacial adhesion and flexural and thermal properties of jute/poly lactic acid biocomposites. *J. Biobased Mater. Bioenergy* **2007**, *1*, 331–340.

44. Aziz, S.H.; Ansell, M.P. The effect of alkalization and fibre alignment on the mechanical and thermal properties of kenaf and hemp bast fibre composites: Part 1—Polyester resin matrix. *Compos. Sci. Technol.* **2004**, *64*, 1219–1230. [CrossRef]

45. Durcilene, A.; Silva, D.; Paula, R.C.M.; Feitosa, J.P.A. Graft copolymerisation of acrylamide onto cashew gum. *Eur. Polym. J.* **2007**, *43*, 2620–2629.

46. Doh, S.J.; Lee, J.Y.; Lim, D.Y.; Im, J.N. Manufacturing and analyses of wet-laid nonwoven consisting of carboxymethyl cellulose fibers. *Fiber. Polm.* **2013**, *14*, 2176–2184. [CrossRef]

47. El-Sakhawy, M.; Tohamy, H.A.S.; Salama, A.; Kamel, S. Thermal properties of carboxymethyl cellulose acetate butrate. *Cellulose Chem. Technol.* **2019**, *53*, 667–675. [CrossRef]

48. Sandler, N.; Reiche, K.; Heinämäki, J.; Yliruusi, J. Effect of Moisture on Powder Flow Properties of Theophylline. *Pharmaceutics* **2010**, *2*, 275–290. [CrossRef]

49. *United States Pharmacopeia and National Formulary (USP 29-NF 24)*; United States Pharmacopeia Convention: Rockville, MD, USA, 2007; Volume 31, p. 1349. Available online: http://www.pharmacopeia.cn/v29240/usp29nf24s0_m13210.html (accessed on 16 May 2020).

50. Carr, R.L. Evaluating flow properties of solids. *Chem. Eng.* **1965**, *72*, 163–168.

51. Fowler, H.W. Powder flow and compaction. In *Cooper and Gunn's Tutorial Pharmacy*, 6th ed.; Carter, S.J., Ed.; CBS Publishers: Delhi, India, 2000.

52. Wells, J.I. *Pharmaceutical Preformulation: The Physicochemical Properties of Drug Substances*, 1st ed.; John Wiley and Sons: New York, NY, USA, 1988.

 *processes*

Article

Isolation, Identification, and Optimization of γ-Aminobutyric Acid (GABA)-Producing *Bacillus cereus* Strain KBC from a Commercial Soy Sauce *moromi* in Submerged-Liquid Fermentation

Wan Abd Al Qadr Imad Wan-Mohtar [1,2], Mohamad Nor Azzimi Sohedein [1], Mohamad Faizal Ibrahim [3,*], Safuan Ab Kadir [1], Ooi Poh Suan [4], Alan Wong Weng Loen [4], Soumaya Sassi [1,5] and Zul Ilham [2,5]

[1] Functional Omics and Bioprocess Development Laboratory, Biotechnology Program, Institute of Biological Sciences, Faculty of Science, University of Malaya, Kuala Lumpur 50603, Malaysia; qadyr@um.edu.my (W.A.A.Q.I.W.-M.); azzimi@um.edu.my (M.N.A.S.); safuankadir1207@gmail.com (S.A.K.); sassi.soumaya94@gmail.com (S.S.)

[2] Bioresources and Bioprocessing Research Group, Institute of Biological Sciences, Faculty of Science, University of Malaya, Kuala Lumpur 50603, Malaysia; ilham@um.edu.my

[3] Department of Bioprocess Technology, Faculty of Biotechnology and Biomolecular Sciences, Universiti Putra Malaysia, Serdang 43400, Malaysia

[4] Lot 3406, Jalan Perusahaan 3, Kamunting Industrial Area, Kwong Bee Chun Sdn. Bhd. Soy Sauce Factory, Kamunting 34600, Malaysia; ps.ooi_06@yahoo.com (O.P.S.); kkkbcfood@yahoo.com (A.W.W.L.)

[5] Biomass Energy Laboratory, Institute of Biological Sciences, Faculty of Science, University of Malaya, Kuala Lumpur 50603, Malaysia

* Correspondence: faizal_ibrahim@upm.edu.my; Tel.: +60-397691936

Received: 24 March 2020; Accepted: 24 April 2020; Published: 30 May 2020

Abstract: A new high γ-aminobutyric acid (GABA) producing strain of *Bacillus cereus* was successfully isolated from soy sauce *moromi*. This *B. cereus* strain named KBC shared similar morphological characteristics (Gram-positive, rod-shaped) with the reference *B. cereus*. 16S rRNA sequence of *B. cereus* KBC was found to be 99% similar with *B. cereus* strain OPWW1 under phylogenetic tree analysis. *B. cereus* KBC cultivated in unoptimized conditions using De Man, Rogosa, Sharpe (MRS) broth was capable of producing 523.74 mg L^{-1} of GABA within five days of the cultivation period. By using response surface methodology (RSM), pH level, monosodium glutamate (MSG) concentration and temperature were optimized for a high concentration of GABA production. The pH level significantly influenced the GABA production by *B. cereus* KBC with p-value = 0.0023. GABA production by *B. cereus* KBC under the optimized condition of pH 7, MSG concentration of 5 g L^{-1} and temperature of 40 °C resulted in GABA production of 3393.02 mg L^{-1}, which is 6.37-fold higher than under unoptimized conditions. Overall, this study has shown that *B. cereus* KBC isolated from soy sauce *moromi* is capable of producing a high concentration of GABA together with the optimal fermentation conditions that have been statistically analysed using RSM.

Keywords: GABA; fermented food; functional food; non-protein amino acid; soy sauce fermentation

---

1. Introduction

Gamma-aminobutyric acid (GABA) is a non-protein amino acid produced by the α-decarboxylation of L-glutamate. The non-protein amino acid is the molecular compound that has the standard structure of an amino acid consisting of N-terminal and C-terminal. Most amino acids have at least one asymmetric carbon and are chiral [1]. Amino acids are classified as non-protein when they are not part of the 22 such molecules that are translated into proteins by the standard genetic code [2]. These non-protein amino acids play important roles as metabolites in the organism, function as allelopathic chemicals, nutrient acquisition and in signalling as well as a stress response. Researchers also reported that they are responsible for significant medical issues in both invertebrate and vertebrate animals [3]. Biosynthesis of non-protein amino acid of GABA is catalysed by the glutamate decarboxylase (GAD) enzyme [4]. GABA is usually produced by microorganisms associated with fermented foods, such as fermented fish [5], fermented cod gut [6], fermented *tempeh* [7], fermented milk [8], fermented *adzuki* beans [9] and recently fermented soybean of soy sauce [10]. GABA plays an important role in the central nervous system (CNS) as the primary inhibitory neurotransmitter. Due to its major inhibitory functions in the brain, GABA is studied as a treatment for various neurological diseases, such as epilepsy, schizophrenia, stiff-person syndrome and anxiety disorders [11]. GABA also demonstrated hypotensive, diuretic and tranquillizer effects [12]. A high concentration of GABA can also be found in pancreatic islets, which was associated with insulin secretion, hence, can be used to treat diabetes [13,14]. Furthermore, the oral administration of GABA can effectively decrease blood pressure in a hypertensive patient [8]. These pharmaceutical applications of GABA have risen the commercial production of functional food with a high concentration of GABA as its bioactive component.

In order to commercially produce GABA, several biosynthetic techniques were employed, such as by using sourdough fermentation, immobilized cell technology and batch fermentation [15]. GABA produced from the fermentation of naturally-occurring microorganisms have higher demand in comparison to chemically-synthesized GABA since customers prefer a nutrient source that is naturally-produced rather than chemically-synthesized [16]. Production of GABA via naturally-occurring microorganisms is preferred as they are naturally present in the food production processes, low-cost, and can be adopted as a functional food [10] rather than painstaking and expensive chemically-synthesized GABA [17], which are more prone to antihypertensive drugs [18]. Naturally, humans have an innate sense toward natural things; therefore, food naturalness is crucial among the majority of consumers [19]. Furthermore, natural GABA also improves the taste of the food while reducing the risk of contamination with pathogenic microorganisms [20]. In order to identify a high GABA-producing strain, numerous microorganisms such as *Lactobacillus paracasei*, *Lactobacillus plantarum*, *Lactobacillus brevis*, *Lactococcus lactis*, *Streptococcus salivarius*, *Uonascus purpureu* and *Streptomyces bacillaris* have been isolated from various type of food like cheeses, kimchi, tea, fresh milk and fermented fish [15]. There are many studies that have reported that the culture conditions play an important role in GABA production. It can be done by (1) finding and/or developing a higher-yield GABA-producing microorganism and (2) optimizing fermentation conditions. GABA produced in fermented food is a good option as it can be directly consumed [21].

Soy sauce is a liquid seasoning used worldwide in cooking and eating [22]. It can be a potential natural functional food for GABA production since the chemically synthesized GABA are expensive compared to naturally produced-GABA [23]. In 2016, several isolated microorganisms from local Malaysian soy sauce producers were reported to be capable of producing GABA due to the abilities of the trio microorganisms, i.e., *Aspergillus oryzae* NSK (from *koji*), *Bacillus* sp. (from *moromi*) and strict anaerobic *Tetragenococcus* sp. (from *moromi*). Based on these findings, the ability of traditional solid-state fermentation of soy sauce production processes using soybean as a substrate has been explored to produce GABA by the microorganism from *moromi*. Previous studies have been done to isolate microorganisms from soy sauce *koji* for GABA production. For example, *Aspergillus oryzae* NSK isolated from the soy sauce *koji* [24] produced 73.13 ± 1.77 mg L^{-1} of GABA. However, there is no conclusive study has been done to determine the capability of microorganisms isolated from the *moromi* stage for GABA production. Therefore, a complete study from isolation, identification and optimization of a newly isolated microorganism from commercial soy sauce *moromi* for GABA production has been conducted.

2. Materials and Methods

2.1. Isolation of B. cereus KBC

Three *moromi* samples ranging from 10, 25 and 80 days were obtained from an established commercial soy sauce factory in Perak, Malaysia. Soy sauce production consists of several stages of fermentation (Figure 1). The first stage is a solid-state fermentation, in which soybeans and wheat flour are mixed with fungal species, such as *Aspergillus sojae* or *Aspergillus oryzae*, as the starter and left fermenting for several days to form *koji* [25]. During the first stage of making *koji*, the soybean was soaked overnight using tap water to allow it to be softened. The dehulled soybean was boiled at 100 °C for 3 h until the colour appeared to be slightly golden. Meanwhile, 1 kg of dried soybean was mixed with 500 g of wheat flour and 0.0005 g of *A. oryzae* that works as a starter. The mixture was loaded into four standard steel trays evenly, each loaded with 4 cm thickness. The *koji* trays were covered thoroughly by the parchment paper and incubated at around 27–35 °C for two weeks. After the incubation in a humid and aerated incubation chamber was completed, a greenish-yellow mash could be observed, known as *koji*. The second stage involves the mixing of *koji* with 20% (*w/v*) of sodium chloride solution (brine) and incubating for six months in a tank to produce *moromi*. *Moromi* is pressed in the final stage to collect the liquid product, which is the soy sauce.

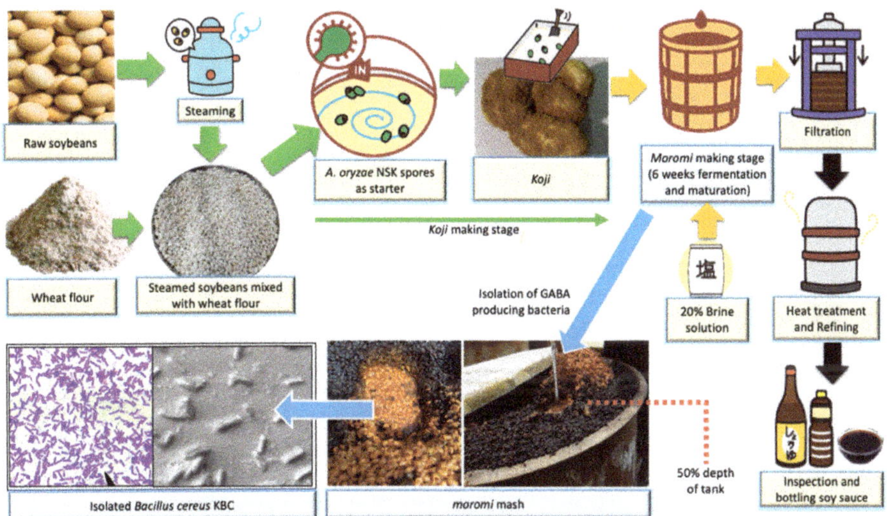

Figure 1. Soy sauce production blueprint and isolation point for *Bacillus cereus* strain KBC.

The collected *moromi* samples were stored in 500 mL Schott bottles in the cold temperature of less than 4 °C. Brine sample at the bottom of the *moromi* sample was pipetted and serial dilution was conducted. The isolation was conducted by spreading and streaking the *moromi* sample on de Man, Rogosa, Sharpe (MRS) (69964-500G, Sigma-Aldrich, Dorset, UK) agar plates. The agar plates were incubated at 30 °C for 5 days under anaerobic conditions using an anaerobic jar supplied with *Anaerocult* A (Merck) following the methods by Liu et al. [26]. After incubation, the plates were screened for colony growth and a single colony was streaked on a new MRS agar plate to obtain a pure isolate. The pure colonies were transferred to MRS media slants and kept as stock cultures at −4 °C. This pure isolate was used for morphological observation and identification. For long term preservation, the identified master strain was cultured overnight in MRS broth and 500 μL of the overnight culture was inoculated into a 2 mL glass vial containing 500 μL of 50% *v/v* glycerol. The glass vials containing the master strains were stored in −80 °C for future use.

2.2. Morphological Analysis

The isolated pure culture was Gram-stained and observed under a light microscope [27] to identify their morphological characteristics. This isolated pure culture was also observed under 1000× magnification using a Scanning Electron Microscope (SEM) (Brand ZEISS, Model MERLIN Compact, Oberkcohen, Germany) [28] to confirm its morphological structure.

2.3. Identification Using 16S rRNA

Identification of *Bacillus cereus* strain KBC was conducted by Apical Scientific Sdn. Bhd. (Seri Kembangan, Selangor, Malaysia) using 16S rRNA sequencing with 1.5 kb full length based on Tamura et al. [29]. The extracted bacterial DNA was amplified using universal primers 27F and 1492R, as mentioned by Jawan et al. [30]. The procedures consisted of 25 µL of total reaction volume, which consisted of genomic DNA purified using in-house extraction blueprint (0.3 pmol of each primer, 0.5 U DNA Taq polymerase deoxynucleotides triphosphates (dNTPs, 400 µM each), supplied PCR buffer and deionised water). The PCR was done strictly according to Jawan et al. [30]: 1 cycle for initial denaturation (94 °C for 2 min); 25 cycles (98 °C for 10 s; 53 °C for 30 s; 68 °C for 1 min) for annealing-amplified DNA extension via Eppendorf Mastercycler gradient (Eppendorf, Hamburg, Germany). The PCR products were purified and directly sequenced with primers 785F and 907R using BigDye® Terminator v3.1 Cycle Sequencing Kit (Applied Biosystems Co., Foster City, MA, USA). Finally, the sequences fragment (1414 bp) were compared with those deposited in the GenBank DNA database using Basic Local Alignment Search Tool (BLAST; https://www.ncbi.nlm.nih.gov/nuccore/NR_074540.1,NR_113266.1,NR_115714.1,NR_112630.1, NR_114582.1,NR_115526.1,NR_152692.1,NR_113991.1,NR_114581.1,NR_043403.1). A phylogenetic tree of Neighbour Joining (Unrooted Tree) by NCBl Blast Tree Method based on 16S rRNA 137 genes (bacteria only) was, excluding uncultured bacteria (taxid: 77133), constructed to determine the closest bacterial species by using Molecular Evolutionary Genetics Analysis (MEGA) software (Version 6.0, https://www.megasoftware.net, the software is online by Tamura et al. [29]). Distance-clustering was generated using bootstrap values based on 1000 replications. The resulting phylogenetic tree was re-run in Mega X software, as described by Wan-Mohtar et al. [10], and the isolated bacteria was identified as the same species by the closest K_{nuc}.

2.4. Inoculum Preparation

The stock culture of *B. cereus* KBC in an MRS media slant was thawed to room temperature before being streaked on MRS agar plates under sterile conditions. The *B. cereus* KBC on the MRS agar plates was cultured for two days to check their viability, then, a loop of the bacterial colony was transferred into 100-mL Erlenmeyer flask containing 50 mL of MRS broth. The MRS broth culture was incubated for 24 h and 1% inoculum with the cell's viability count of 10^6 CFU mL^{-1} was used for the fermentation to produce GABA.

2.5. Medium Preparation

MRS agar and broth (Sigma-Aldrich, Dorset, UK) were used in these experiments. This media composed of (in g/L) peptone (10.0), yeast extract (4.0), glucose (20.0), dipotassium hydrogen phosphate (2.0), sodium acetate (5.0), triammonium citrate (2.0), magnesium sulphate (0.2), manganese sulphate (0.05), and meat extract (8.0). The MRS agar and broth were prepared by measuring a suitable amount of media (31 g of MRS powder in 500 mL distilled water) and autoclaved at 121 °C for 15 min. Monosodium glutamate (MSG) was added into the media for the optimisation of GABA production according to the amount specified by the response surface methodology (RSM).

2.6. Production of GABA by an Isolated Strain

GABA production was conducted by transferring 150 mL of MRS broth into 250-mL Erlenmeyer flasks. The flasks were then inoculated with 1.5 mL of 1% inoculum with the cells viability count of 10^6 CFU mL^{-1}. The inoculated flasks were incubated at 30 °C, 100 rpm for 7 days using a shaker incubator (Binder, Bohemia, NY, USA).

2.7. Optimisation of GABA Production Using RSM

The GABA production by the isolated strain was optimised using RSM with central composite design (CCD) in Design Expert 7.0 software (Version 7, Godward St NE, Suit 6400, Minneapolis, MN, USA). The fermentation for optimisation was conducted in a 100-mL Erlenmeyer flask containing 50 mL MRS broth. The variables being tested are temperature, pH and MSG concentrations with the ranges shown in Table 1. The experimental design constructed by CCD with the α-value set at 1.0 generated a total of twenty runs as listed in Table 2. All the experimental runs were conducted accordingly, and the GABA concentration was measured and inserted as a response for the RSM analysis.

Table 1. The selected variables, range and levels inputted for optimisation study.

Variables	Range and Levels		
	−1	0	1
pH	3.0	5.0	7.0
MSG (g L^{-1})	1.0	3.0	5.0
Temperature (°C)	20	30	40

Table 2. Central composite design (CCD) design with studied variables and actual responses for the gamma-aminobutyric acid (GABA) production (mg L^{-1}) of *B. cereus* KBC.

Run No.	Variables			Actual Response
	pH	MSG (g L^{-1})	Temperature (°C)	GABA (mg L^{-1})
1	3	1	40	1989.54
2	5	3	20	3303.81
3	5	3	30	2178.9
4	7	5	40	3393.02
5	5	3	30	1990.2
6	5	5	30	2141.45
7	3	3	30	1857.36
8	5	3	30	2181.97
9	3	5	20	1965.97
10	5	3	30	2063.18
11	3	5	40	2060.37
12	5	3	30	2146.49
13	7	5	20	3585.29
14	7	1	40	3018.52
15	5	3	40	2379.37
16	5	1	30	2520.9
17	7	3	30	3303.19
18	3	1	20	1969.95
19	7	1	20	1766.99
20	5	3	30	2047.25

2.8. Analytical Procedures

2.8.1. Determination of Dry Cell Weight

Centrifuge tubes of 1.5-mL were pre-dried in a dehydrator overnight and weighed. A 1 mL of *Bacillus* strain culture was collected aseptically and centrifuged at 4000 rpm for 15 min. The supernatant was collected and stored in another 1.5 mL tube for the determination of GABA. The pellets were washed with distilled water and centrifuged again. The water was discarded, and the pellets were dried in a dehydrator overnight before the biomass was weighed.

2.8.2. Determination of GABA

The supernatant was filtered through a 0.22-μm pore-size nylon filter (Fisher Scientific, Brecon, UK). The supernatant was injected into an HPLC equipped with a Hypersil Gold C-18 column (250 × 4.6 mm I.D., particle size 5/um; Thermo Scientific, Meadow, UK). Mobile phase (a mixture of 60% solution A (aqueous solution of 100.02 millimolar (mM) sodium acetate, 3.59 mM triethylamine and 12.49 mM acetic acid in 1000 mL) adjusted to pH 5.8, 28% solution B (deionized water), and 12% solution C (acetonitrile) was used during the separation processes. The flow rate of the mobile phase was set at 0.6 mL/min. The separation process was conducted at room temperature, and the detection was monitored at 254 nm. Gradient HPLC separations were performed on a Shimadzu LC 20AT apparatus, consisting of a pump system, a CT0-10ASVP model oven with 20-μL injection loop injector, and a Model SPD-M20A PDA detector in conjunction with a DELL model DELL Optiplex integrator.

2.9. Statistical Analysis

All experimental runs were done in triplicates and the respective mean ± SD was determined using the software, GraphPad Prism 7 (GraphPad Software Inc., 2016, San Diego, CA, USA) and presented as error bars in the graph or ± symbol in the table. It should be noted the error bars might not appear in the graph if SD is smaller than the size of the symbol for the data point.

For the optimisation of GABA production by *B. cereus* KBC using RSM, the statistical tool in Design Expert 7.0 software was used to conduct an analysis of variance (ANOVA) for the experimental responses. The significance of the model and variables being studied was determined based on p-value < 0.05.

3. Results and Discussion

3.1. Identification of B. cereus KBC

3.1.1. Morphological Characteristics of B. cereus KBC

Morphological characteristics of the bacterial strain isolated from the soy sauce *moromi* are shown in Figure S1. It can be observed that the isolated strain exhibited round, opaque and milky-coloured colonies when grown on MRS agar plates. Gram staining also clearly showed that this strain is a Gram-positive with rod-shaped morphology and has been confirmed under 1000× magnification using an SEM.

3.1.2. 16S rRNA Identification of B. cereus KBC

Molecular identification was performed on the microorganism isolated from *moromi* in order to identify the species of the strain. The base pairs of the isolated strain were estimated using agarose gel electrophoresis under ultraviolet (UV) light (Figure S2). The marker in Lane 1 was used to construct the standard curve for base pairs length determination. The base pairs of BCSNSK were estimated to be 1414 bp. The phylogenetic tree (Figure 2) confirmed the first isolated strain to be closely related with the species *Bacillus cereus* and thus will be designated as *Bacillus cereus* strain KBC (*B. cereus* KBC) in further discussion.

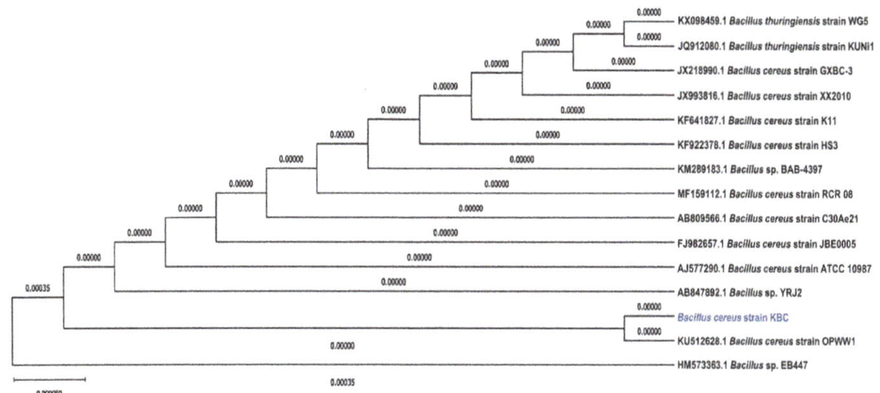

Figure 2. Phylogenetic tree of *B. cereus* strain KBC with evolutionary distance. The *B. cereus* KBC was closely related to *B. cereus* strain OPWW1.

3.2. GABA Production by B. cereus KBC

The biomass and GABA-producing capability of *Bacillus cereus* strain KBC was studied by measuring its biomass and GABA content in MRS broth for seven days. The graph in Figure 3 shows biomass and GABA production during the fermentation period. The biomass increased until it reached the highest value of 0.57 ± 0.07 g L^{-1} on day two, then decreased until day seven. The biomass concentration could demonstrate the viability of the cells. *B. cereus* KBC displayed rapid growth and reached maximum cell concentration on day two. Nutrient limitation, such as a limited amount of carbon source, might cause a decrease in the biomass of *B. cereus* KBC. The GABA content increased each day until day five in which it reached the maximum value of 532.74 ± 5.89 mg L^{-1} before decreasing. GABA could improve the growth of bacteria by acting as a growth factor. The previous study has shown the importance of GABA as a growth factor for a gut microorganism known as KLE1738 [31], as the isolated *B. cereus* strain KBC may also produce GABA to stimulate their growth. On day five of fermentation, the GABA production was the highest while the biomass is the lowest. On the next day, the biomass suddenly increases while the GABA content decreases. *Bacillus cereus* strain KBC could utilise the GABA for survivability and result in this unbalanced growth. More research is needed to be done to study the existent of GABA-dependent metabolism mechanism in *B. cereus* strain KBC. In comparison with a study by Ab Kadir et al. [24], GABA is produced mostly during the log and stationary phase, while, in this present study, GABA was produced during stationary phase. This situation might be caused by nutrient limitation such as peptone (nitrogen source) in the MRS broth. These components are vital elements for the microorganism in order to synthesize amino acids for growth and secondary metabolite production such as GABA [24]. The result also showed GABA concentration of 286 mg L^{-1} at day zero, which could be originated from the inoculum culture as GABA introduced in the fermentation broth, and it was continuously produced by *B. cereus* KBC while growing during the fermentation period. It was observed that there is an increment of 247 mg L^{-1} of GABA from day zero until day five (maximum amount of GABA production). In the study by Ab Kadir et al. [24], maximum GABA concentrations of 194 mg L^{-1} were achieved around day two to day four, using *Aspergillus oryzae* strain NSK isolated from soy sauce *koji*, which grows and behaves differently from *B. cereus* KBC. Technically both *A. oryzae* strains NSK (from *koji*) and *B. cereus* KBC (from *moromi*) are responsible in metabolizing the soybean while producing GABA but at different stages and rates.

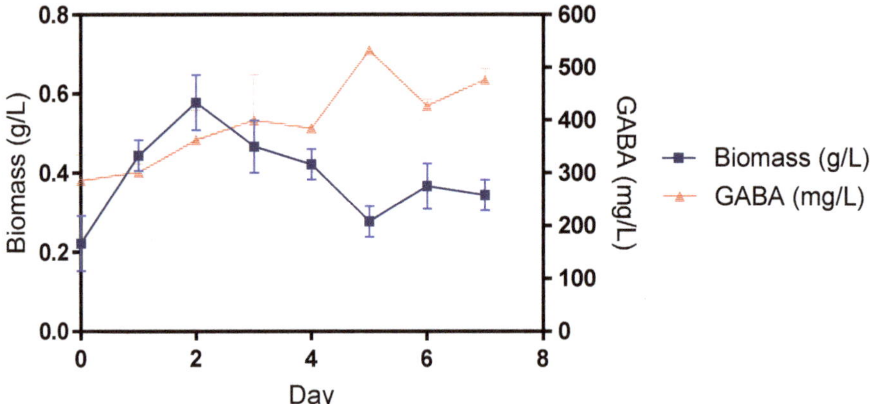

Figure 3. Biomass and GABA production of *Bacillus cereus* strain KBC in a 7-day cultivation period.

3.2.1. Optimisation of GABA Production by *B. cereus* KBC

The ANOVA results for GABA production of *B. cereus* KBC are shown in Table 3. It was found that 48.21% (R^2 = 0.4821) of the variability in the actual response could be described using the CCD linear model. The p-value was 0.0127, indicating that the model was significant ($p < 0.05$). In Figure 4, the effect of a single factor on GABA production was demonstrated. Among the three variables, pH (A, p-value = 0.0023) showed significant effect on the GABA production at $p < 0.05$ (Figure 3a). pH value is an important factor for GABA production [21], which affect the cells growth and glutamate decarboxylase (GAD) activity. GABA production usually being conducted in acidic pH conditions range between 3.5 and 5.0, depending on the types of microorganism and different properties of GADs. However, higher initial pH could trigger the cells to produce more GABA into the system.

Table 3. Analysis of variance (ANOVA) results for the actual responses using the CCD linear model for GABA production of *B. cereus* KBC.

Source	Sum of Squares	Mean Square	DF	F Value	Prob > F	
Model	3,088,531.300	1,029,510.433	3	4.965	0.0127 *	significant
A: pH	2,728,832.702	2,728,832.702	1	13.161	0.0023 *	significant
B: MSG	353,508.535	353,508.535	1	1.705	0.2101	
C: Temperature	6190.063	6190.063	1	0.030	0.8650	
Residual	3,317,449.610	207,340.600	16			
Pure Error	31,289.248	6257.850	5			
Lack of Fit	3,286,160.362		11	47.739	0.0002 *	significant
Cor Total	6,405,980.910	298,741.851	19			

Standard Deviation = 455.35	Mean = 2393.19	Adequate Precision = 7.221
R^2 = 0.4821		Adjusted R^2 = 0.3850

* Significant value.

In addition, optimising MSG concentration is aimed to stimulate the production of GABA by GAD via the GABA shunt way. However, it was reported that excessive MSG could inhibit cell growth and decrease GABA production [32,33]. In this present study, MSG (B, p-value = 0.2101) and temperature (C, p-value = 0.8650) were observed to be less influential on the GABA production of *B. cereus* KBC (Figure 4b,c, respectively). It should be noted that the optimal concentrations of MSG are different for various microorganisms in GABA production. Figure 5 shows the merged effect of pH, MSG concentration, and temperature on GABA production displayed as the response surface profiles. Figure 5a shows the effect of pH (A) and MSG concentration (B), Figure 5b shows the effect of MSG concentration (B) and temperature (C), and Figure 5c shows the effect of pH (A) and temperature (C) on GABA production of *B. cereus* KBC. The GABA production can be seen heavily influenced by the pH level, while the temperature and the MSG concentration only provide small effects on the GABA production of BSCKBC. Although some studies reported that temperature is a crucial parameter in GABA production [21], maintaining the temperature at the range of that *B. cereus* KBC could sustain its metabolic pathway could retain GABA production at a high level.

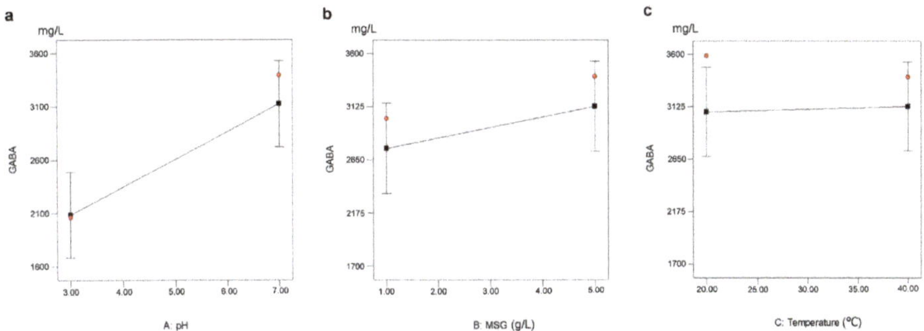

Figure 4. One factor profiles, which demonstrated the effects of (**a**) pH, (**b**) MSG concentration (g L^{-1}), and (**c**) temperature (°C) on GABA production of *B. cereus* KBC.

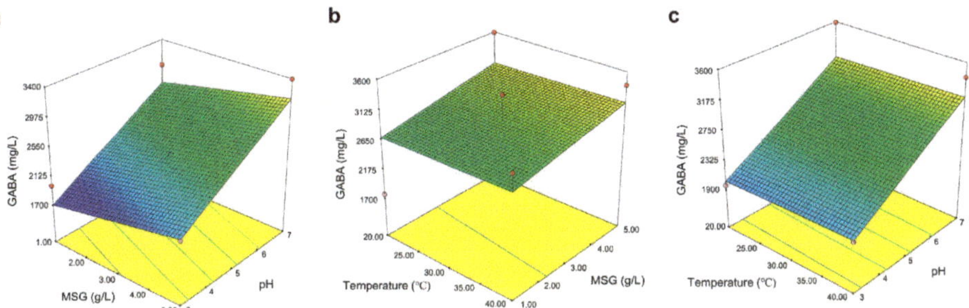

Figure 5. Response surface profile of GABA production from *B. cereus* KBC indicated the effects between (**a**) MSG and pH, (**b**) temperature and MSG, and (**c**) temperature and pH.

3.2.2. Validation of the Optimised Conditions

After the construction of the RSM linear model for GABA production by *B. cereus* KBC, the Design Expert 7.0 software was used to conduct optimisation for high GABA production. The software calculated the optimum conditions to be at pH 7, MSG concentration of 5 g L^{-1} and temperature of 40 °C. The predicted response was 3128.42 mg L^{-1}, while the actual response generated from the experiment was 3393.02 mg L^{-1}. The actual response was comparatively higher than the predicted response calculated by the software. Interestingly, the GABA production by *B. cereus* KBC under optimised conditions appeared to be 6.37-fold higher than the GABA production by *B. cereus* KBC under unoptimised conditions (532.74 mg L^{-1}). By just specifically controlling the pH, temperature and MSG concentration, six times more GABA can be produced from the same microorganism. In functional food production, this could mean more GABA can be produced in each fermentation run and save more production cost. The final product will also contain a much higher concentration of GABA and provide more benefit to the consumer.

Comparison between GABA production from different microorganisms isolated from various food sources is shown in Table 4. The GABA concentration produced by *B. cereus* strain KBC (532.74 mg L^{-1})

isolated from Malaysian soy sauce *moromi* was relatively higher than the GABA concentration (73.13 mg L^{-1}) produced by *A. oryzae* NSK isolated from the Malaysian soy sauce *koji* [24]. Under optimized conditions, *B. cereus* strain KBC demonstrated slightly higher GABA concentration (3393.02 mg L^{-1}) than *A. oryzae* NSK (3278.31 mg L^{-1}) [10]. In addition, the GABA production by several lactic acid bacteria was several folds higher than GABA produced by *B. cereus* strain KBC. Examples of lactic acid bacteria with high GABA production are *Lactobacillus paracasei* NFRI 7415 isolated from fermented fish (31145.3 mg L^{-1}), *Lactobacillus brevis* BJ20 isolated from kimchi (2465 mg L^{-1}), and *Lactococcus lactis* subsp. *lactis* B isolated from kimchi and yoghurt (6410 mg L^{-1}). However, GABA production by *B. cereus* strain KBC was slightly higher than the GABA produced by *Lactobacillus plantarum* DSM19463 isolated from cheeses (498.1 mg L^{-1}).

Table 4. GABA concentration by microorganisms isolated from various food sources.

Isolated Strain	Source	GABA Concentration (mg L^{-1})	Reference
Lactobacillus paracasei NFRI 7415	Fermented fish	31,145.30	[5]
Lactobacillus brevis BJ20	Kimchi	2465.00	[6]
Lactococcus lactis subsp. *lactis* B	Kimchi and yoghurt	6410.00	[34]
Lactobacillus plantarum DSM19463	Cheeses	498.10	[35]
Aspergillus oryzae NSK	Soy sauce *koji*	73.13	[24]
Aspergillus oryzae NSK	Soy sauce *koji*	354.08	[36]
Aspergillus oryzae NSK (unoptimized)	Soy sauce *koji*	3278.31	[10]
Bacillus cereus strain KBC (unoptimized)	Soy sauce *moromi*	532.74	This study
Bacillus cereus strain KBC (optimized)	Soy sauce *moromi*	3393.02	This study

4. Conclusions

A novel bacteria strain was successfully isolated from the soy sauce *moromi* by using MRS media under anaerobic conditions. The 16S rRNA sequencing and phylogenetic tree analysis revealed that the strain belongs to *B. cereus* species. The GABA-producing capability of this strain was studied, and the results demonstrated that *B. cereus* KBC managed to produce 532.74 mg L^{-1} GABA. The optimisation of GABA production using RSM also demonstrated that pH significantly influenced the GABA production of *B. cereus* KBC. The optimized conditions for high GABA production of *B. cereus* KBC were found to be at pH 7, 5 g L^{-1} MSG concentration and temperature of 40 °C. These optimized conditions resulted in a GABA production of 3393.02 mg L^{-1}, which is 6.37-fold higher than in unoptimised condition (532.74 mg L^{-1}). This study will be useful in isolating and optimizing potential GABA-producing strains for the production of soy sauce with a high concentration of GABA in the future.

Supplementary Materials: The following are available online at http://www.mdpi.com/2227-9717/8/6/652/s1, Figure S1: (a) Fermentation tank containing soy sauce *moromi* at a commercial soy sauce factory in Perak, Malaysia, (b) 80 days soy sauce *moromi* sample in which the bacterial strain was isolated, (c) Morphologies of *B. cereus* KBC on MRS agar plate, (d) *B. cereus* KBC under light microscope (400x magnification) after Gram staining, (e) *B. cereus* KBC at 1000x magnification under Scanning Electron Microscope (Bar = 10 μm) and Figure S2: Agarose gel electrophoresis of 16S rRNA isolated from *B. cereus* KBC culture plate. Lane 1 and 3 correspond to 10kb marker. Lane 2 corresponds to the sample (*B. cereus* KBC). Lane 4 corresponds to negative control (-ve) and Lane 5 corresponds to positive control (+ve).

Author Contributions: Conceptualization, W.A.A.Q.I.W.-M.; data curation, M.N.A.S.; funding acquisition, W.A.A.Q.I.W.-M. and M.F.I.; investigation, M.N.A.S. and S.A.K.; methodology, M.N.A.S.; project administration, Z.I.; resources, W.A.A.Q.I.W.-M. and M.F.I.; software, Z.I.; supervision, W.A.A.Q.I.W.-M.; validation, W.A.A.Q.I.W.-M., M.F.I. and O.P.S.; visualization, S.A.K., A.W.W.L. and S.S.; writing—original draft, M.N.A.S.; writing—review and editing, W.A.A.Q.I.W.-M. and M.F.I. All authors have read and agreed to the published version of the manuscript.

Funding: This research was funded by the Ministry of Education Malaysia under the Fundamental Research Grant Scheme (FRGS: FP066-2018A), SATU Joint Research Scheme Program Universiti Malaya (ST006-2019) and Geran Putra (9559300) Universiti Putra Malaysia.

Acknowledgments: All the authors would like to acknowledge all the financial supports for this research project and the collaboration works between the University of Malaya and the Universiti Putra Malaysia.

Conflicts of Interest: The authors declare no conflict of interest.

References

1. Pizzarello, S. Non-Protein Amino Acids. In *Encyclopedia of Astrobiology*; Gargaud, M., Amils, R., Cleaves, H.J., Eds.; Springer: Berlin/Heidelber, Germany, 2015.
2. Mander, L.; Liu, H. *Comprehensive Natural Products II: Chemistry and Biology*; Elsevier: Amsterdam, The Netherlands, 2010. [CrossRef]
3. Vranova, V.; Rejsek, K.; Skene, K.R.; Formanek, P. Non-protein amino acids: Plant, soil and ecosystem interactions. *Plant Soil* **2011**, *342*, 31–48. [CrossRef]
4. Petroff, O.A.C. GABA and glutamate in the human brain. *Neuroscientist* **2002**, *8*, 562–573. [CrossRef] [PubMed]
5. Komatsuzaki, N.; Shima, J.; Kawamoto, S.; Momose, H.; Kimura, T. Production of γ-aminobutyric acid (GABA) by Lactobacillus paracasei isolated from traditional fermented foods. *Food Microbiol.* **2005**. [CrossRef]
6. Lee, B.J.; Kim, J.S.; Kang, Y.M.; Lim, J.H.; Kim, Y.M.; Lee, M.S.; Jeong, M.H.; Ahn, C.B.; Je, J.Y. Antioxidant activity and γ-aminobutyric acid (GABA) content in sea tangle fermented by Lactobacillus brevis BJ20 isolated from traditional fermented foods. *Food Chem.* **2010**. [CrossRef]
7. Aoki, H.; Furuya, Y.; Endo, Y.; Fujimoto, K. Effect of γ-aminobutyric acid-enriched tempeh-like fermented soybean (gaba-tempeh) on the blood pressure of spontaneously hypertensive rats. *Biosci. Biotechnol. Biochem.* **2003**. [CrossRef] [PubMed]
8. Inoue, K.; Shirai, T.; Ochiai, H.; Kasao, M.; Hayakawa, K.; Kimura, M.; Sansawa, H. Blood-pressure-lowering effect of a novel fermented milk containing γ-aminobutyric acid (GABA) in mild hypertensives. *Eur. J. Clin. Nutr.* **2003**. [CrossRef] [PubMed]
9. Liao, W.C.; Wang, C.Y.; Shyu, Y.T.; Yu, R.C.; Ho, K.C. Influence of preprocessing methods and fermentation of adzuki beans on γ-aminobutyric acid (GABA) accumulation by lactic acid bacteria. *J. Funct. Foods* **2013**. [CrossRef]
10. Wan-Mohtar, W.A.A.Q.I.; Ab Kadir, S.; Halim-Lim, S.A.; Ilham, Z.; Hajar-Azhari, S.; Saari, N. Vital parameters for high gamma-aminobutyric acid (GABA) production by an industrial soy sauce koji Aspergillus oryzae NSK in submerged-liquid fermentation. *Food Sci. Biotechnol.* **2019**. [CrossRef]
11. Wong, C.G.T.; Bottiglieri, T.; Snead, O.C. GABA, γ-hydroxybutyric acid, and neurological disease. *Ann. Neurol.* **2003**. [CrossRef]
12. Capitani, G.; De Biase, D.; Aurizi, C.; Gut, H.; Bossa, F.; Grütter, M.G. Crystal structure and functional analysis of Escherichia coli glutamate decarboxylase. *EMBO J.* **2003**. [CrossRef]
13. Adeghate, E.; Ponery, A.S. GABA in the endocrine pancreas: Cellular localization and function in normal and diabetic rats. *Tissue Cell* **2002**. [CrossRef] [PubMed]
14. Cavagnini, F.; Pinto, M.; Dubini, A.; Invitti, C.; Cappelletti, G.; Polli, E.E. Effects of gamma aminobutyric acid (GABA) and muscimol on endocrine pancreatic function in man. *Metabolism* **1982**. [CrossRef]
15. Dhakal, R.; Bajpai, V.K.; Baek, K.H. Production of GABA (γ-aminobutyric acid) by microorganisms: A review. *Braz. J. Microbiol.* **2012**. [CrossRef]
16. Li, H.; Cao, Y. Lactic acid bacterial cell factories for gamma-aminobutyric acid. *Amino Acids* **2010**, *39*, 1107–1116. [CrossRef] [PubMed]
17. Lin, S.; Brower, P.L.; Winkle, D.D. Chiral HPLC separations for process development of S-(+)-isobutyl GABA, a potential anti-epileptic agent. *J. Liq. Chromatogr. Relat. Technol.* **1996**. [CrossRef]
18. Venkatesh, T.; Mainkar, P.S.; Chandrasekhar, S. Total synthesis of (±)-galanthamine from GABA through regioselective aryne insertion. *Org. Biomol. Chem.* **2019**. [CrossRef] [PubMed]
19. Román, S.; Sánchez-Siles, L.M.; Siegrist, M. The importance of food naturalness for consumers: Results of a systematic review. *Trends Food Sci. Technol.* **2017**, *67*, 44–57. [CrossRef]
20. Holzapfel, W.H.; Geisen, R.; Schillinger, U. Biological preservation of foods with reference to protective cultures, bacteriocins and food-grade enzymes. *Int. J. Food Microbiol.* **1995**, *24*, 343–362. [CrossRef]
21. Cui, Y.; Miao, K.; Niyaphorn, S.; Qu, X. Production of gamma-aminobutyric acid from lactic acid bacteria: A systematic review. *Int. J. Mol. Sci.* **2020**, *21*, 995. [CrossRef]

22. Murooka, Y.; Yamshita, M. Traditional healthful fermented products of Japan. *J. Ind. Microbiol. Biotechnol.* **2008**, *35*, 791. [CrossRef]

23. Sarasa, S.B.; Mahendran, R.; Muthusamy, G.; Thankappan, B.; Selta, D.R.F.; Angayarkanni, J. A Brief Review on the Non-protein Amino Acid, Gamma-amino Butyric Acid (GABA): Its Production and Role in Microbes. *Curr. Microbiol.* **2020**, *77*, 534–544. [CrossRef] [PubMed]

24. Ab Kadir, S.; Wan-Mohtar, W.A.A.Q.I.; Mohammad, R.; Abdul Halim Lim, S.; Sabo Mohammed, A.; Saari, N. Evaluation of commercial soy sauce koji strains of Aspergillus oryzae for γ-aminobutyric acid (GABA) production. *J. Ind. Microbiol. Biotechnol.* **2016**. [CrossRef] [PubMed]

25. Fukushima, D. *Industrialization of Fermented Soy Sauce Production Centering around Japanese Shoyu*; Food Science and Technology (Marcel Dekker, Inc.): New York, NY, USA, 2004; pp. 1–88.

26. Liu, W.; Bao, Q.; Qing, M.; Chen, X.; Sun, T.; Li, M.; Zhang, J.; Yu, J.; Bilige, M.; Sun, T.; et al. Isolation and identification of lactic acid bacteria from Tarag in Eastern Inner Mongolia of China by 16S rRNA sequences and DGGE analysis. *Microbiol. Res.* **2012**. [CrossRef] [PubMed]

27. Sohedein, M.N.A.; Wan-Mohtar, W.A.A.Q.I.; Hui-Yin, Y.; Ilham, Z.; Chang, J.S.; Supramani, S.; Siew-Moi, P. Optimisation of biomass and lipid production of a tropical thraustochytrid Aurantiochytrium sp. UMACC-T023 in submerged-liquid fermentation for large-scale biodiesel production. *Biocatal. Agric. Biotechnol.* **2020**. [CrossRef]

28. Mohd Hanafiah, Z.; Wan Mohtar, W.H.M.; Abu Hasan, H.; Jensen, H.S.; Klaus, A.; Wan-Mohtar, W.A.A.Q.I. Performance of wild-Serbian Ganoderma lucidum mycelium in treating synthetic sewage loading using batch bioreactor. *Sci. Rep.* **2019**. [CrossRef] [PubMed]

29. Tamura, K.; Stecher, G.; Peterson, D.; Filipski, A.; Kumar, S. MEGA6: Molecular evolutionary genetics analysis version 6.0. *Mol. Biol. Evol.* **2013**. [CrossRef] [PubMed]

30. Jawan, R.; Kasimin, M.E.; Jalal, S.N.; Mohd Faik, A.A.; Abbasiliasi, S.; Ariff, A. Isolation, characterisation and in vitro evaluation of bacteriocins-producing lactic acid bacteria from fermented products of Northern Borneo for their beneficial roles in food industry. *J. Phys. Conf. Ser.* **2019**, *1358*, 012020. [CrossRef]

31. Strandwitz, P.; Kim, K.H.; Terekhova, D.; Liu, J.K.; Sharma, A.; Levering, J.; McDonald, D.; Dietrich, D.; Ramadhar, T.R.; Lekbua, A.; et al. GABA-modulating bacteria of the human gut microbiota. *Nat. Microbiol.* **2019**. [CrossRef]

32. Zhuang, K.; Jiang, Y.; Feng, X.; Li, L.; Dang, F.; Zhang, W.; Man, C. Transcriptomic response to GABA-producing Lactobacillus plantarum CGMCC 1.2437T induced by L-MSG. *PLoS ONE* **2018**. [CrossRef]

33. Yang, S.Y.; Lü, F.X.; Lu, Z.X.; Bie, X.M.; Jiao, Y.; Sun, L.J.; Yu, B. Production of γ-aminobutyric acid by Streptococcus salivarius subsp. thermophilus Y2 under submerged fermentation. *Amino Acids* **2008**. [CrossRef]

34. Lu, X.; Chen, Z.; Gu, Z.; Han, Y. Isolation of γ-aminobutyric acid-producing bacteria and optimization of fermentative medium. *Biochem. Eng. J.* **2008**. [CrossRef]

35. Di Cagno, R.; Mazzacane, F.; Rizzello, C.G.; De Angelis, M.; Giuliani, G.; Meloni, M.; De Servi, B.; Gobbetti, M. Synthesis of γ-aminobutyric acid (GABA) by Lactobacillus plantarum DSM19463: Functional grape must beverage and dermatological applications. *Appl. Microbiol. Biotechnol.* **2010**. [CrossRef] [PubMed]

36. Hajar-Azhari, S.; Wan-Mohtar, W.A.A.Q.I.; Ab Kadir, S.; Rahim, M.H.A.; Saari, N. Evaluation of a Malaysian soy sauce koji strain Aspergillus oryzae NSK for γ-aminobutyric acid (GABA) production using different native sugars. *Food Sci. Biotechnol.* **2018**. [CrossRef] [PubMed]

 *processes*

Article

Investigation of the Thermal Properties of Electrodes on the Film and Its Heating Behavior Induced by Microwave Irradiation in Mounting Processes

Kenji Kanazawa [1], Takashi Nakamura [2], Masateru Nishioka [2] and Sei Uemura [1],*

[1] Sensing System Research Center (SSRC), National Institute of Advanced Industrial Science and Technology (AIST), 1-1-1 Higashi, Tsukuba, Ibaraki 305-8565, Japan; kenji.kanazawa@aist.go.jp
[2] Research Institute for Chemical Process Technology, National Institute of Advanced Industrial Science and Technology (AIST), 4-2-1, Nigatake, Miyagino-ku, Sendai 983-8551, Japan; nakamura-mw@aist.go.jp (T.N.); m-nishioka@aist.go.jp (M.N.)
* Correspondence: sei-uemura@aist.go.jp

Received: 2 April 2020; Accepted: 5 May 2020; Published: 9 May 2020

Abstract: We have developed a novel microwave (MW) soldering system using a cylindrical single-mode TM_{110} MW cavity that spatially separates the electric fields at the top and bottom of the cavity and the magnetic field at the center of the cavity. This MW reactor system automatically detects the suitable resonance frequency and provides the optimum MW irradiation conditions in the cylindrical cavity via a power feedback loop. Furthermore, we investigated the temperature properties of electrodes by MW heating with the simulation of a magnetic field in the TM_{110} cavity toward the mounting of electronic components by MW heating. We also developed a short-time melting technology for solder paste on polyimide substrate using MW heating and succeeded in mounting a temperature sensor using the novel MW heating system without damaging the electronic components, electronic circuits, and the substrate.

Keywords: electronics package; induction heating; magnetic field; electric field; TM_{110} single-mode cavity; solder; eddy current

---

1. Introduction

The Internet of Things (IoT) applications include smart grids, smart homes [1], sensors [2,3], and actuators [4]. Their applications are remotely accessible and enable the exchange of information, allowing for the control of electrical devices and monitoring healthcare from anywhere on the network. This multifunctionality has seen the IoT receive considerable attention in an attempt to revolutionize our lifestyles [5,6]. In smart sensors and actuators, technological developments in mounting electronic components on a flexible substrate (such as textile, plastic substrates, and stretchable thermoplastic elastomer resins) are urgently required [2,3]. However, the heating process used in conventional industrial solder mounting (i.e., heating by reflow oven) requires temperatures of more than 170 °C for about 300 s [7,8], which means that the glass transition temperature (T_g) of flexible substrates such as polyethylene terephthalate (PET, T_g = 120 °C) and polyethylene naphthalate (PEN, T_g = 140 °C) should be higher than this if conventional industrial solder mounting systems are to be used. Hence, it is impossible to mount components on flexible and/or stretchable substrates that have low thermostability using solder as a connection material.

To resolve the temperature problems, electrically conductive adhesives and heating processes (electronic packaging technology), which serve as electric and thermal conductors as well as mechanical connectors without heat damage, are important (Figure 1). Despite the development of polymer-based conductive adhesives based on polymers with a lower melting point than solder paste [9], the bond strength and conductivity are lower than those of solder paste, and so this approach has not been adopted. Thus, a new solder paste with low melting point, high conductivity, and strong adhesion force between the electronic components and substrates is strongly required.

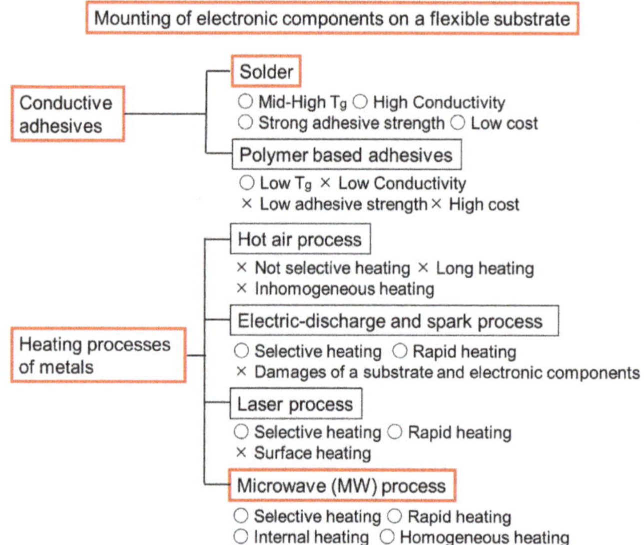

Figure 1. Candidates for conductive adhesives and heating processes of metal in mounting of electronic components on flexible substrate. Circle marks (○) and cross marks (×) mean desirable and undesirable properties in the mounting processes of electronic components on flexible substrate, respectively. Red squares highlight the focuses of this study.

Regarding the heating process of metal, various methods using lasers [10,11], electric-discharge [12], and microwave (MW) [13,14] techniques have been reported. Although selective laser sintering (SLS) was developed by Dechkard [15] and has been widely utilized by industry, understanding chemical, mechanical, and material metallurgical phenomena in SLS process was difficult. Electric-discharge techniques enable rapid heating. Chaim reported that the spark plasma sintering (SPS) process demonstrates sintering, consolidation and crystal growth by spark plasma systems [16]. Among the methods of metal sintering, MW-based techniques particularly represent a powerful approach because MW can selectively heat materials. Heating by MW is based upon the ability of materials to absorb and transform electromagnetic energy into heat. Materials are classified into three categories, with respect to their interaction with MW. (a) MW reflectors (high conductive materials such as metals); the material is not effectively heated by MW. (b) MW transmitters (low loss insulator materials such as Teflon and quartz); MW can penetrate through the material without any absorption, losses or heat generation. (c) MW absorbers (high loss insulator such as dielectric materials); they take up the energy from the MW field and get heated up rapidly [14,17,18]. So far, MW dielectric heating has been used for inorganic/organic syntheses [19–21] and metal nanoparticle synthesis [22,23]. Bulk metals cannot be heated by an electric field because the spark phenomenon occurs as follows. When MW interacts with a bulk metal, the electrons on the material's surface get sloshed around. If the metal has an edge, the charges can pile up and result in a high concentration of voltage. If the voltage is high enough, it can rip an electron off a molecule in the air, leading to creating a spark. Therefore, the electric field and magnetic field are separated from each other, and the heating process by a magnetic field is needed to heat metal without spark generation. Various MW-assisted reactors are commercially available for application; however, the most commercial MW reactor is a multi-mode reactor and the electromagnetic distributions dramatically changes when materials are set in the reactor and metals cannot be heated during controlling the electromagnetic distributions, causing the inhomogeneous heating of the reactor [22,23]. Hence, an applicator with a multi-mode reactor is generally used for the continuous vulcanization of rubber, heating food, and heating of dielectric materials. On the other hand, an applicator with a single-mode is suitable for the heating of a metal because electromagnetic distributions can be controlled during MW irradiation. However, MW reactor systems with a cylindrical single mode cavity are scarce.

In this study, we focused on an MW sintering system to mount electronic components on substrates with low thermostability. We originally developed an MW reactor system that separates the electric fields at the top and bottom and the magnetic field at the center in a cylindrical single-mode MW cavity. This MW reactor system automatically detects the suitable resonance frequency and provides the optimum MW irradiation conditions in the cylindrical cavity via a power feedback loop, leading to the generation of standing wave in the center of the MW cavity. Consequently, a homogeneous heating zone can be created along the central line of the MW cavity. In this paper, we describe the heating property of the electrode pad shape effect by MW that was prepared by silver paste screen printing and the mounting of a thermo-sensor device on polyimide substrate using MW in a magnetic field without damaging the sensor or the substrate. Furthermore, the results of thermogravimetric (TG), differential thermal analysis (DTA), and differential scanning calorimetry (DSC) are performed to evaluate the decomposition behavior of the solder paste.

2. Experimental Section

2.1. Materials

Conductive silver paste (REXALPHA RA FS 074, Toyochem Co. Ltd., Tokyo, Japan) and solder paste (L20-BLT5-T7F, Sn42.0/Bi58.0 (in wt.%), melting point: 138 °C, Senjyu Metal Industry Co., Ltd., Tokyo, Japan) were used as received without further addition of solvent and flux. Scotch tape (standard 3 M brand transparent tape with a thickness of 52 μm) was used to prepare the solder paste samples. A thermo-sensor (SHT31-DIS-B, size: 2.5 mm × 2.5 mm × 0.9 mm, SENSIRION Corp., Zurich, Switzerland) for mounting by MW irradiation was used. As references, SHTDA2 (SENSIRION Corp., Zurich, Switzerland) with an SHT35 thermo-sensor (SENSIRION Corp., Zurich, Switzerland) and a digital thermo-meter (Vaisala, Vantaa, Finland) were used.

2.2. Preparation of Samples and the Induction Soldering Method by MW

2.2.1. Preparation of Silver Paste Pad for the Investigation of the Pad Shape on Induction Soldering by MW and the Induction Soldering Method by MW

Samples for the investigation of pad shape effect with silver paste on induction soldering by MW were prepared by screen printing machine (NT-15TVA, Neotechno Japan Corp., Tokyo, Japan) with a squeeze speed of 20 mm/s on polyimide sheet substrate, followed by heating with a hot plate at 150 °C for 20 min (the sizes given in Table S1).

To investigate the pad shape effect on induction soldering by MW, the prepared pads were mounted coaxially at the center of the TM_{110} cavity and heated by MW at 1, 5, 10, 20, and 30 W of output powers for 30 s under the atmosphere, respectively.

2.2.2. Preparation of Samples for Investigation of Solder Heating Behavior by the Induction Soldering of MW and Induction Soldering by MW

The samples of solder paste for MW sintering tests were prepared using the blade cote method. First, a 1 cm × 1 cm square of scotch tape with a thickness of about 60 μm was formed on a 3 cm × 3 cm polyimide sheet. Second, solder paste was spread using a slide glass and the scotch tape was removed (Figure S1).

To investigate solder heating behavior by MW, the prepared samples were mounted coaxially at the center of the TM_{110} cavity and heated by MW with 1, 5, 10, 20, and 30 W output powers for 70 s under the atmosphere. The cooling times of the samples were less than 20 s.

2.2.3. Preparation of Samples for Mounting Electric Components by the Induction Soldering of MW and Induction Soldering by MW

The circuit patterns of silver paste were prepared by a screen printing machine (NT-15TVA, Neotechno Japan Corp., Tokyo, Japan) with a squeeze speed of 20 mm/s on polyimide sheet substrate, followed by heating with hot plate at 150 °C for 20 min. Solder paste was also printed on the circuit patten. After this, thermo-sensor was placed on the solder paste.

To package the electronic component by MW, the prepared samples were mounted coaxially at the center of the TM_{110} cavity and heated by MW with 1 W of output power for 40 s, 5 W of output power for 22 s, 10 W of output power for 11 s, 20 W of output power for 8 s, and 30 W of output power for 4 s under the atmosphere, respectively. The cooling time of samples were less than 20 s.

2.3. MW Sintering System

The MW reactor system (Ryowa Electronics Co., Ltd, Miyagi, Japan) consists of a variable-frequency microwave generator (2.45 ± 0.05 GHz, 200 W) and a cylindrical single-mode cavity (Figure 2). As shown in Figure 2b, the inner diameter of the TM_{110} single-mode cavity was designed based on the incident electromagnetic wave frequency (2.45 GHz). The inner diameter and height of the cavity were 14.6 cm and 11.0 cm. Samples were mounted coaxially at the center of the TM_{110} cavity. The distance between the sample and microwave antenna was 13.45 cm. The oscillation frequency was monitored for matching with the resonance frequency. The applied power was controlled by the temperature feedback module. The surface temperature of samples was measured using an infrared (IR) thermal camera (Model PI-200, Optris GmbH, Berlin, Germany) placed through the open circular slit (1.5 cm diameter) of the cavity.

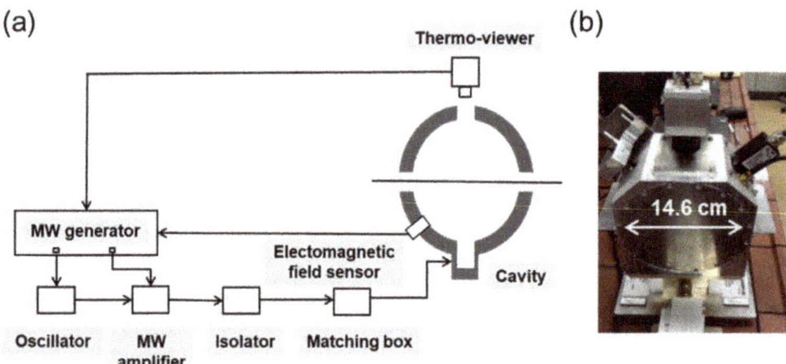

Figure 2. (a) Induction soldering system of microwave (MW) and (b) TM_{110} single-mode cavity.

2.4. Electromagnetic Simulation in TM_{110} Mode Cavity with and without Bulk Silver

The electromagnetic field distribution in cavity was determined by the finite element method (FEM) simulations performed using COMSOL Multiphysics® software (version 5.4, 2018, COMSOL Inc., Boston, MA, USA). The simulation module included the MW antenna and the elliptical cavity with 15.8-cm inner diameter and an 8.0-cm height. The base of the antenna was set to the lumped port that introduced the electromagnetic wave. The walls of the elliptical cavity and the MW antenna were made of aluminum and used as the impedance boundary condition. The numerical model for the simulation of the electromagnetic field is described below:

$$\nabla \times \left(\mu_r^{-1} \times E \right) - k_0^2 \left(\varepsilon_r - \frac{j\sigma}{\omega \varepsilon_0} \right) E = 0$$

where E denotes the electric field vector inside the cavity, μ_r the relative permeability, j the imaginary unit, σ the conductivity, ω the angular frequency, and ε_0 the permittivity. The model used material parameters for air: $\sigma = 0$, $\mu_r = \varepsilon_r = 1$. In the bulk silver with a thickness of 8 μm, the same parameters were used except for the conductivity, which was 61.6×10^6 S/m. The FEM mesh configurations, such as the number of elements and time of solving, are summarized in Table S2.

2.5. Instruments

The thermal characteristics of the solder paste were measured by differential thermal analysis (DTA, TG/DTA6200, Seiko Instrument Inc., Chiba, Japan), thermogravimetry (TG, TG/DTA6200, Seiko Instrument Inc., Chiba, Japan), and differential scanning calorimetry (DSC, EXSTAR DSC 7020, Seiko Instrument Inc., Chiba, Japan) at a heating rate of 5 °C/min in air. Optical images of the sample were recorded on a VW-9000/VW-600c digital microscope (Keyence Corp., Osaka, Japan).

2.6. Method of Measuring Room Temperature

The room temperature was measured using the thermo-sensor mounted by the induction soldering of MW. To convert from voltage to temperature, the following expression, which is defined by supplier of a temperature sensor (SysCom. Inc., New York, NY, USA), was used:

$$\text{Temperature (\%)} = (\text{output voltage} \times 1000/25) - 40$$

The input voltage was applied using a direct current (DC) signal source (SS7012, Hioki E.E. Corp., Nagano, Japan), and the output voltage was obtained using a multi-input data recording system (NR-600, Keyence Corp., Osaka, Japan). Furthermore, a temperature sensor (SHTDA-2, SysCom. Inc., New York, NY, USA) was used as a reference, because the sensor mounted by MW was a type of SHTDA-2 sensor.

In addition, a digital thermo-meter (HM41, Vaisara, Vantaa, Finland) was employed as a reference to check the room temperature.

3. Results and Discussion

3.1. Induction Soldering System by MW

Metals cannot be heated by MW in an electric field because sparks occur. Therefore, we originally developed a frequency-controlled MW system that can compensate for shifts in the resonance frequency through automatic detection and tracking, and which can then feedback the best condition for MW irradiation. Consequently, a strong magnetic field was uniformly formed along the central line of the cavity, where 90° is the angle direction of sample conveying direction (Figure 3). To confirm whether the spark phenomenon occurs or not using the developed induction soldering system of MW, we tried to heat metals by MW in a magnetic field. As the result, metal was successfully heated without causing the spark phenomenon. From this result, we found that the sample could be heated by the magnetic field of MW.

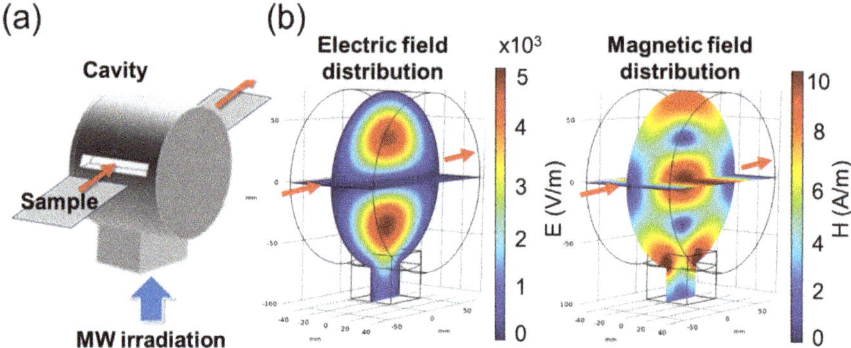

Figure 3. (**a**) Induction soldering system of MW and (**b**) the simulated distribution of electric and magnetic fields in the TM$_{110}$ cavity. The frequencies of the simulations are 2.4446 GHz. Red arrows are sample conveying direction.

3.2. MW Sintering Properties and Morphology of Solder Paste

Figure 4 shows the temperature dependency with respect to output power of MW. Furthermore, infrared (IR) camera images are shown in Figure 5. The temperatures of the samples under MW irradiation at 1 and 5 W for 70 s were less than 100 °C and the solder paste did not melt (Figure S1a,b). In contrast, the temperatures of the samples heated at 10, 20, and 30 W exceeded 140 °C and the samples melted, leading to the formation of balls, as shown in Figure S1c,d. The formation of balls would mean insufficient wetting of the polyimide substrate by the molten solder paste. The sample heated by a 10-W output power took 15 s to melt, whereas the samples subjected to 20 and 30 W of output power took about 10 s to melt. In the process of induction soldering by MW at 10, 20, and 30 W output powers, the temperature increased dramatically from around 110 °C. This would be because of the evaporation process of the solvent contained in the solder paste, because no endothermic process was observed at around 110 °C in the TG/TDA and DSC measurements (Figure S2). After reaching around 150 °C, the temperature decreased. This decrease in temperature after the melting process would be because (i) melted balls were formed and the balls would reflect MW [14,17,18] and/or (ii) penetration depth decreased with an increase in conductivity [24], and eddy current was not generated.

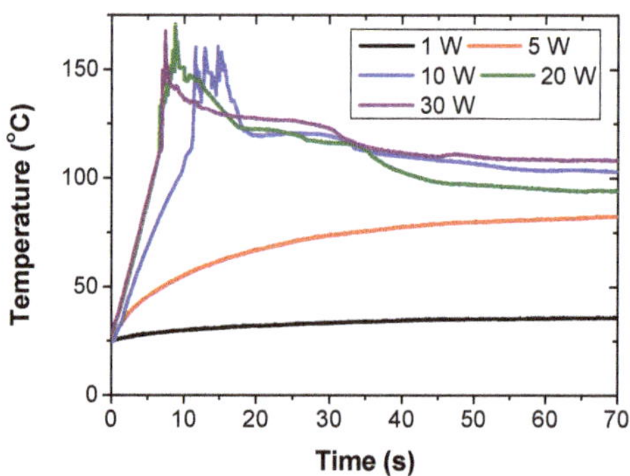

Figure 4. Changes in temperature of solder paste on a polyimide sheet by MW irradiation at 1, 5, 10, 20, and 30 W output powers for 70 s.

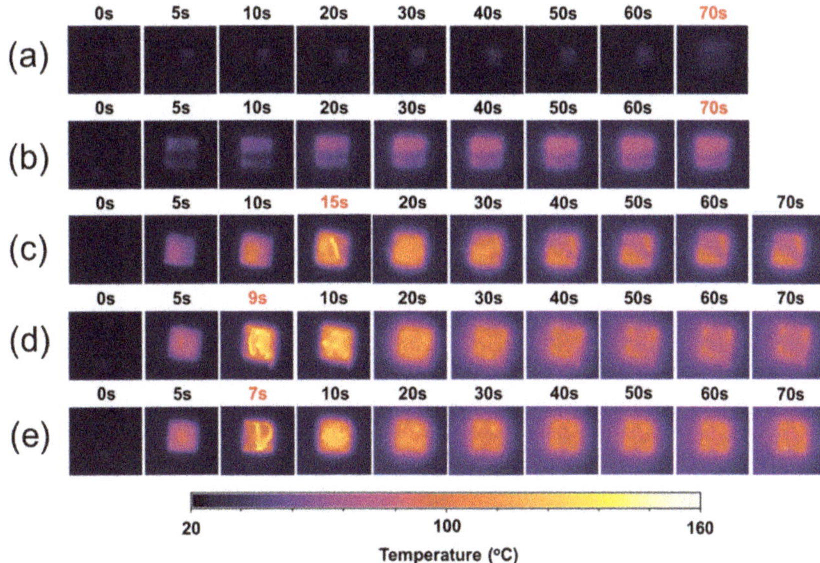

Figure 5. Infrared (IR) thermal imaging of solder paste on a polyimide sheet by MW irradiation with (**a**) 1, (**b**) 5, (**c**) 10, (**d**) 20, and (**e**) 30 W output powers for 70 s. Red characters indicate the times of maximum temperature during the heating process.

3.3. Thermal Property of Silver Pad Patterns by the Induction Soldering of MW

Figure 6 and Table S1 show the temperature difference (ΔT) between initial temperature (T_0) and arrived temperature (T) of the samples by the induction soldering of MW and ΔT, as defined in the below equation:

$$\Delta T = T - T_0$$

ΔT of (1 mm, 0.5–20 mm) samples ($\Delta T_{(1,\ 0.5-20)}$) by the induction soldering of MW at 1, 5, 10, 20, 30 W of output power for 30 s were less than 10 °C, as shown in Figure 6a and Table S1; $\Delta T_{(1,\ 0.5-20)}$ = 0.2–0.8 °C for 1 W, 1.1–1.5 °C for 5 W, 1.7–2.6 °C for 10 W, 3.3–5.5 °C for 20 W, and 5.3–8.6 °C for 30 W, respectively. These results suggest that ΔT is not mainly dependent on the length of samples. To elucidate the lowΔT reason, the simulation of magnetic field distribution around bulk silver in the cavity was performed by using COMSOL® software (Figure 7). Figure 7a shows the magnetic field distribution of conveying sample which is z axis = 0 in the cavity without sample, while Figure 7b,c show magnetic field distribution in the cavity with samples whose thickness are 8 μm. A square size drawn with a solid line in Figure 7a is 20 mm in width × 20 mm in length whose length is within the sample size. In particular, the penetration depth of bulk silver, which was estimated by the equation [24] was 1.3 μm. The magnetic field distribution intensities around the bulk silver samples, which are (width, length) = (1 mm, 5 mm), (1 mm, 10 mm), (1 mm, 15 mm), and (1 mm, 20 mm), are almost the same as those shown in Figure 7b. This is because the direction of magnetic field vibration is the same with the direction of length and the magnetic field intensity decreases from the center to length direction [25–27]. Therefore, the samples do not have the possibility to absorb MW effectively, even if the sample length increases. In general, magnetic field induces the circular electrical current, which called eddy current and the eddy current also induces Joule heat, leading to a temperature increase in the sample [24]. Due to low eddy current, which is induced by low magnetic field intensity, the generation of Joule heat is also low. Therefore, the $\Delta T_{(1,\ 5-20)}$ would be also low.

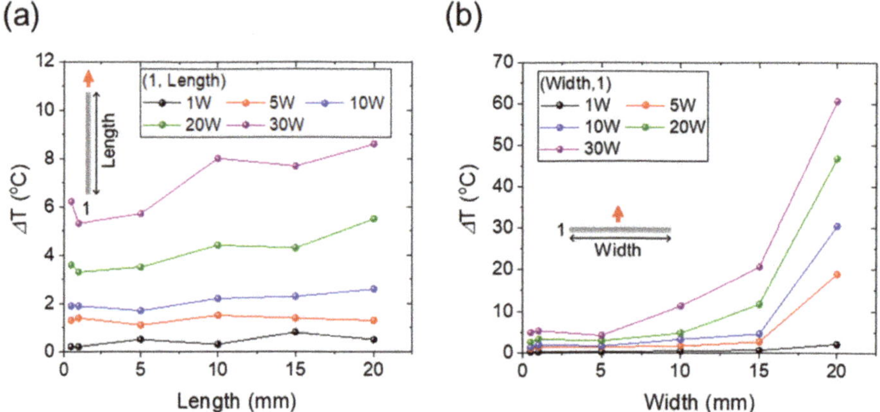

Figure 6. (**a**) $\Delta T_{(1, 0.5)}$, $\Delta T_{(1, 1)}$, $\Delta T_{(1, 5)}$, $\Delta T_{(1, 10)}$, $\Delta T_{(1, 15)}$, $\Delta T_{(1, 20)}$, (**b**) $\Delta T_{(0.5, 1)}$, $\Delta T_{(1, 1)}$, $\Delta T_{(5, 1)}$, $\Delta T_{(10, 1)}$, $\Delta T_{(15, 1)}$, and $\Delta T_{(20, 1)}$ by MW irradiation at 1, 5, 10, 20, 30 W output powers for 30 s, respectively. Inset images are the sample shape. The red arrows of inset are the direction of conveying the samples.

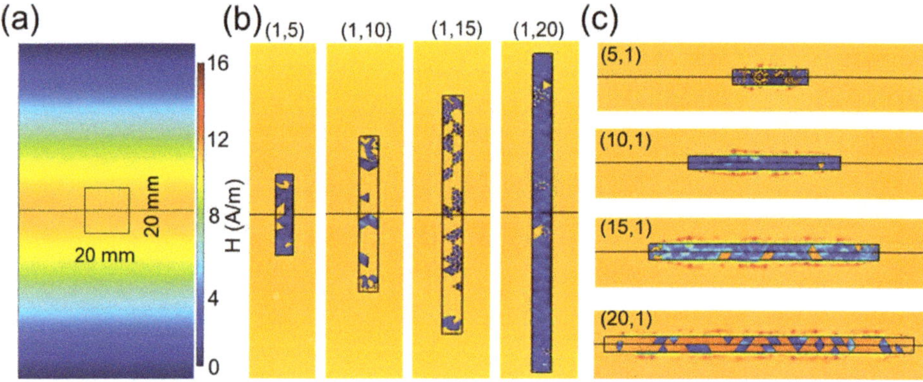

Figure 7. Simulation of magnetic field distribution around bulk silver against the direction of conveying the samples in the TM_{110} cavity: (**a**) empty cavity, (**b**) (width, length) = (1 mm, 5 mm), (1 mm, 10 mm), (1 mm, 15 mm), (1 mm, 20 mm), (**c**) (5 mm, 1 mm), (10 mm, 1 mm), (15 mm, 1 mm), and (20 mm, 1 mm) bulk silver with 8 μm of thickness, respectively.

On the other hand, the ΔT of samples ($\Delta T_{(0.5-20, 1)}$) by the induction soldering of MW at 1, 5, 10, 20, 30 W output powers for 30 s increased as an increase in both width and MW output power, as shown in Figure 6b and Table S1; $\Delta T_{(0.5-20, 1)} = 0.1–2.1$ °C for 1 W, $\Delta T_{(0.5-20, 1)} = 1.0–19$ °C for 5 W, $\Delta T_{(0.5-20, 1)} = 1.4–30.5$ °C for 10 W, and $\Delta T_{(0.5-20, 1)} = 2.6–46.8$ °C for 20 W, and $\Delta T_{(0.5-20, 1)} = 4.2–60.6$ °C for 30 W, respectively. As shown in Figure 7a,c, the magnetic field in the cavity was uniformly widespread at the 90° angle direction against the sample conveying direction, and the magnetic field intensity around the bulk silver also increased with an increase in width [25–27]. The increase in magnetic field intensity around bulk silver accompanied with an increase in width would be caused by a uniform strong magnetic field along width direction and 90° angle against the sample conveying direction [25–27]. Due to generation of higher eddy current associated with an increase in magnetic field intensity resulting from an increase in width, higher Joule heat, which is induced by eddy current, is generated with an increase in width and consequently ΔT would increase [24]. To confirm this consideration, simulations of resistive loss around bulk silver with 8 μm of thickness were carried out. As shown in

Figure 8, resistive loss around the sample was concentrated with the edge of bulk silver. Furthermore, the resistive loss also increased with an increase in width, which also supports the notion that eddy current would increase with an increase in width. From the results of the magnetic field and resistive loss simulations of the bulk silver, an increase in temperature would be caused by an increase in magnetic field intensity and eddy current with an increase in width.

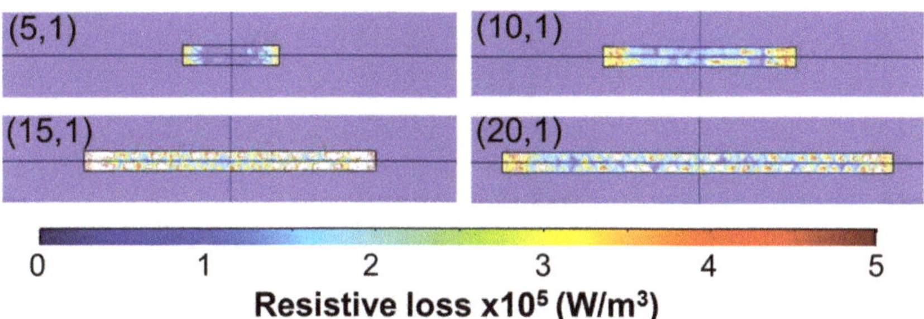

Figure 8. Simulation of resistive loss around bulk silver against the direction of conveying the samples in the TM$_{110}$ cavity: (width, length) = (5 mm, 1 mm), (10 mm, 1 mm), (15 mm, 1 mm), and (20 mm, 1 mm) bulk silver with 8 μm of thickness, respectively.

3.4. Mounting a Device on a Polyimide Sheet Using the MW Method

We mounted a temperature sensor (SHT-31-DIS-B) on a polyimide substrate by MW irradiation (Figure 9). First, silver paste was spread on the polyimide sheet as an electric circuit using the screen-printing method, and the pattern was heated at 100 °C for 1 h on a hot plate. Second, solder paste was spread on the silver circuit to form a connection between the circuit and the sensor, and then a thermo-sensor was mounted on the silver pattern (Figure 9a). Figure 9b,c show the sensor and connection part between the silver circuit and sensor before and after MW irradiation with 10 W of output power for 10 s, confirming that the solder paste melted and the sensor was connected to the electric circuit through the melted solder paste without damaging the substrate. In addition, as shown in the thermal image of the sample under MW irradiation with 10 W of output power for 10 s (Figure 9d), the temperature of the sensor part increased rapidly and reached ~160 °C for 11 s, whereas the temperature of the silver electric circuit did not increase during the MW irradiation. This is because the reaction area (area of solder paste containing sensor unit) is larger than the silver electric circuit and the width of the electric circuit was less than 6 mm against magnetic field direction, leading to effective solder heating and melting by MW. Figure 9e shows the temperature dependency with respect to various output powers of MW irradiation. The results confirm that the solder melted under 5–30 W of output power, similar to the results of Figure 9c, after the temperature had exceeded 150 °C. In the process of heating only solder paste by MW, the temperature of the solder paste dramatically decreased after reaching around 150 °C during the induction soldering by MW (Figure 5), while, in the process of mounting a device by MW, the temperature decrease of the solder paste was not observed after reaching 160 °C (Figure 9e). The reasons would be as follows. The melting and adherence between the solder and the sensor (Figure 9c) would occur without the formation of a small ball, leading to the generation of higher eddy current than the process of heating only solder. In addition to this, dielectric substances and metals in the sensor would also absorb MW and heat up with the generation of eddy current.

Finally, we checked the room temperature using the thermo-sensor mounted by MW irradiation to confirm that the device works well and that neither the sensor nor the silver electric circuit had been damaged by MW irradiation. The room temperature values measured by the commercial digital thermometer and the SHTDA-2 sensor were 23.4 and 24.6 °C, as summarized in Table 1. The room temperature measured by the sensor mounted using MW was 24.8 °C. These results indicate that the thermo-sensor mounted by MW works well and did not suffer any damage from induction soldering by MW. On the other hand, we also succeeded in mounting a thermo-sensor on polyethylene terephthalate film (PET) by the induction soldering of MW without any damage to the sensor and PET. Their detail results would be report elsewhere.

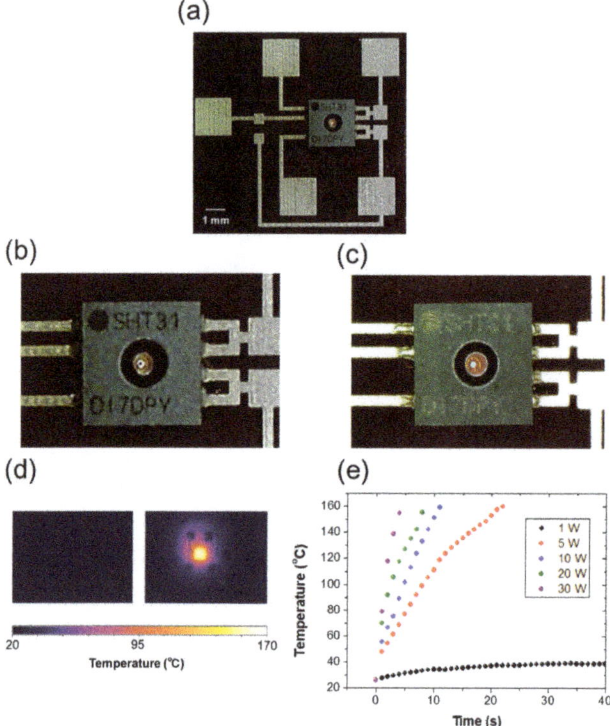

Figure 9. (**a**) Image of a thermo-sensor. Photographs of a sensor and connection part (**b**) before and (**c**) after MW irradiation at 10 W of output power. (**d**) Thermography of samples by MW irradiation with 10 W of output power for 10 s. (**e**) Temperature properties by MW irradiation with 1 (black circle), 5 (red circle), 10 (blue circle), 20 (green circle), and 30 W (purple circle) output powers for 40, 22, 11, 8, and 4 s, respectively.

Table 1. Room temperate measured by digital meter and sensors.

Temperature Sensor	Voltage (V)	Temperature [b] (°C)
Digital meter	-	23.4
SHTDA-2	1.6158	24.6
SHT-31-DIS-B [a]	1.6210	24.8

[a] A sensor which was mounted by induction soldering of MW. [b] Calculated as (output voltage × 1000/25) − 40.

4. Conclusions

We have developed an MW reactor system with a cylindrical single mode TM_{110} cavity that separates the electric fields at the top and bottom and a magnetic field at the center in the cavity. The system enables the heating of electrodes with silver paste and the melting of solder paste. In the investigation of the temperature properties of the electrodes by MW, we found that temperature increase was mainly dependent on not the length but width of the electrode and was higher if there was an increase in the width of electrodes. In the simulation of magnetic field distribution by using COMSOL® software (version 5.4, COMSOL Inc., Boston, MA, USA, 2018), magnetic field intensity around samples also increased with an increase in width, while magnetic field intensity around samples did not increase when the length was large. When the width of the sample was large, the increase in magnetic field around the sample in simulation was because magnetic field distribution is uniformly widespread at the 90° direction of the conveying the samples. For this reason, we found that the temperature increase of the samples highly depends on the width of the sample, which corelated with magnetic field intensity and eddy current value. Furthermore, we successfully mounted a temperature sensor on a polyimide substrate, without damaging the sensor device or the substrate, using MW for a short time period (10 s) and a low output power (10 W); the temperature of the substrate was less than 100 °C during induction soldering by MW and the substrate surface in a microscope image was not changed by MW. Comparisons with a commercial digital thermometer and thermo-sensor show that the mounted temperature sensor device works well, suggesting that induction heating by MW can be applied during the mounting process instead of conventional reflow heating. The developed MW system, the investigation of electrode pad shape effect on induction heating by MW, and the mounting process of electric components by MW should be further developed for use in electronics' packaging.

Supplementary Materials: The following are available online at http://www.mdpi.com/2227-9717/8/5/557/s1, Figure S1: (**a**) A photograph and (**b**) an optical image of solder paste before MW heating process on polyimide sheet. (**c**) A photograph and (**d**) an optical image of solder paste after MW heating process on polyimide sheet, Figure S2: (**a**)TG/DTA and (**b**) DSC curves of solder paste at a scan rate of 5 °C/min in air, Table S1: $\Delta T_{(1, 0.5)}$, $\Delta T_{(1, 1)}$, $\Delta T_{(1, 5)}$, $\Delta T_{(1, 10)}$, $\Delta T_{(1, 15)}$, $\Delta T_{(1, 20)}$, $\Delta T_{(0.5, 1)}$, $\Delta T_{(5, 1)}$, $\Delta T_{(10, 1)}$, $\Delta T_{(15, 1)}$, and $\Delta T_{(20, 1)}$ by MW irradiating at 1, 5, 10, 20, 30 W output powers for 30 s. The thickness of the samples is 12 mm. The thickness of the samples is 12 mm, Table S2: Information about number of elements, meshfree time, and solution time of magnetic field distribution in simulation using Comsol Multiphysics® software.

Author Contributions: K.K., T.N., M.N., and S.U. conceived and designed the experiments; K.K. performed the experiments; K.K. and S.U. analyzed the data; K.K. wrote the paper; K.K., T.N., M.N., and S.U. reviewed and edited the paper; S.U. supervised the research. All authors have read and agreed to the published version of the manuscript.

Funding: This research received no external funding.

Conflicts of Interest: The authors declare no conflict of interest.

References

1. Alaa, M.; Zaidan, A.; Zaidan, B.; Talal, M.; Kiah, M. A review of smart home applications based on Internet of Things. *J. Netw. Comput. Appl.* **2017**, *97*, 48–65. [CrossRef]
2. Liu, Y.; Wang, H.; Zhao, W.; Zhang, M.; Qin, H.; Xie, Y. Flexible, Stretchable Sensors for Wearable Health Monitoring: Sensing Mechanisms, Materials, Fabrication Strategies and Features. *Sensors* **2018**, *18*, 645. [CrossRef] [PubMed]
3. Parrilla, M.; Cuartero, M.; Crespo, G.A. Wearable potentiometric ion sensors. *TrAC Trebds Anal. Chem.* **2019**, *110*, 303–320. [CrossRef]
4. Acome, E.; Mitchell, S.K.; Morrissey, T.G.; Emmett, M.B.; Benjamin, C.; King, M.; Radakovitz, M.; Keplinger, C. Hydraulically amplified self-healing electrostatic actuators with muscle-like performance. *Science* **2018**, *359*, 61–65. [CrossRef] [PubMed]
5. Ammar, M.; Russello, G.; Crispo, B. Internet of Things: A survey on the security of IoT frameworks. *J. Inf. Secur. Appl.* **2018**, *38*, 8–27. [CrossRef]
6. Pan, J.; McElhannon, J. Future Edge Cloud and Edge Computing for Internet of Things Applications. *IEEE Internet Things J.* **2018**, *5*, 439–449. [CrossRef]
7. Abtew, M.; Selvaduray, G. Lead-free solders in microelectronics. *Mater. Sci. Eng. Rep.* **2000**, *27*, 95–141. [CrossRef]
8. Suganuma, K.; Kim, S.J.; Kim, K.S. High-temperature lead-free solders: Properties and possibilities. *JOM* **2009**, *61*, 64–71. [CrossRef]
9. Li, Y.; Wong, C.P. Recent advances of conductive adhesives as a lead-free alternative in electronic packaging: Materials, processing, reliability and applications. *Mater. Sci. Eng. R* **2006**, *51*, 1–35. [CrossRef]

10. Santos, E.C.; Shiomi, M.; Osakada, K.; Laoui, T. Rapid manufacturing of metal components by laser forming. *Int. J. Mach. Tools Manuf.* **2006**, *46*, 1459–1468. [CrossRef]

11. Gu, D.D.; Meiners, W.; Wissenbach, K.; Poprawe, R. Laser additive manufacturing of metallic components: Materials, processes and mechanisms. *Int. Mater. Rev.* **2012**, *57*, 133–164. [CrossRef]

12. Lee, W.H.; Seong, J.G.; Yoon, Y.H.; Jeong, C.H.; Van Tyne, C.J.; Lee, H.G.; Chang, S.Y. Synthesis of TiC reinforced Ti matrix composites by spark plasma sintering and electric discharge sintering; a comparative assessment of microstructural and mechanical properties. *Ceram. Int.* **2019**, *45*, 8108–8114. [CrossRef]

13. Mishra, R.R.; Sharma, A.K. Microwave-material interaction phenomena: Heating mechanisms, challenges and opportunities in material processing. *Compos. Part A* **2016**, *81*, 78–97. [CrossRef]

14. Khaled, D.E.; Novas, N.; Gazquez, J.A.; Manzano-Agugliaro, F. Microwave dielectric heating: Applications on metals processing. *Renew. Sustain. Energ. Rev.* **2018**, *82*, 2880–2892. [CrossRef]

15. Zou, Y.; Li, C.H.; Liu, J.A.; Wu, J.M.; Hu, L.; Gui, R.F.; Shi, Y.S. Towards fabrication of high-performance Al_2O_3 ceramics by indirect selective laser sintering based on particle packing optimization. *Ceram. Int.* **2019**, *45*, 12654–12662. [CrossRef]

16. Chaim, R.; Levin, M.; Shlayer, A.; Estournes, C. Sintering and densification of nanocrystalline ceramic oxide powders: A review. *Adv. Appl. Ceram.* **2008**, *107*, 159–169. [CrossRef]

17. Mirzaei, A.; Neri, G. Microwave-assisted synthesis of metal oxide nanostructures for gas sensing application: A review. *Sens. Actuators B Chem.* **2016**, *237*, 749–775. [CrossRef]

18. Loharkar, P.K.; Ingle, A.; Jhavar, S. Parametric review of microwave-based materials processing and its applications. *J. Mater. Res. Technol.* **2019**, *8*, 3306–3326. [CrossRef]

19. Levin, E.E.; Grebenkemper, J.H.; Pollock, T.M.; Seshadri, R. Protocols for High Temperature Assisted-Microwave Preparation of Inorganic Compounds. *Chem. Mater.* **2019**, *31*, 7151–7159. [CrossRef]

20. Thomas-Hillman, I.; Laybourn, A.; Dodds, C.; Kingman, S.W. Realising the environmental benefits of metal–organic frameworks: Recent advances in microwave synthesis. *J. Mater. Chem. A* **2018**, *6*, 11564–11581. [CrossRef]

21. Horikoshi, S.; Osawa, A.; Abe, M.; Serpone, N. On the generation of hot-spots by microwave electric and magnetic fields and their impact on a microwave-assisted reaction in the presence of metallic Pd nanoparticles on an activated carbon support. *J. Phys. Chem. C* **2011**, *115*, 23030–23035. [CrossRef]

22. Nishioka, M.; Miyakawa, M.; Kataoka, H.; Koda, H.; Sato, K.; Suzuki, T.M. Continuous synthesis of monodispersed silver nanoparticles using a homogeneous heating microwave reactor system. *Nanoscale* **2011**, *3*, 2621–2626. [CrossRef] [PubMed]

23. Miyakawa, M.; Hiyoshi, N.; Koda, H.; Watanabe, K.; Kunigami, H.; Kunigami, H.; Miyazawa, A.; Nishioka, M. Continuous syntheses of carbon-supported Pd and Pd@Pt core-shell nanoparticles using a flow-type single-mode microwave reactor. *RSC Adv.* **2020**, *10*, 6571–6575. [CrossRef]

24. Morgan, S.P., Jr. Effect of surface roughness on eddy current losses at microwave frequencies. *J. Appl. Pnys.* **1949**, *20*, 352–362. [CrossRef]

25. Webb, A. Cavity- and waveguide-resonators in electron paramagnetic resonance, nuclear magnetic resonance, and magnetic resonance imaging. *Prog. Nucl. Magn. Reson. Spectrosc.* **2014**, *83*, 1–20. [CrossRef]

26. Verhoeven, W.; van Rens, J.F.M.; Kemper, A.H.; Rietman, E.H.; van Doorn, H.A.; Koole, I.E.; Kieft, R.; Mutsaers, P.H.A.; Luiten, O.J. Design and characterization of dielectric filled TM_{110} micraowave cavities for ultrafast electron microscopy. *Rev. Sci. Instrum.* **2019**, *90*, 080370. [CrossRef]

27. Cao, Z.; Yoshikawa, N.; Taniguchi, S. Directional selectivity of microwave H field heating of Au thin films and non-doped Si plates. *Mater. Chem. Phys.* **2009**, *117*, 14–17. [CrossRef]

Communication

Concentration of Lipase from *Aspergillus oryzae* Expressing *Fusarium heterosporum* by Nanofiltration to Enhance Transesterification

Hans Wijaya [1,2], Kengo Sasaki [3], Prihardi Kahar [3], Emmanuel Quayson [1], Nova Rachmadona [1], Jerome Amoah [3], Shinji Hama [4], Chiaki Ogino [1,*] and Akihiko Kondo [1,3]

[1] Department of Chemical Science and Engineering, Graduate School of Engineering, Kobe University, 1-1 Rokkodaicho, Nada-ku, Kobe 657-8501, Japan; hanswijayalipi2014@gmail.com (H.W.); quaysonemmanuel45@gmail.com (E.Q.); nvarach@yahoo.co.id (N.R.); akondo@kobe-u.ac.jp (A.K.)
[2] Research Center for Biotechnology, Indonesian Institute of Sciences (LIPI), Jl. Raya Bogor Km 46, Cibinong, Bogor 16911, West Java, Indonesia
[3] Graduate School of Science, Technology and Innovation, Kobe University, 1-1 Rokkodaicho, Nada-ku, Kobe 657-8501, Japan; sikengo@people.kobe-u.ac.jp (K.S.); pri@port.kobe-u.ac.jp (P.K.); asbe.71@gmail.com (J.A.)
[4] Bio-energy Corporation, Research and Development Laboratory, 2-9-7 Minaminanamatsucho, Amagasaki 660-0053, Japan; hama@bio-energy.jp
* Correspondence: ochiaki@port.kobe-u.ac.jp; Tel./Fax: +81-78-803-6193

Received: 12 March 2020; Accepted: 7 April 2020; Published: 11 April 2020

Abstract: Nanofiltration membrane separation is an energy-saving technology that was used in this study to concentrate extracellular lipase and increase its total activity for biodiesel production. Lipase was produced by recombinant *Aspergillus oryzae* expressing *Fusarium heterosporum* lipase (FHL). A sulfonated polyethersulfone nanofiltration membrane, NTR-7410, with a molecular weight cut-off of 3 kDa was used for the separation, because recombinant lipase has a molecular weight of approximately 20 kDa, which differs from commercial lipase at around 30 kDa for Callera™ Trans L (CalT). After concentration via nanofiltration, recombinant lipase achieved a 96.8% yield of fatty acid methyl ester (FAME) from unrefined palm oil, compared to 50.2% for CalT in 24 h. Meanwhile, the initial lipase activity (32.6 U/mL) of recombinant lipase was similar to that of CalT. The composition of FAME produced from recombinant concentrated lipase, i.e., C14:1, C16:0, C18:0, C18:1 cis, and C18:2 cis were 0.79%, 34.46%, 5.41%, 45.90%, and 12.46%, respectively, after transesterification. This FAME composition, even after being subjected to nanofiltration, was not significantly different from that produced from CalT. This study reveals the applicability of a simple and scalable nanofiltration membrane technology that can enhance enzymatic biodiesel production.

Keywords: nanofiltration; lipase; *Fusarium heterosporum*; fatty acid methyl ester

---

1. Introduction

Over the past decade, interest in biodiesel as an alternative to diesel fuel continues to increase throughout the world due to concerns about global climate change. Increasingly, there is a desire for renewable/sustainable energy sources, and an interest in developing domestic supplies of fuel that are more secure [1]. Biodiesel (fatty acid alkyl esters) is produced from renewable natural sources such as vegetable oils (e.g., palm oil), animal fats, and microalgal oil [2,3]. Biodiesel can be used directly in existing diesel engines without major modifications, or as a mixture with petroleum diesel, and the burning of biodiesel produces gas emissions such as sulfur oxide, which are less harmful than those emitted by the burning of petroleum-based fuels [4].

The viscosity of vegetable oils is improved via a transesterification pathway which involves triglycerides and alcohols of lower molecular weights and homogenous or heterogenous substances that are used as catalysts to yield biodiesel and glycerol [4]. Transesterification via enzymatic catalysis has attracted much attention because it is an eco-friendly process that produces no by-products, features easy product recovery, and requires a low reaction temperature [5]. However, the process is expensive, and has a relatively slow reaction rate [6,7]. A variety of lipases (EC 3.1.1.3) from various microorganisms (*Candida antarctica*, *Rhizopus oryzae*, *Pseudomonas cepacia*, *Thermomyces lanuginosus*, etc.) have been used to accomplish both transesterification and esterification [8–10]. Many researchers have attempted to solve the limitations of lipase-catalyzed biodiesel production by immobilizing the enzymes or cells on a suitable matrix [8,11] or via the use of a lipase cocktail [12]. In contrast, at least one previous study has successfully conducted biodiesel production using recombinant *Aspergillus oryzae* that expresses *Fusarium heterosporum* lipase (FHL), which has demonstrated a high level of tolerance to water [13]. With the use of that particular enzyme, however, the conversion of oil to fatty acid methyl ester (FAME) remained low. To increase the conversion rate, the total activity of this enzyme was increased using a simple process such as a concentrating method. In general, the conversion increases proportionally with the increase of lipase concentration [14]. Concentration methods such as precipitation [15,16] require costly chemicals such as ammonium sulfate. Another concentration method is membrane separation technology, which has advantages that include energy savings, selectivity, no chemical requirement, and simplicity of operation and scale-up [17].

Reinehr et al. [17] previously reported a membrane concentration of lipase that could be produced from *Aspergillus niger* by using microfiltration and ultrafiltration separation processes. The present study is the first to apply a nanofiltration membrane to simply concentrate lipase produced from recombinant *A. oryzae* (expressing FHL) and enable a high level of transesterification compared to a commercially available lipase, Callera™ Trans L (CalT) (Novozymes, Bagsvaerd, Denmark). In this study, we use unrefined palm oil as a model substrate for transesterification. Palm oil is well known as one of the most suitable sources for biodiesel production. Indonesia and Malaysia produce approximately 85% of global crude palm oil, which is likely to increase in the future [3,18]. The aim of the present study was to efficiently produce FAME from unrefined palm oil using membrane-concentrated lipase.

2. Materials and Methods

2.1. Materials and Microorganisms

Unrefined palm oil was purchased as a substrate from Malang, East Java, Indonesia. *Aspergillus oryzae* expressing FHL used in this study was obtained as described previously [19,20].

2.2. Lipase Production

Sakaguchi flasks (500 mL) containing 100 mL of DP medium (2% glucose, 2% polypeptone, 1% KH_2PO_4, 0.2% $NaNO_3$, 0.05% $MgSO_4 \cdot 7H_2O$) were aseptically inoculated with spores from *A. oryzae* expressing FHL in Czapek-Dox (CD)-NO_2-methionine selection plate agar [20]. The flasks were incubated at 30 °C and shaken at 150 rpm for 96 h on a reciprocal shaker. The culture broth was collected and then centrifuged at 6000× *g* for 15 min at 4 °C to recover the supernatant. The culture supernatant was dialyzed in MEMBRA-CEL® dialysis tubing with a molecular weight cut-off (MWCO) of 3500 Da (RC, SERVA Electrophoresis GmbH, Heidelberg, Germany), followed by filtrations through different filter papers in the following order: (1) a polycarbonate filter (0.8 μm pore size); (2) a polycarbonate filter (0.5 μm pore size); and (3) a polystyrene filter (0.22 μm pore size). The supernatant was then subjected to nanofiltration-membrane concentration. Lipase from Callera™ Trans L, a liquid lipase from *Thermomyces lanuginosus* lipase (CalT) (Novozymes, Bagsvaerd, Denmark), was used as a control.

2.3. Nanofiltration Membrane Separation Processes

A sulfonated polyethersulfone nanofiltration membrane, NTR-7410, with a 3000 Da MWCO was obtained from the Nitto Denko Corporation (Osaka, Japan). The membrane was cut into a circle (diameter: 7.5 cm; effective area: 32 cm^2). The nanofiltration (NF) process was carried out at room temperature using a flat membrane test cell (model C40-B, Nitto Denko Corporation, Osaka, Japan) [21]. The lipase supernatant was then subjected to the test cell. The inside of the test cell was stirred at 300 rpm at a pressure of 2.5 MPa under nitrogen gas for one hour.

2.4. Sodium Dodecyl Sulfate-Polyacrylamide Gel Electrophoresis (SDS-PAGE) and Zymography

The *A. oryzae* supernatant expressing FHL was either directly applied to polyacrylamide gel electrophoresis in the presence of SDS-PAGE, or applied after filtration through a 5K MWCO Spin-X Ultrafiltration Concentrator (Corning, UK) to concentrate the lipase. CalT was applied to SDS-PAGE after dilution. The proteins were then stained with Coomassie brilliant blue R-250. Zymography analysis was carried out to detect lipase activity, as described previously [22]. The SDS-PAGE gels were incubated for one hour at room temperature in developing solution consisting of 3 mM α-naphthyl acetate, 1 mM Fast Red TR (Sigma, St. Louis, MO, USA), and 100 mM sodium phosphate buffer, pH 8.0. Precision Plus Protein™ Dual Color Standards (Bio-Rad, Hercules, CA, USA) were used as a standard marker.

2.5. Measurement of Lipase Activity and Protein Assay

The hydrolytic activities of lipase were tested using *p*-nitrophenyl butyrate (pNPB) as a chromogenic substrate. A stock solution was prepared by dissolving 5 µL of pNPB in 250 µL of ethanol, with a further dilution to 50 mL using distilled water. The stock solution was then incubated in a Bioshaker (Taitec, Saitama, Japan) for 10 min at 30 °C to allow the development of lipase-hydrolytic activity. After incubation, 5% trichloroacetate was added to terminate the reaction. The absorbance of para nitrophenol (pNP) that was produced was measured at 400 nm (UV-Vis spectrophotometer, Shimadzu, Kyoto, Japan). One unit (U$_2$) of lipase activity was defined as the amount of lipase that liberates 1 µmol of pNP from pNPB per minute [12]. Protein concentrations were measured using a Pierce™ BCA Protein Assay Kit (Thermo Scientific™, Rockford, IL, USA).

2.6. FAME Production by Enzyme

FAME production was carried out in triplicate using a 10 mL glass tube with a silicon cap equipped with a stirrer (Thermo Scientific VARIOMAG Magnetic Stirrers, Waltham, MA, USA) for circulation at 800 rpm. The glass tube was then immersed in a heated water bath Thermo Minder EX TAITEC (Taitec, Saitama, Japan) that was set to 30 °C. The reaction mixture consisted of 4 g of unrefined palm oil and 1.2 mL of lipase (unconcentrated and concentrated lipases from recombinant *A. oryzae* and CalT diluted 245-fold). The transesterifications were carried out at 32.6 U/mL lipase activity for concentrated lipase from recombinant *A. oryzae* and for that from CalT diluted 245-fold (Table 1). To avoid deactivation of the lipase, 186 µL of methanol (corresponding to a 1:1 molar ratio of unrefined palm oil to methanol) was added step-wise at 0, 2, 4, and 6 h. Samples were taken at 0, 2, 4, 6, 9, and 24 h [13].

Table 1. Comparison of lipase activity and protein content.

Process	Volume (mL)	Lipase Activity (U/mL)	Protein (mg/mL)
Recombinant lipase before concentration by NTR7410	350	6.4 ± 0.1	0.7 ± 0.0
Recombinant lipase after concentration by NTR7410	65	32.6 ± 3.1	4.0 ± 0.2
Permeate from NTR7410	280	0	0
Callera™ Trans L (CalT)		8396.7 ± 378.4	25.1 ± 1.4

2.7. Analytical Methods

Fatty acid methyl ester (FAME) produced during the course of the transesterification reaction was measured via gas chromatography. Samples taken at specified times were centrifuged at 12,000× *g* for 5 min at 5 °C, and the upper layer was analyzed via a GC-17A (Shimadzu, Kyoto, Japan) equipped with a ZB-5HT capillary column (0.25 mm × 15 m) (Phenomenex, Torrance, CA, USA), an auto-sampler, and a flame ionization detector, as previously described [12]. During the analysis, the temperatures of the injector and detector were set at 320 °C and 380 °C, respectively, using helium as a carrier gas at a flow rate of 58 mL/min. The column was configured with an initial temperature of 130 °C for 2 min, which was raised to 350 °C at 10 °C/min, and then to 370 °C at 7 °C/min. The FAME composition in each reaction mixture was reported as the percentage of the oil in the reaction mixture using tricaprylin as an internal standard [11].

Transesterification was conducted by following the protocol from the fatty acid methylation kit (Nacalai Tesque Inc., Kyoto, Japan). The FAME composition was analyzed using a gas chromatography-mass spectrometer (GC-MS) (Shimadzu, Kyoto, Japan). The GC-MS was equipped with a 0.25 mm × 30 m DB-23 capillary column (J&W Scientific, Folsom, CA, USA). The carrier was helium gas with a flow rate of 0.8 mL/min at 1:5 split ratios. The initial column temperature was 250 °C, which was increased to 50 °C for 1 min and then increased 25 °C/min to 190 °C and 5 °C/min to 235 °C for 4 min. An internal standard C8:0 (octanoic acid) was included in each sample and FAME was detected at the provided retention time (Supplementary Table S1). The amount of FAME (%) was calculated as the percentage of each fatty acid to the total weight of fatty acids produced [23].

3. Results and Discussion

3.1. Characterization of Lipases before and after Membrane Concentration

As described previously, the molecular weights of lipases ranged from 20 to 80 kDa [17,24] or up to 150 kDa [25]. Thus, NTR-7410 with a MWCO of 3 kDa [26] was selected as the membrane that would best concentrate lipase. Nanofiltration concentration was performed for one hour at 2.5 MPa.

Both concentrated and unconcentrated lipases produced from recombinant *A. oryzae* were characterized and compared with commercial lipase, CalT. At first, the molecular weight of the lipase produced by recombinant *A. oryzae* was determined via SDS-PAGE. Then, lipase enzyme activity was detected using the Zymography technique. As a result, the supernatant of *A. oryzae* contained plural proteins with molecular weights that ranged from 20 to 50 kDa (Figure 1A). However, lipase enzyme was detected as a single band at around 20 kDa (Figure 1B). As expected, the concentration of lipase had definitely increased. By comparison, the CalT contained a major protein at around 30 kDa along with some minor proteins that also showed lipase activity (Figure 1C,D). Therefore, we assumed that the observed bands other than that at around 30 kDa also represented small amounts of lipase (Figure 1D). These results suggest that the FHL produced in *A. oryzae* was a smaller molecule compared with the lipases produced in CalT. In addition, recombinant *A. oryzae* produced some other unknown proteins that were not lipase. Due to this contamination by other enzymes, the lipase produced from recombinant *A. oryzae* showed low specific activity (Table 1) compared with lipase reported elsewhere (more than 66 U/mg) [25]. However, the nanofiltration concentration of lipase produced by recombinant *A. oryzae* successfully increased its total activity from 6.4 U/mL to 32.6 U/mL (five-fold concentration factor) in the short time of one hour. Enzyme activity losses in this study were low (5.5%, Table 2), and may have occurred due to adsorption on the membrane surface as a function of fouling [17]. The activity loss was generally lower, because the use of NTR-7410 (MWCO of 3 kDa) could retain more lipase protein than that produced by recombinant *A. oryzae* at around 20 kDa. The denaturation did not occur due to the pressure applied lower than 400 MPa as used in [27].

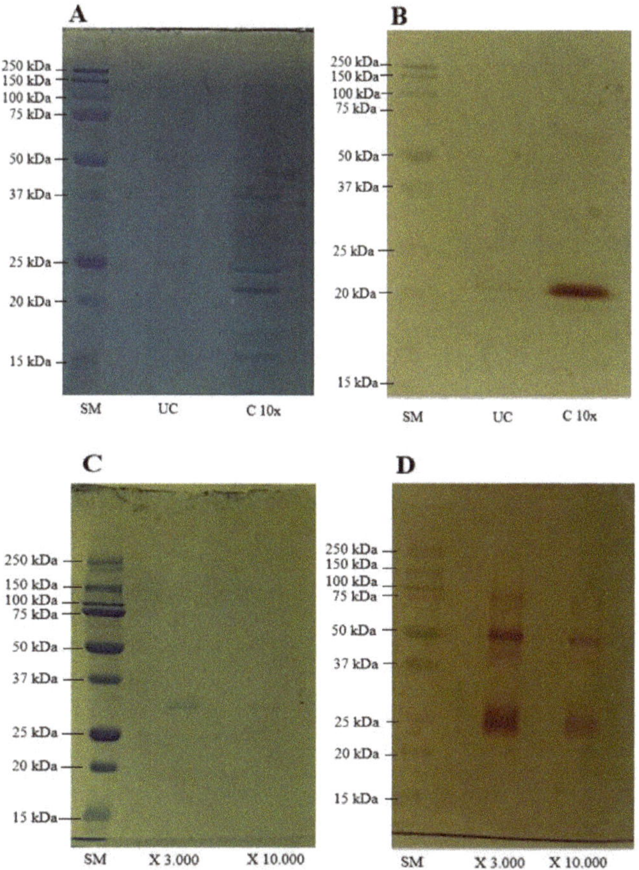

Figure 1. SDS-PAGE and Zymogram. (**A**) SDS-PAGE for the culture supernatant of *Aspergillus oryzae* expressing *Fusarium heterosporum* lipase (FHL) (UC: unconcentrated, C 10×: concentrated 10-fold). (**B**) Zymogram for the detection of lipase produced from recombinant *A. oryzae*. (**C**) SDS-PAGE of the lipase from *Thermomyces lanuginosus* (CalT). (**D**) Zymogram of CalT. (X 3.000 at dilution 3.000-fold, X 10.000 at dilution 10.000-fold). SM: standard marker.

Table 2. Concentration factors and activity loss obtained in previous processes involving the concentration of proteins using membrane separation technologies.

Protein Concentrated	Process Used	Concentration Factor Obtained	Activity Loss	Ref.
Lipase (recombinant *Aspergillus oryzae*)	Nanofiltration (3 kDa)	5	5.5%	This study
Lipase (*Aspergillus niger*)	Sequential micro- and ultrafiltration	3	17% in microfiltration and 22% in ultrafiltration	[17]
Phytase (*Aspergillus niger*)	Ultrafiltration (10 kDa)	4.3	14%	[28]
Inulinase (*Kluyveromyces marxianus*)	Ultrafiltration (100 kDa)	5.5	18.4%	[29]
Lignin-peroxidase (*Streptomyces viridosporus*)	Ultrafiltration (10 kDa)	10	10%	[30]
Phycocyanin (*Spirulina* sp.)	Sequential micro- and ultrafiltration	2	13.6% in 1 µm pore size	[31]

3.2. Efficient FAME Production by Concentrated Lipase

FAME yield was compared for concentrated and unconcentrated lipases produced from recombinant *A. oryzae* and commercial lipase (CalT). In these reactions, the total lipase activity of concentrated lipase produced by recombinant *A. oryzae* was arranged to be nearly the same as that of diluted CalT. The results for the FAME yield are shown in Figure 2. Interestingly, FAME production was enhanced by using concentrated lipase produced by recombinant *A. oryzae* (designated as AC), compared with diluted CalT (designated as CalT), and unconcentrated lipase (designated as BC). Other controls were 0.5 AC + 0.5 CalT, which contained half of the concentrated lipase produced by recombinant *A. oryzae* and half of diluted CalT, and 0.5 BC + 0.5 CalT, which contained half of the unconcentrated lipase produced by recombinant *A. oryzae* and half of diluted CalT. By comparing these data, it was apparent that activity for lipase produced by recombinant *A. oryzae* to convert unrefined palm oil to FAME was significantly enhanced by nanofiltration concentration, compared with that produced by CalT. The reason for this remains unclear. We hypothesized that the FHL produced by recombinant *A. oryzae* would have different characteristics from commercial lipase (CalT).

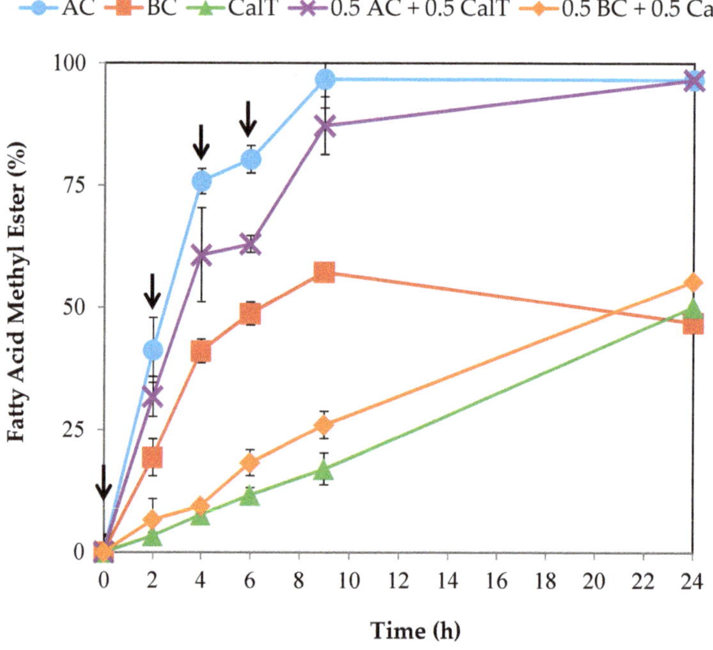

Figure 2. Fatty acid methyl ester (FAME) production by lipase enzymes. AC: lipase produced from recombinant *A. oryzae* after nanofiltration (NF) concentration; BC: lipase produced from recombinant *A. oryzae* before NF concentration; CalT: lipase from *Thermomyces lanuginosus* showing lipase activity similar to AC (diluted around 245-fold); 0.5 AC + 0.5 CalT: half AC and half CalT; and 0.5 BC + 0.5 CalT: half BC and half CalT. The arrows indicate the time to add methanol.

The FAME compositions after 24 h of enzymatic reaction were analyzed by GC-MS, as shown in Table 3. In general, there were no major differences in the FAME composition converted from unrefined palm oil by both concentrated and unconcentrated lipases, produced either by recombinant *A. oryzae* or CalT. These results suggest that the biodiesel produced using concentrated lipase from recombinant *A. oryzae* has high potential to be used in the same manner as biodiesel produced from

CalT. Furthermore, nanofiltration concentration can be used to enhance the quality of lipase used in biodiesel production without a loss of FAME quantity.

Previously, most of the research focusing on enzymatic catalysis has employed lipase immobilized on polymer support as a catalyst. However, the immobilization process is neither simple nor inexpensive [32]. Using the suggested membrane separation technology to concentrate lipase therefore simplifies the process and reduces the cost. In addition, concentrated lipase produced from recombinant *A. oryzae* can be used as a sole lipase or as a supplement to other commercially available lipases to reduce costs and improve biodiesel conversion yields from unconventional feedstock.

Table 3. FAME profiles of transesterification results.

Lipase Variations	FAME Compositions (%)				
	C14:1	C16:0	C18:0	C18:1 cis	C18:2 cis
AC	0.79 ± 0.15	34.46 ± 0.37	5.41 ± 1.01	45.90 ± 1.71	12.46 ± 0.29
BC	0.89 ± 0.04	35.45 ± 0.12	5.84 ± 0.08	44.88 ± 0.03	12.32 ± 0.09
CalT	0.81 ± 0.04	34.75 ± 0.44	5.85 ± 0.18	45.06 ± 0.12	12.48 ± 0.19
0.5 AC + 0.5 CalT	0.79 ± 0.01	34.46 ± 0.03	5.75 ± 0.08	45.14 ± 0.09	12.94 ± 0.12
0.5 BC + 0.5 CalT	0.76 ± 0.08	34.71 ± 0.57	5.76 ± 0.11	45.24 ± 0.44	12.66 ± 0.56

AC: lipase produced from *A. oryzae* expressing FHL after NF concentration, BC: lipase produced from *A. oryzae* expressing FHL before NF concentration, CalT: lipase from *Thermomyces lanuginosus* showing lipase activity similar to AC (diluted around 245-fold), 0.5 AC + 0.5 CalT: half AC and half CalT, and 0.5 BC + 0.5 CalT: half BC and half CalT.

4. Conclusions

In this study, the total activity of lipase produced by recombinant *A. oryzae* was successfully increased at about 5-fold using nanofiltration membrane separation technology. Concentrated lipase produced from recombinant *A. oryzae* showed a higher FAME yield of 96.8% from unrefined palm oil, compared to 50.2% for CalT in a 24 h period, although the lipase activity (32.6 U/mL) was nearly the same between concentrated lipase and CalT. FAME composition using concentrated lipase was unchanged, compared with that using CalT. The FAME, C14:1, C16:0, C18:0, C18:1 cis, and C18:2 cis were produced, respectively, at 0.79%, 34.46%, 5.41%, 45.90%, and 12.46% after transesterification at 30 °C for 24 h. In this study, a simple and inexpensive process was developed using a nanofiltration membrane that is expected to improve enzymatic and industrial biodiesel production.

Supplementary Materials: The following are available online at http://www.mdpi.com/2227-9717/8/4/450/s1, Table S1: Retention time of target compound of fatty acid methyl ester obtained in Gas Chromatography-Mass Spectrometer analysis.

Author Contributions: Conceptualization, H.W., K.S., P.K., J.A., E.Q., N.R.; experiment, H.W., E.Q., N.R.; methodology, K.S., P.K., J.A., S.H., C.O., A.K.; validation, K.S., P.K., J.A.; data curation, H.W., K.S., P.K., J.A.; formal analysis, H.W., K.S., J.A., P.K.; writing—original draft preparation, H.W., K.S., P.K., J.A., E.Q.; writing—review and editing, K.S., P.K., J.A., S.H., C.O., A.K. All authors have read and agreed to the published version of the manuscript.

Funding: This research was funded by the International Joint Program, Science and Technology Research Partnership for Sustainable Development (SATREPS), Innovative Bioproduction Kobe (iBioK) from the Japan Science and Technology Agency (JST), the Japan International Cooperation Agency (JICA), and The Ministry of Education, Culture, Sports, Science and Technology.

Acknowledgments: We appreciate the help provided by Ayami Fujino and Yasunobu Takeshima.

Conflicts of Interest: The authors declare no conflicts of interest.

References

1. Hoekman, S.K.; Broch, A.; Robbins, C.; Ceniceros, E.; Natarajan, M. Review of biodiesel composition, properties, and specifications. *Renew. Sustain. Energy Rev.* **2012**, *16*, 143–169. [CrossRef]
2. Ge, J.C.; Kim, H.Y.; Yoon, S.K.; Choi, N.J. Optimization of palm oil biodiesel blends and engine operating parameters to improve performance and PM morphology in a common rail direct injection diesel engine. *Fuel* **2020**, *260*, 116326. [CrossRef]
3. Lam, M.K.; Jamalluddin, N.A.; Lee, K.T. *Production of Biodiesel Using Palm Oil*, 2nd ed.; Elsevier Inc.: Amsterdam, The Netherlands, 2019; ISBN 9780128168561.
4. Atadashi, I.M.; Aroua, M.K.; Aziz, A.A. Biodiesel separation and purification: A review. *Renew. Energy* **2011**, *36*, 437–443. [CrossRef]
5. Bhangu, S.K.; Gupta, S.; Ashokkumar, M. Ultrasonic enhancement of lipase-catalysed transesterification for biodiesel synthesis. *Ultrason. Sonochem.* **2017**, *34*, 305–309. [CrossRef]
6. Gog, A.; Roman, M.; Toşa, M.; Paizs, C.; Irimie, F.D. Biodiesel production using enzymatic transesterification—Current state and perspectives. *Renew. Energy* **2012**, *39*, 10–16. [CrossRef]
7. Amini, Z.; Ilham, Z.; Ong, H.C.; Mazaheri, H.; Chen, W.H. State of the art and prospective of lipase-catalyzed transesterification reaction for biodiesel production. *Energy Convers. Manag.* **2017**, *141*, 339–353. [CrossRef]
8. Narwal, S.K.; Gupta, R. Biodiesel production by transesterification using immobilized lipase. *Biotechnol. Lett.* **2013**, *35*, 479–490. [CrossRef]
9. Hama, S.; Noda, H.; Kondo, A. How lipase technology contributes to evolution of biodiesel production using multiple feedstocks. *Curr. Opin. Biotechnol.* **2018**, *50*, 57–64. [CrossRef]
10. Bajaj, A.; Lohan, P.; Jha, P.N.; Mehrotra, R. Biodiesel production through lipase catalyzed transesterification: An overview. *J. Mol. Catal. B Enzym.* **2010**, *62*, 9–14. [CrossRef]
11. Amoah, J.; Quayson, E.; Hama, S.; Yoshida, A.; Hasunuma, T.; Ogino, C.; Kondo, A. Simultaneous conversion of free fatty acids and triglycerides to biodiesel by immobilized *Aspergillus oryzae* expressing *Fusarium heterosporum* lipase. *Biotechnol. J.* **2017**, *12*. [CrossRef]
12. Amoah, J.; Ho, S.H.; Hama, S.; Yoshida, A.; Nakanishi, A.; Hasunuma, T.; Ogino, C.; Kondo, A. Lipase cocktail for efficient conversion of oils containing phospholipids to biodiesel. *Bioresour. Technol.* **2016**, *211*, 224–230. [CrossRef] [PubMed]
13. Amoah, J.; Ho, S.H.; Hama, S.; Yoshida, A.; Nakanishi, A.; Hasunuma, T.; Ogino, C.; Kondo, A. Converting oils high in phospholipids to biodiesel using immobilized *Aspergillus oryzae* whole-cell biocatalysts expressing *Fusarium heterosporum* lipase. *Biochem. Eng. J.* **2016**, *105*, 10–15. [CrossRef]
14. He, Q.; Xu, Y.; Teng, Y.; Wang, D. Biodiesel Production Catalyzed by Whole-Cell Lipase from *Rhizopus chinensis*. *Chin. J. Catal.* **2008**, *29*, 41–46. [CrossRef]
15. Menoncin, S.; Domingues, N.M.; Freire, D.M.G.; Toniazzo, G.; Cansian, R.L.; Oliveira, J.V.; Di Luccio, M.; de Oliveira, D.; Treichel, H. Study of the extraction, concentration, and partial characterization of lipases obtained from *Penicillium verrucosum* using solid-state fermentation of soybean bran. *Food Bioprocess Technol.* **2010**, *3*, 537–544. [CrossRef]
16. Preczeski, K.P.; Kamanski, A.B.; Scapini, T.; Camargo, A.F.; Modkoski, T.A.; Rossetto, V.; Venturin, B.; Mulinari, J.; Golunski, S.M.; Mossi, A.J.; et al. Efficient and low-cost alternative of lipase concentration aiming at the application in the treatment of waste cooking oils. *Bioprocess Biosyst. Eng.* **2018**, *41*, 851–857. [CrossRef]
17. Reinehr, C.O.; Treichel, H.; Tres, M.V.; Steffens, J.; Brião, V.B.; Colla, L.M. Successive membrane separation processes simplify concentration of lipases produced by *Aspergillus niger* by solid-state fermentation. *Bioprocess Biosyst. Eng.* **2017**, *40*, 843–855. [CrossRef]
18. Varkkey, H.; Tyson, A.; Choiruzzad, S.A.B. Palm oil intensification and expansion in Indonesia and Malaysia: Environmental and socio-political factors influencing policy. *For. Policy Econ.* **2018**, *92*, 148–159. [CrossRef]
19. Takaya, T.; Koda, R.; Adachi, D.; Nakashima, K.; Wada, J.; Bogaki, T.; Ogino, C.; Kondo, A. Highly efficient biodiesel production by a whole-cell biocatalyst employing a system with high lipase expression in *Aspergillus oryzae*. *Appl. Microbiol. Biotechnol.* **2011**, *90*, 1171–1177. [CrossRef]
20. Hama, S.; Tamalampudi, S.; Suzuki, Y.; Yoshida, A.; Fukuda, H.; Kondo, A. Preparation and comparative characterization of immobilized *Aspergillus oryzae* expressing *Fusarium heterosporum* lipase for enzymatic biodiesel production. *Appl. Microbiol. Biotechnol.* **2008**, *81*, 637–645. [CrossRef]

21. Sasaki, K.; Tsuge, Y.; Sasaki, D.; Teramura, H.; Inokuma, K.; Hasunuma, T.; Ogino, C.; Kondo, A. Mechanical milling and membrane separation for increased ethanol production during simultaneous saccharification and co-fermentation of rice straw by xylose-fermenting *Saccharomyces cerevisiae*. *Bioresour. Technol.* **2015**, *185*, 263–268. [CrossRef]

22. Febriani; Hertadi, R.; Kahar, P.; Akhmaloka; Madayanti, F. Isolation and purification of novel thermostable alkaline lipase from local thermophilic microorganism. *Biosci. Biotechnol. Res. Asia* **2010**, *7*, 617–622.

23. Amza, R.L.; Kahar, P.; Juanssilfero, A.B.; Miyamoto, N.; Otsuka, H.; Kihira, C.; Ogino, C.; Kondo, A. High cell density cultivation of *Lipomyces starkeyi* for achieving highly efficient lipid production from sugar under low C/N ratio. *Biochem. Eng. J.* **2019**, *149*, 107236. [CrossRef]

24. Priji, P.; Sajith, S.; Faisal, P.A.; Benjamin, S. *Pseudomonas* sp. BUP6 produces a thermotolerant alkaline lipase with trans-esterification efficiency in producing biodiesel. *3 Biotech* **2017**, *7*, 369. [CrossRef]

25. Sharma, A.; Meena, K.R.; Kanwar, S.S. Molecular characterization and bioinformatics studies of a lipase from *Bacillus thermoamylovorans* BHK67. *Int. J. Biol. Macromol.* **2018**, *107*, 2131–2140. [CrossRef] [PubMed]

26. Sasaki, K.; Okamoto, M.; Shirai, T.; Tsuge, Y.; Fujino, A.; Sasaki, D.; Morita, M.; Matsuda, F.; Kikuchi, J.; Kondo, A. Toward the complete utilization of rice straw: Methane fermentation and lignin recovery by a combinational process involving mechanical milling, supporting material and nanofiltration. *Bioresour. Technol.* **2016**, *216*, 830–837. [CrossRef]

27. Marie-Olive, M.N.; Athes, V.; Combes, D. Combined effects of pressure and temperature on enzyme stability. *High Press. Res.* **2000**, *19*, 317–322. [CrossRef]

28. Fernández-Lorente, G.; Ortiz, C.; Segura, R.L.; Fernández-Lafuente, R.; Guisán, J.M.; Palomo, J.M. Purification of different lipases from *Aspergillus niger* by using a highly selective adsorption on hydrophobic supports. *Biotechnol. Bioeng.* **2005**, *92*, 773–779. [CrossRef]

29. Golunski, S.; Astolfi, V.; Carniel, N.; De Oliveira, D.; Di Luccio, M.; Mazutti, M.A.; Treichel, H. Ethanol precipitation and ultrafiltration of inulinases from *Kluyveromyces marxianus*. *Sep. Purif. Technol.* **2011**, *78*, 261–265. [CrossRef]

30. Gottschalk, L.M.F.; Bon, E.P.S.; Nobrega, R. Lignin Peroxidase from *Streptomyces viridosporus* T7A: Enzyme Concentration Using Ultrafiltration. *Biotechnol. Appl. Biochem.* **2008**, *147*, 23–32. [CrossRef]

31. Chaiklahan, R.; Chirasuwan, N.; Loha, V.; Tia, S.; Bunnag, B. Separation and purification of phycocyanin from *Spirulina* sp. using a membrane process. *Bioresour. Technol.* **2011**, *102*, 7159–7164. [CrossRef]

32. Wancura, J.H.C.; Tres, M.V.; Jahn, S.L.; de Oliveira, J.V. Lipases in liquid formulation for biodiesel production: Current status and challenges. *Biotechnol. Appl. Biochem.* **2019**, 1–20. [CrossRef] [PubMed]

Article

Biotechnology and Bioprocesses: Their Contribution to Sustainability

Alejandro Barragán-Ocaña [1,*], Paz Silva-Borjas [1], Samuel Olmos-Peña [2] and Mirtza Polanco-Olguín [1]

[1] Instituto Politécnico Nacional (IPN), Centro de Investigaciones Económicas, Administrativas y Sociales (CIECAS), Lauro Aguirre 120. Col. Agricultura, Alcaldía. Miguel Hidalgo, Ciudad de México C. P. 11360, Mexico; psilva@ipn.mx (P.S.-B.); mpolancoo@ipn.mx (M.P.-O.)

[2] Universidad Autónoma del Estado de México (UAEM), Centro Universitario UAEM Valle de Chalco, Hermenegildo Galena No.3, Colonia María Isabel, Valle de Chalco CP, Estado de México 56615, Mexico; solmosp@uaemex.mx

* Correspondence: abarragano@ipn.mx; Tel.: +52-5557-296-000

Received: 21 March 2020; Accepted: 2 April 2020; Published: 7 April 2020

Abstract: Significant advancements in biotechnology have resulted in the development of numerous fundamental bioprocesses, which have consolidated research and development and industrial progress in the field. These bioprocesses are used in medical therapies, diagnostic and immunization procedures, agriculture, food production, biofuel production, and environmental solutions (to address water-, soil-, and air-related problems), among other areas. The present study is a first approach toward the identification of scientific and technological bioprocess trajectories within the framework of sustainability. The method included a literature search (Scopus), a patent search (Patentscope), and a network analysis for the period from 2010 to 2019. Our results highlight the main technological sectors, countries, institutions, and academic publications that carry out work or publish literature related to sustainability and bioprocesses. The network analysis allowed for the identification of thematic clusters associated with sustainability and bioprocesses, revealing different related scientific topics. Our conclusions confirm that biotechnology is firmly positioned as an emerging knowledge area. Its dynamics, development, and outcomes during the study period reflect a substantial number of studies and technologies focused on the creation of knowledge aimed at improving economic development, environmental protection, and social welfare.

Keywords: bioeconomy; bioprocesses; applications; policy; social welfare; sustainability

---

1. Introduction

Biotechnology and bioprocesses are two important tools for economic progress and social welfare. The industrial, academic, and government sectors are bound to face technical problems as they develop competitive biotechnological products and processes using synthetic biology, genetics, and molecular biology as alternatives to chemical-based applications. In this regard, the biological control of microbial consortia based on synthetic biology solutions and the regulation and optimization of the migration from batch production to continuous production are ongoing tasks [1,2]. In the biopharmaceutical industry, improved bioprocesses are always in demand to address new regulatory requirements, quality control needs, and production problems in biological products, cell culture titration, and the production of biosimilars [3–5].

Applications derived from biotechnology are very diverse, including food design, processing, and optimization to improve nutritional intake [6]; the optimization of processes to purify monoclonal antibodies for the treatment of different conditions, the analysis of host cell proteins (HCPs), and the production of hematopoietic stem cells (HSCs) for therapeutic purposes [7–9]; the development of microorganisms for the processing and transformation of biomass into fuels [10–12]; the production of raw materials based on fermentation processes, such as ethanol, butanol [13–15], and other products traditionally derived from chemical sources, such as aliphatic, aromatic, and other macromolecules using bioprocesses, such as (a) separate hydrolysis and fermentation (SHF), (b) simultaneous saccharification and fermentation (SSF), and (c) consolidated bioprocessing (CBP) [16]; and the construction of bioelectronic devices for applications in multivariate data analysis, experiment design, mathematical models, sensors, and biosensors whose data are processed by software to monitor and optimize processes [17–22].

Other important bioprocesses involve the large-scale production of secondary metabolites relevant for the food, cosmetics, pharmaceutical [23], wastewater treatment, and bioremediation industries (all of these are high-value processes) using bacteria and plant cells produced in vitro to protect endangered or scarce plants or to obtain metabolites [24,25] and enzymes produced by filamentous fungi, leveraging the advances of genetic engineering and molecular biology [26]; the development of cells that can be used in the production of new drugs [27]; the application of enzymatic processes to treat textiles [28]; the use of bacteria for the production of enzymes and various chemical products [29]; the use of nanotechnology, for instance, the nano-encapsulation of bioactive compounds, intelligent packaging systems in food production, biocatalysts and biosensors, and microbiological identification [30,31]; the collection and commercialization of recyclable and biodegradable biopolymers such as PLA (polylactide) [32]; and the development of regenerative medicine solutions [33].

Culture collections (CCs) and microbial biological resources centers (mBRCs) are two critical elements during the microorganism characterization and preservation process. In the second case, in spite of pending challenges, Europe has achieved substantial progress in the areas of databases, quality, infrastructure, legislation, and project development. This progress contributes to the preservation of biodiversity and ecosystems and, certainly, to stimulating the innovation, research, and development of biotechnology-based applications [34,35]. Molecular and genetic characterizations of living collections of biological resources provide added value to these biorepositories. As a consequence, their development has technical, financial, and regulatory implications to address depending on the type of collection (microbial cultures and animal or plant germplasm) [36].

Acetic acid bacteria are an interesting group of organisms with potential for the generation of diverse metabolites of industrial application based on sustainable processes; however, current processes still have limitations to address large-scale industrial demand [37]. Another example is the development of yeasts that produce high amounts of glutathione to be used in drugs, cosmetics, and foods; in the wine production industry, the antioxidant effect of glutathione and its action against unwanted aromatic compounds are particularly adequate [38].

As can be appreciated, the impact of biotechnology on social welfare is evident and has been widely discussed. Areas of focus such as sustainability, the decrease in CO_2 emissions, technological change, and the bioeconomy are associated with this field of study, whose potential is vital for the development of numerous products based on inputs derived from agriculture or other renewable sources of biological origin [39–41]. For example, marine algae can be used to produce biofuels and bioenergy as a substitute for fossil fuels [42,43]. In the case of genetically modified and improved seeds, potential risks and benefits are the subject of heated debate, especially around the ethics of their development and use, and the issues related to economic and environmental impact have yet to be addressed [44,45]. It is also possible to use materials created by biological agents or to use these agents in environmental remediation applications [46]. This new reality underscores the need to analyze the peculiarities of these inventions in order to address their resulting intellectual property rights adequately [47].

Thus, under an full (economic, social, and environmental) approach, concerned with social welfare and the development of the bioeconomy, sustainability is closely oriented toward achieving objectives and promoting economic growth [48,49]. Although all of these biological resources are renewable and the solutions that they provide are socially valuable, it is essential to respect and preserve their biological sources and optimize the use of water and energy to carry out the bioprocesses. As a consequence, the use of environmental indicators (climate change, water, energy, land use, chemical risk) is necessary to manage resources sustainably [50]. However, evaluating the social impact of these advances is one of the most neglected tasks in the field of bioeconomics because the attention has been focused on environmental and techno-economic elements [51]. Additionally, the adverse effects of globalization on economic equality and the preservation of biodiversity must be considered in each case and context and paying attention to social indicators related to health, food, and employment, among others, must be paramount [52–54].

Biotechnology uses bioprocesses as an operating mechanism, and the development and improvement of these processes provide technological alternatives to solve myriad problems in the health, food, energy, agriculture, and many other industrial sectors. As a result, academia, the business sector, governments, non-governmental organizations, and the societies in which all these applications have a positive economic, social, or environmental impact have taken an interest in bioprocesses. Nevertheless, the alternatives brought forth by the bioeconomy to promote economic development should not be limited to technological advancement per se but include other aspects of interest to different actors.

Specifically, the emphasis must be placed on developments that support social welfare and mitigate environmental impact. Although technologically and economically efficient solutions to address environmental and other types of technical issues that represent a significant social impact are already being developed, it is also true that many challenges have to do with the creation of policies, regulations, and ethical guidelines concerning biosecurity, as well as with technical and risk assessments, industrial scale-up, the efficient use of renewable resources, and industry-driven ad hoc mechanisms to address specific problems derived from this area of application (see Figure 1).

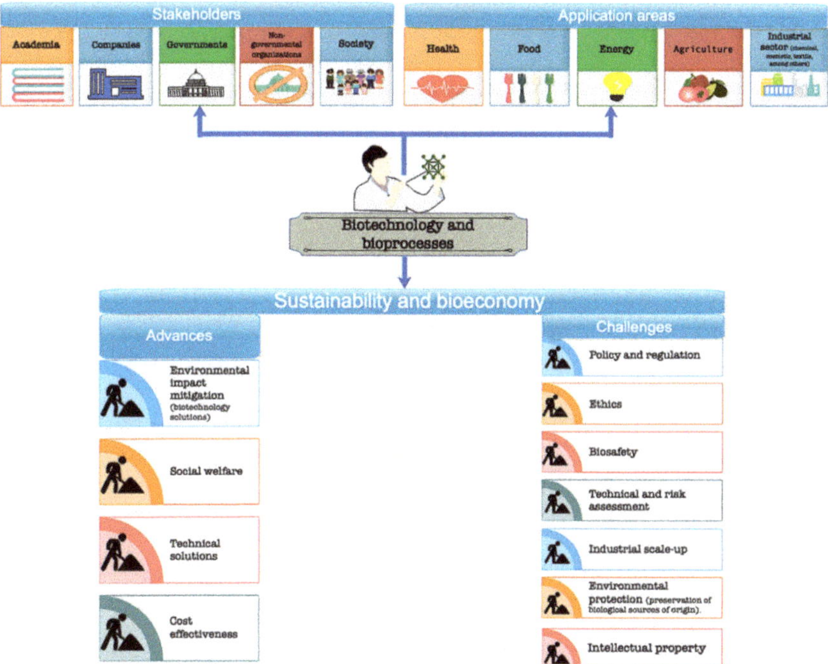

Figure 1. Bioeconomy, biotechnology, and sustainability. Source: elaborated by the authors.

2. Method

A literature search and a patent application search were carried out using the Scopus (in the case of Scopus, the search criterion was as follows: (TITLE-ABS-KEY (bioproce*) AND TITLE-ABS-KEY (sustainab*)) AND PUBYEAR > 2009) [55] and Patentscope (in the case of Patentscope, the advanced search criteria for English language in all offices was: bioproce* AND sustainab*) [56] databases, respectively, for the period from 2010 to 2019, as a first approach to understand the relationship between sustainability and bioprocess. The purpose of the present study was to identify documents and patent applications related to the development and analysis of sustainability involving bioprocesses to approach our object of study along the economic, social, and environmental axes. In addition, the search sought to reveal an initial outline of the scientific and technological trajectories around these terms during the study period.

Thus, the first of these databases identified the 676 most relevant publications by country, knowledge area, institution, and source; the lowest number of published documents (29) corresponded to 2010, whereas the highest (103) corresponded to 2018. Concerning the patent search, the data considered were the number of applications per country, per institution, and technology area; 1233 applications were found, and 2013 was the year with the most significant number of applications (156). The data obtained from Scopus were subjected to network analysis, including co-occurrence, using the authors' keywords as a unit of analysis and full counting. Additionally, a co-authorship study was included in the analysis, considering the country as a unit of analysis and full counting.

3. Results

Figure 2 shows that the United States was the leader in bioprocess and sustainability-related research during the study period. Among the ten leading countries, India, China, and Germany published more than 50 documents each. The most developed areas, those with more than 100 documents, were biochemistry, genetics, and molecular biology; chemical engineering; immunology and microbiology; environmental science; energy and engineering. These data highlight the need to increase the number of basic research projects in disciplines focused on the development or improvement of new bioprocesses and their industrial scale-up and the creation of technological applications for the medical, food, environmental, and power generation sectors.

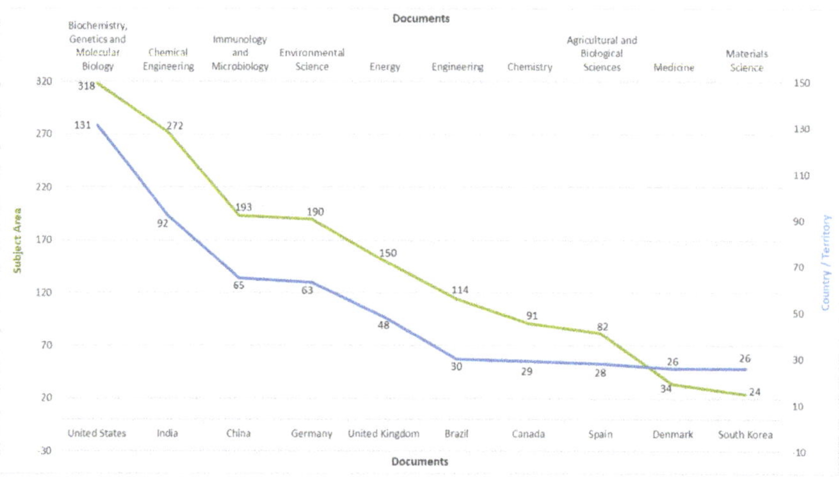

Figure 2. Documents by area of knowledge and country (2010–2019). Source: elaborated by the authors based on Scopus [55].

According to Journal Citation Reports (JCR) [57] 2018, two important academic journals led the list with more than 30 published documents: (1) Bioresource Technology, with an impact factor of 6.669 and ranked Q1 in the following three areas: (a) Agricultural Engineering, (b) Biotechnology and Applied Microbiology, and (c) Energy and Fuels; and (2) Applied Microbiology and Biotechnology, with an impact factor of 3.670 and ranked Q2 in Biotechnology and Applied Microbiology. They are followed by journals with ten or more published documents: Biotechnology for Biofuels (impact factor: 5.452); Biotechnology and Bioengineering (impact factor: 4.260); Current Opinion in Biotechnology (impact factor: 8.083); Renewable and Sustainable Energy Reviews (impact factor: 10.556); Journal of Chemical Technology and Biotechnology (impact factor: 2.659); and Biotechnology Journal (impact factor: 3.543) (the quartiles for the rest of the journals with at least 10 documents are distributed according to different categories as follows: (1) Biotechnology for Biofuels: Biotechnology and Applied Microbiology (Q1) and Energy and Fuels (Q1); (2) Biotechnology and Bioengineering: Biotechnology and Applied Microbiology (Q1); (3)- Current Opinion in Biotechnology: Biochemical Research Methods (Q1) and Biotechnology and Applied Microbiology (Q1); (4) Renewable and Sustainable Energy Reviews: Green and Sustainable Science and Technology (Q1) and Energy and Fuels (Q1); (5) Journal of Chemical Technology and Biotechnology: Biotechnology and Applied Microbiology (Q2), Chemistry, Multidisciplinary (Q2), Engineering, Environmental (Q3), Engineering, Chemical (Q2); and (6) Biotechnology Journal: Biochemical Research Methods (Q1) and Biotechnology and Applied Microbiology (Q2).) These top ten journals include the participation of institutions from Denmark, the Netherlands, England, France, the United States, Korea, India, and Brazil (see Figure 3).

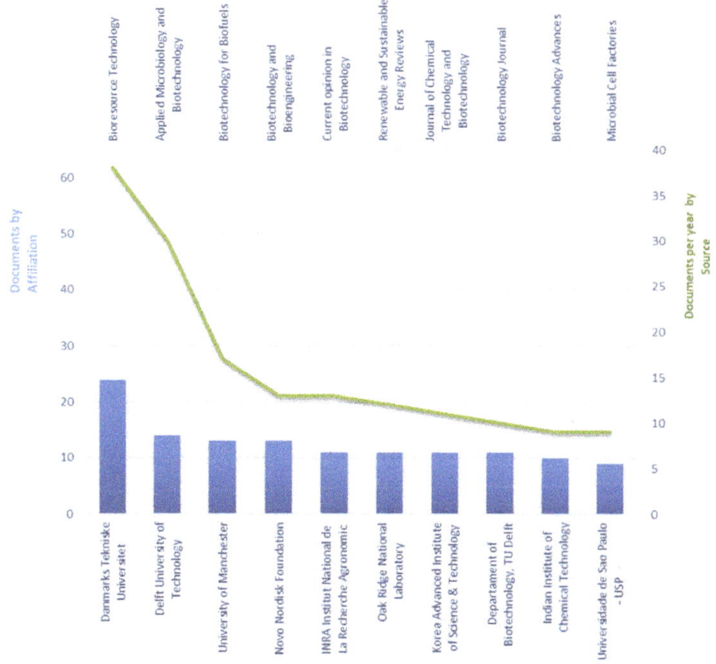

Figure 3. Documents by affiliation and source of publication (2010–2019). Source: elaborated by the authors based on Scopus [55].

In regard to technological development, patent applications were clearly dominated by the United States, where 552 applications were filed. This country was followed by the Patent Cooperation Treaty (PCT), Australia, and the European Patent Office, with more than 100 applications each. By institution,

Genomatica, Inc. ranked first with 118 applications, followed by Regents of the University of California with 44, and the Massachusetts Institute of Technology with 35; all of these institutions are based in the United States. The participation of independent inventors and another American company are noteworthy. These data show that the United States, Australia, and European countries are the target market for this type of invention, according to the data obtained by searching in English (see Figure 4).

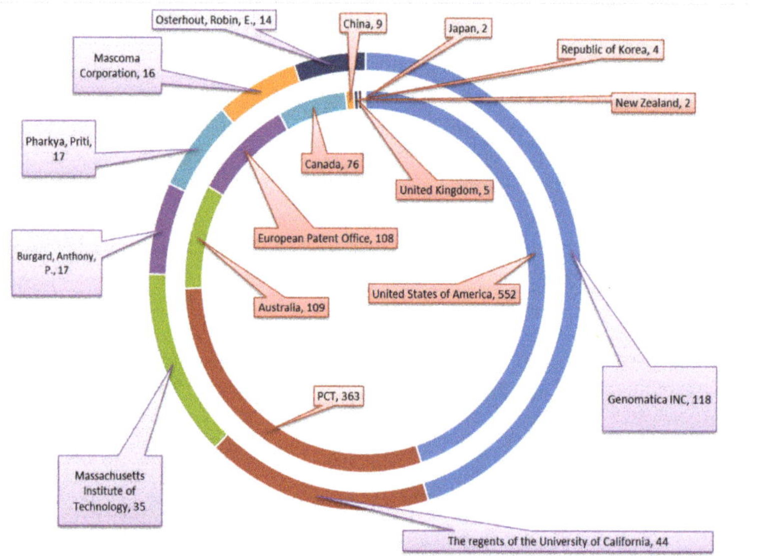

Figure 4. Patent applications by country and institution (2010–2019). Source: elaborated by the authors based on World Intellectual Property Organization (WIPO) [56].

Most of these applications are related to scientific and technological areas of traditional expertise in the development of bioprocesses, fermentation processes, the use of enzymes to obtain various chemical compounds, and other formulations and applications involving microorganisms or enzymes, in addition to compositions of microorganisms and enzymes that are essential in many applications and processes. The rest of the categories are related to the development of new devices; the manufacture of organic compounds and pharmaceutical products; medical appliances; the production of deodorization, disinfection, and sterilization materials; indexing associated with other microorganism subclasses; applications related to water treatment, wastewater, sewage, and sludges; peptide generation processes; separation methods, and applications associated with sugars, sugar derivatives, nucleosides, nucleic acids, and nucleotides, among other examples (see Table 1).

The network analysis based on the literature search showed that sustainability and bioprocesses are central points in this search; both keywords are grouped in the largest cluster, together with smaller nodes such as green chemistry, enzymes, biocatalysis, and industrial biotechnology, among others. Around this cluster, the figure shows three smaller groupings, whose main nodes are as follows: (1) fermentation, biofuels, bioethanol, biorefining, and lignocellulose; (2) microalgae, anaerobic digestion, bioenergy, biodiesel, biohydrogen, and dark fermentation; and (3) consolidated bioprocessing, biofuel, biobutanol, and clostridium. Finally, a smaller cluster, whose main component is associated with synthetic biology, can also be appreciated. These links showcase the vigorous dynamics of bioprocess design research in connection with sustainability. They outline where most of the scientific research is taking place and where the new areas of opportunity originating around this activity can be found (see Figure 5).

Table 1. Patent applications by technological field (2010–2019).

No.	Code	Classification	Number
1	C12P	Fermentation or enzyme-using processes to synthesize a desired chemical compound or composition or to separate optical isomers from a racemic mixture.	735
2	C12N	Microorganisms or enzymes … ; compositions thereof … propagating, preserving, or maintaining microorganisms; mutation or genetic engineering; culture media …	623
3	C12M	Apparatus for enzymology or microbiology …	114
4	C07C	Acyclic or carbocyclic compounds …	82
5	A61K	Preparations for medical, dental, or toilet purposes …	81
6	C12R	Indexing scheme associated with subclasses … relating to microorganisms	78
7	C02F	Treatment of water, wastewater, sewage, or sludge …	76
8	C07K	Peptides …	54
9	B01D	Separation	51
10	C07H	Sugars; derivatives thereof; nucleosides; nucleotides; nucleic acids …	47

Source: elaborated by the authors based on WIPO [56,58].

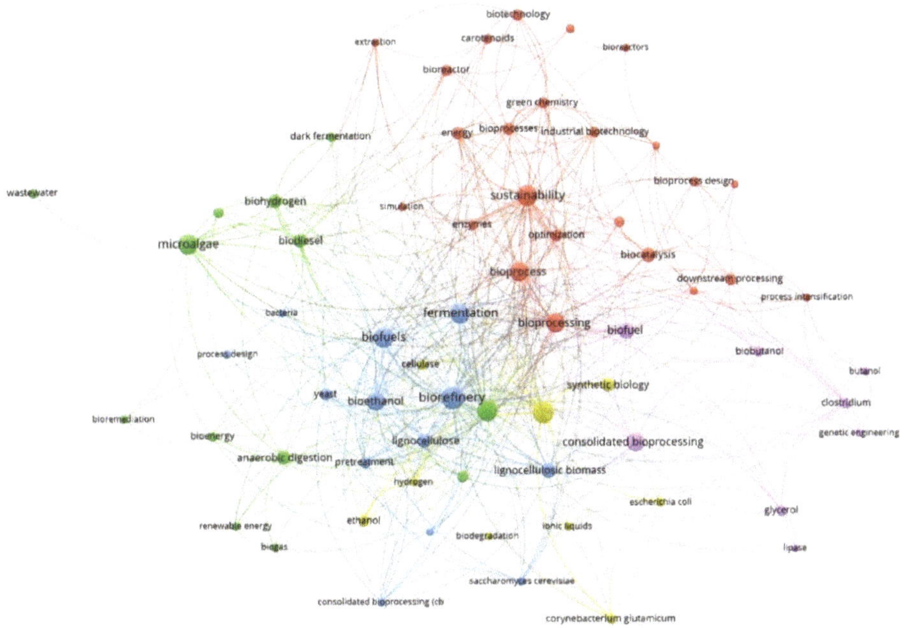

Figure 5. Keyword co-occurrence analysis: (author)-full counting. Source: elaborated by the authors based on Scopus [55].

As mentioned before, the United States was determined to be the primary originator of publications focused on sustainability and bioprocesses, and the same was valid for institutional collaboration. However, although with lower frequency and proximity, the presence of other economies such as India, Germany, China, England, South Africa, Mexico, Chile, Malaysia, Australia, Turkey, Italy, and Singapore, among others, could also be observed in the network; individually, these countries connected to different clusters. Therefore, it is necessary to increase our efforts to generate new research focused on the sustainability of bioprocesses in local environments in collaboration with scientists from institutions in different parts of the world, since biotechnological solutions cannot always be applied globally, hence the need for ad hoc solutions to specific problems, especially in developing countries (see Figure 6).

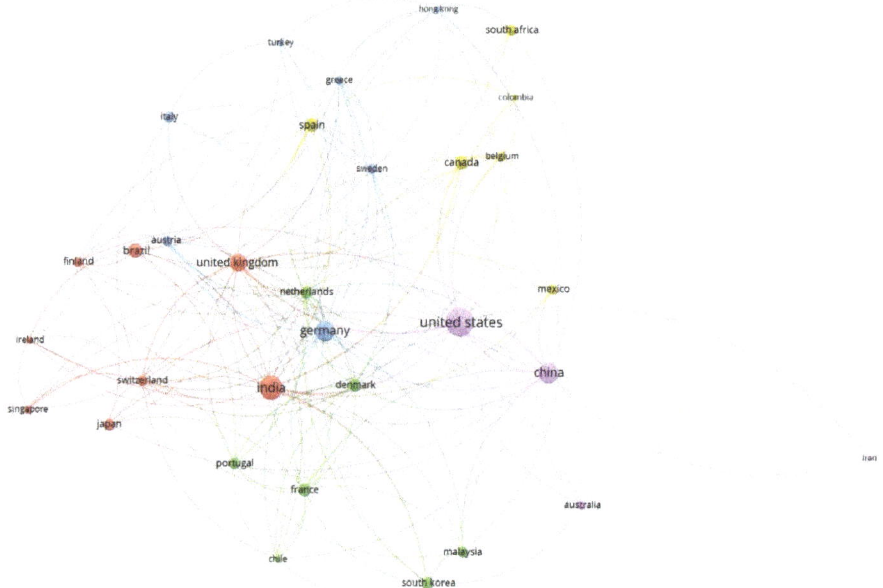

Figure 6. Co-authorship analysis: country-full counting. Source: elaborated by the authors based on Scopus [55].

4. Conclusions

Biotechnology has provided society with thousands of bioprocesses to address diseases and food demands, to develop petroleum product substitutes, to provide alternatives for energy production, and to solve agricultural problems, among other benefits. Applications and products based on biological sources are the framework of a bioeconomy that contributes to the economic development of regions and countries. However, social welfare and care for the environment must be inherent to these applications, which is why the generation of ad hoc indicators of these two areas is necessary to monitor these areas.

The scientific and technological trajectory shows how sustainability and bioprocesses are topics of great interest and constantly growing, although further efforts are still needed to move toward an integrated sustainability framework. Microbiology and enzymology are often prevalent in this technological field, although new areas of opportunity are emerging around the new demand for sustainable solutions to support economic growth and industrial development, especially basic science projects, which need to be explored and exploited in greater depth. New bioprocesses based on biorefining, bioethanol, consolidated bioprocessing, microalgae, lignocellulose, biocatalysis, and biohydrogen are among the products and technologies of the future.

In the following decades, actors from the academic, business, and government sectors, in addition to non-governmental organizations and society in general, will have to intensify their collaboration mechanisms, especially in developing economies, where the challenge to move forward sustainably is harder and problems associated with poverty and inequality tend to be more serious. Although biotechnology has shown unprecedented progress, it is also true that the design of bioprocesses must be geared to sustainability criteria, in which social impact must be a priority. Additional aspects to take into account to guarantee the successful development of future biotechnological applications for the benefit of economic development, environmental protection, and social welfare are legislative, normative, and ethical considerations; the optimization of resources and the conservation of biological sources; technical and risk assessment; biosecurity, and intellectual property.

Author Contributions: Conceptualization and methodology: A.B.-O. and P.S.-B.; Investigation: A.B.-O., P.S.-B., S.O.-P. and M.P.-O.; Formal Analysis: A.B.-O., P.S.-B. and S.O.-P.; Data curation: P.S.-B., S.O.-P. and M.P.-O. All authors have read and agreed to the published version of the manuscript.

Funding: SIP grant numbers 20195587 and 20200773.

Acknowledgments: We wish to acknowledge the support provided by the National Polytechnic Institute (Instituto Politécnico Nacional—IPN) and the Secretariat for Research and Postgraduate Studies (Secretaría de Investigación y Posgrado—SIP).

Conflicts of Interest: The authors declare no conflict of interest.

References

1. Croughan, M.S.; Konstantinov, K.B.; Cooney, C. The Future of Industrial Bioprocessing: Batch or Continuous? *Biotechnol. Bioeng.* **2015**, *112*, 648–651. [CrossRef]
2. Shong, J.; Jimenez Diaz, M.R.; Collins, C.H. Towards synthetic microbial consortia for bioprocessing. *Curr. Opin. Biotechnol.* **2012**, *23*, 798–802. [CrossRef] [PubMed]
3. Whitford, W.G. Single-use technology supporting the comeback of continuous bioprocessing. *Pharm. Bioprocess.* **2013**, *1*, 249–253. [CrossRef]
4. Cramer, S.M.; Holstein, M.A. Downstream bioprocessing: Recent advances and future promise. *Curr. Opin. Chem. Eng.* **2011**, *1*, 27–37. [CrossRef]
5. Wuest, D.M.; Harcum, S.W.; Lee, K.H. Genomics in mammalian cell culture bioprocessing. *Biotechnol. Adv.* **2012**, *30*, 629–638. [CrossRef]
6. Jägerstad, M.; Piironen, V.; Walker, C.; Ros, G.; Carnovale, E.; Holasova, M.; Nau, H. Increasing natural food folates through bioprocessing and biotechnology. *Trends Food Sci. Technol.* **2005**, *16*, 298–306. [CrossRef]
7. Levy, N.E.; Valente, K.N.; Choe, L.H.; Lee, K.H.; Lenhoff, A.M. Identification and characterization of host cell protein product-associated impurities in monoclonal antibody bioprocessing. *Biotechnol. Bioeng.* **2014**, *111*, 904–912. [CrossRef]
8. Tscheliessnig, A.L.; Konrath, J.; Bates, R.; Jungbauer, A. Host cell protein analysis in therapeutic protein bioprocessing—Methods and applications. *Biotechnol. J.* **2013**, *8*, 655–670. [CrossRef]
9. Kelley, B. Industrialization of mAb production technology: The bioprocessing industry at a crossroads. *mAbs* **2009**, *1*, 443–452. [CrossRef]
10. Lynd, L.E.; van Zyl, W.H.; McBride, J.E.; Laser, M. Consolidated bioprocessing of cellulosic biomass: An update. *Curr. Opin. Biotechnol.* **2005**, *16*, 577–583. [CrossRef]
11. Jin, M.; Balan, V.; Gunawan, C.; Dale, B.E. Consolidated bioprocessing (CBP) performance of clostridium phytofermentans on AFEX-treated corn stover for ethanol production. *Biotechnol. Bioeng.* **2011**, *108*, 1290–1297. [CrossRef] [PubMed]
12. Parisutham, V.; Kim, T.H.; Lee, S.K. Feasibilities of consolidated bioprocessing microbes: From pretreatment to biofuel production. *Bioresour. Technol.* **2014**, *161*, 431–440. [CrossRef] [PubMed]
13. Blanch, H.W. Bioprocessing for biofuels. *Curr. Opin. Biotechnol.* **2012**, *23*, 390–395. [CrossRef] [PubMed]
14. Yuan, W.J.; Chang, B.L.; Ren, J.G.; Liu, J.P.; Bai, F.W.; Li, Y.Y. Consolidated bioprocessing strategy for ethanol production from Jerusalem artichoke tubers by *Kluyveromyces marxianus* under high gravity conditions. *J. Appl. Microbiol.* **2012**, *112*, 38–44. [CrossRef]
15. Favaro, L.; Jansen, T.; Van Zyl, W.H. Exploring industrial and natural Saccharomyces cerevisiae strains for the bio-based economy from biomass: The case of bioethanol. *Crit. Rev. Biotechnol.* **2019**, *39*, 800–816. [CrossRef]

16. Kawaguchi, H.; Hasunuma, T.; Ogino, C.; Kondo, A. Bioprocessing of bio-based chemicals produced from lignocellulosic feedstocks. *Curr. Opin. Biotechnol.* **2016**, *42*, 30–39. [CrossRef]

17. Yang, B.X.; Kim, E.; Liu, Y.; Shi, X.W.; Rubloff, G.W.; Ghodssi, R.; Bentley, W.E.; Pancer, Z.; Payne, G.F. In-film bioprocessing and immunoanalysis with electroaddressable stimuli-responsive polysaccharides. *Adv. Funct. Mater.* **2010**, *20*, 1645–1652. [CrossRef]

18. Lim, M.; Ye, H.; Panoskaltsis, N.; Drakakis, E.M.; Yue, X.; Cass, A.E.G.; Radomska, A.; Mantalaris, A. Intelligent bioprocessing for haemotopoietic cell cultures using monitoring and design of experiments. *Biotechnol. Adv.* **2007**, *25*, 353–368. [CrossRef]

19. Kirdar, A.O.; Green, K.D.; Rathore, A.S. Application of multivariate data analysis for identification and successful resolution of a root cause for a bioprocessing application. *Biotechnol. Prog.* **2008**, *24*, 720–726. [CrossRef]

20. Kumar, V.; Bhalla, A.; Rathore, A.S. Design of experiments applications in bioprocessing: Concepts and approach. *Biotechnol. Prog.* **2014**, *30*, 86–99. [CrossRef]

21. Randek, J.; Mandenius, C.F. On-line soft sensing in upstream bioprocessing. *Crit. Rev. Biotechnol.* **2018**, *38*, 106–121. [CrossRef] [PubMed]

22. Dekker, L.; Polizzi, K.M. Sense and sensitivity in bioprocessing—Detecting cellular metabolites with biosensors. *Curr. Opin. Chem. Biol.* **2017**, *40*, 31–36. [CrossRef] [PubMed]

23. Steingroewer, J.; Bley, T.; Georgiev, V.; Ivanov, I.; Lenk, F.; Marchev, A.; Pavlov, A. Bioprocessing of differentiated plant in vitro systems. *Eng. Life Sci.* **2013**, *13*, 26–38. [CrossRef]

24. Georgiev, M.I.; Weber, J.; Maciuk, A. Bioprocessing of plant cell cultures for mass production of targeted compounds. *Appl. Microbiol. Biotechnol.* **2009**, *83*, 809–823. [CrossRef] [PubMed]

25. Zaidi, K.U.; Ali, A.S.; Ali, S.A.; Naaz, I. Microbial tyrosinases: Promising enzymes for pharmaceutical, food bioprocessing, and environmental industry. *Biochem. Res. Int.* **2014**, *2014*, 854687. [CrossRef] [PubMed]

26. Wang, L.; Ridgway, D.; Gu, T.; Moo-Young, M. Bioprocessing strategies to improve heterologous protein production in filamentous fungal fermentations. *Biotechnol. Adv.* **2005**, *23*, 115–129. [CrossRef]

27. Thomson, H. Bioprocessing of embryonic stem cells for drug discovery. *Trends Biotechnol.* **2007**, *25*, 224–230. [CrossRef] [PubMed]

28. Mojsov, K.D. Trends in bio-processing of textiles: A review. *Adv. Technol.* **2014**, *3*, 135–138. [CrossRef]

29. Pfeifenschneider, J.; Brautaset, T.; Wendisch, V.F. Methanol as carbon substrate in the bio-economy: Metabolic engineering of aerobic methylotrophic bacteria for production of value-added chemicals. *Biofuels Bioprod. Bioref.* **2017**, *11*, 719–731. [CrossRef]

30. Neethirajan, S.; Jayas, D.S. Nanotechnology for the food and bioprocessing industries. *Food Bioprocess. Technol.* **2011**, *4*, 39–47. [CrossRef]

31. Trujillo, L.E.; Ávalos, R.; Granda, S.; Guerra, L.S.; País-Chanfrau, J.M. Nanotechnology applications for food and bioprocessing industries. *Biol. Med.* **2016**, *8*, 1–6. [CrossRef]

32. Sawyer, D.J. Bioprocessing—No longer a field of dreams. *Macromol. Symp.* **2003**, *201*, 271–281. [CrossRef]

33. Manson, C. Regenerative medicine. The industry comes of age. *Med. Device Technol.* **2007**, *18*, 25–30.

34. De Vero, L.; Boniotti, M.B.; Budroni, M.; Buzzini, P.; Cassanelli, S.; Comunian, R.; Gullo, M.; Logrieco, A.F.; Mannazzu, I.; Perugini, I. Preservation, characterization and exploitation of microbial biodiversity: The perspective of the Italian network of culture collections. *Microorganisms* **2019**, *7*, 685. [CrossRef] [PubMed]

35. Schüngel, M.; Stackebrandt, E. Microbial Resource Research Infrastructure (MIRRI): Infrastructure to foster academic research and biotechnological innovation. *Biotechnol. J.* **2015**, *10*, 17–19. [CrossRef] [PubMed]

36. McCluskey, K. A review of living collections with special emphasis on sustainability and its impact on research across multiple disciplines. *Biopreserv. Biobank.* **2017**, *15*, 20–30. [CrossRef] [PubMed]

37. La China, S.; Zanichelli, G.; De Vero, L.; Gullo, M. Oxidative fermentations and exopolysaccharides Production by acetic acid bacteria: A mini review. *Biotechnol. Lett.* **2018**, *40*, 1289–1302. [CrossRef]

38. De Vero, L.; Bonciani, T.; Verspohl, A.; Mezzetti, F.; Giudici, P. High-glutathione producing yeasts obtained by genetic improvement strategies: A focus on adaptive evolution approaches for novel wine strains. *AIMS Microbiol.* **2017**, *3*, 155–170. [CrossRef]

39. McLaren, J.S. Crop biotechnology provides an opportunity to develop a sustainable future. *Trends Biotechnol.* **2005**, *23*, 339–342. [CrossRef]

40. Jordan, N.; Boody, G.; Broussard, W.; Glover, J.D.; Keeney, D.; McCown, B.H.; McIsaac, G.; Muller, M.; Murray, H.; Neal, J.; et al. Sustainable development of the agricultural bio-economy. *Science* **2007**, *316*, 1570–1571. [CrossRef]

41. Jenkins, T. Toward a biobased economy: Examples from the UK. *Biofuels Bioprod. Bioref.* **2008**, *2*, 133–143. [CrossRef]

42. Van Hal, J.W.; Huijgen, W.J.J.; López-Contreras, A.M. Opportunities and challenges for seaweed in the biobased economy. *Trends Biotechnol.* **2014**, *32*, 231–233. [CrossRef] [PubMed]

43. Kircher, M. The transition to a bio-economy: National perspectives. *Biofuels Bioprod. Bioref.* **2012**, *6*, 240–245. [CrossRef]

44. Burton, D.M.; Love, H.A.; Ozertan, G.; Taylor, C.R. Property rights protection of biotechnology innovations. *J. Econ. Manag. Strat.* **2005**, *14*, 779–812. [CrossRef]

45. Bennett, A.B.; Chi-Ham, C.; Barrows, G.; Sexton, S.; Zilberman, D. Agricultural biotechnology: Economics, environment, ethics, and the future. *Annu. Rev. Environ. Resour.* **2013**, *38*, 249–279. [CrossRef]

46. Gavrilescu, M.; Chisti, Y. Biotechnology-a sustainable alternative for chemical industry. *Biotechnol. Adv.* **2005**, *23*, 471–499. [CrossRef]

47. Straus, J. Intellectual property rights and bioeconomy. *JIPLP* **2017**, *12*, 576–590. [CrossRef]

48. Aguilar, A.; Bochereau, L.; Matthiessen, L. Biotechnology as the engine for the Knowledge-Based Bio-Economy. *Biotechnol. Genet. Eng.* **2009**, *26*, 371–388. [CrossRef]

49. Hediger, W. Sustainable development and social welfare. *Ecol. Econ.* **2000**, *32*, 481–492. [CrossRef]

50. Pursula, T.; Aho, M.; Rönnlund, I.; Päällysaho, M. Environmental sustainability indicators for the bioeconomy. In *Towards a Sustainable Bioeconomy: Principles, Challenges and Perspectives*; Filho, W.L., Pociovălişteanu, D.M., Borges de Brito, P.R., Borges de Lima, I., Eds.; World Sustainability Series; Springer: Cham, Switzerland, 2018; pp. 43–61.

51. Rafiaani, P.; Kuppens, T.; Van Dael, M.; Azadi, H.; Lebailly, P.; Van Passel, S. Social sustainability assessments in the biobased economy: Towards a systemic approach. *Renew. Sustain. Energy Rev.* **2018**, *82*, 1839–1853. [CrossRef]

52. Goeschl, T.; Swanson, T. The Social Value of Biodiversity for R&D. *Environ. Resour. Econ.* **2002**, *22*, 477–504.

53. Nissanke, M.; Thorbecke, E. Globalization, poverty, and inequality in Latin America: Findings from case studies. *World Dev.* **2010**, *38*, 797–802. [CrossRef]

54. Adams, W.M.; Aveling, R.; Brockington, D.; Dickson, B.; Elliott, J.; Hutton, J.; Roe, D.; Vira, B.; Wolmer, W. Biodiversity conservation and the eradication of poverty. *Science* **2004**, *36*, 1146–1149. [CrossRef] [PubMed]

55. Scopus. Abstract and Citation Database. 2019. Available online: https://www.scopus.com (accessed on 27 September 2019).

56. WIPO. Patentscope. Search International and National Patent Collections. 2019. Available online: https://patentscope.wipo.int/search/en/search.jsf (accessed on 27 September 2019).

57. Journal Citation Reports -JCR-. Incites Journal Citation Reports. 2019. Available online: https://jcr.clarivate.com (accessed on 16 December 2019).

58. WIPO. International Patent Classification (IPC). 2019. Available online: https://www.wipo.int/classifications/ipc/en/ (accessed on 19 December 2019).

Article

Sperm Proteomics Analysis of Diabetic Induced Male Rats as Influenced by *Ficus carica* Leaf Extract

Umarqayum Abu Bakar [1], Puvaratnesh Subramaniam [1], Nurul Ain Kamar Bashah [1],
Amira Kamalrudin [1], Khaidatul Akmar Kamaruzaman [1], Malina Jasamai [2], Wan Mohd Aizat [3],
M. Shahinuzzaman [4] and Mahanem Mat Noor [1,*]

[1] Centre for Biotechnology and Functional Food, Faculty of Science and Technology, Universiti Kebangsaan
    Malaysia, Bangi 43600, Malaysia; umarqayum92@gmail.com (U.A.B.); puvaratnesh@hotmail.com (P.S.);
    nurulain_0917@yahoo.com (N.A.K.B.); amirakamalrudin@gmail.com (A.K.);
    khaidatulakmar89@gmail.com (K.A.K.)
[2] Faculty of Pharmacy, Universiti Kebangsaan Malaysia, Kuala Lumpur 50300, Malaysia; malina@ukm.edu.my
[3] Institute of Systems Biology (INBIOSIS), Universiti Kebangsaan Malaysia, Bangi 43600, Malaysia;
    wma@ukm.edu.my
[4] School of Chemical Sciences and Food Technology, Faculty of Science and Technology, Universiti
    Kebangsaan Malaysia, Bangi 43600, Malaysia; shahinchmiu@gmail.com
*   Correspondence: mahanem@ukm.edu.my; Tel.: +603-8921-5193

Received: 1 February 2020; Accepted: 25 March 2020; Published: 28 March 2020

Abstract: Diabetes mellitus is shown to bring negative effects on male reproductive health due to
long-term effects of insulin deficiency or resistance and increased oxidative stress. *Ficus carica* (FC),
an herbal plant, known to have high antioxidant activity and antidiabetic properties, has been used
traditionally to treat diabetes. The objective of this study is to determine the potential of the FC
leaf extract in improving sperm quality of streptozotocin (STZ) induced diabetic male rats from
proteomics perspective. A total of 20 male rats were divided into four groups; normal (nondiabetic
rats), negative control (diabetic rats without treatment), positive control (diabetic rats treated with
300 mg/kg metformin), and FC group (diabetic rats treated with 400 mg/kg FC extract). The treatments
were given via oral gavage for 21 consecutive days. The fasting blood glucose (FBG) level of FC
treated group demonstrated a significant ($p < 0.05$) decrease compared to negative group after 21 days
of treatment, as well as a significant ($p < 0.05$) increase in the sperm quality parameters compared
to negative group. Sperm proteomics analysis on FC treated group also exhibited the increase
of total protein expression especially the proteins related to fertility compared to negative group.
In conclusion, this study clearly justified that FC extract has good potential as antihyperglycemic and
profertility agent that may be beneficial for male diabetic patients who have fertility problems.

Keywords: *Ficus carica*; diabetes mellitus; proteomics; sperm quality

---

1. Introduction

According to the World Health Organization [1], there were 422 million people with diabetes
worldwide in 2014. Diabetes causes metabolic disturbances of carbohydrate, fat, and protein that result
in complications of the organ and body systems such as retinopathy, nephropathy, and neuropathy
diabetic [2]. Diabetes also affects the function of the male reproductive system by decreasing sperm
concentration, viability, and increasing sperm apoptosis [3,4]. The male reproductive system is affected
by diabetes probably due to the insulin deficiency or resistance, and the increase of oxidative stress.
Insulin deficiency or resistance is found to directly affect spermatogenesis, testis development, and the
secretion of hormone related to male fertility due to the failure of glucose transportation into sertoli
cells [5]. Meanwhile, Ding et al. [6] reported that the increase of oxidative stress due to hyperglycemia

causes the excessive production of free radicals that eventually induce sperm DNA fragmentation, and reduce the expression of fertility protein.

The emergence of proteomic approach helps researchers discover precisely the root cause of male infertility due to diabetes. Proteomics is the identification and quantification of proteins that give an understanding about protein function and its interactions with other proteins, hence lead to the understanding of certain expressed phenotype [7,8]. Based on a compilation by Amaral et al. [9], there are 6198 different proteins in human spermatozoa, and each protein has its own distinctive role either directly or indirectly in male fertility. In general, diabetes alters the expression of these sperm fertility proteins that finally results in male infertility. An et al. [10] reported that diabetes reduces sperm proteins such as cystatin C and dipeptidyl peptidase 4 which play vital roles in mitochondrial metabolism that affect directly the motility of sperm. While Pavlinkova et al. [4] found that diabetes changes protamine 1/protamine 2 ratio that indicates reduced sperm quality, since protamines are small proteins that bind sperm DNA which are important for DNA stability and sperm maturation. Accurate representation of male infertility due to diabetes by proteomics analysis of sperm protein eases the determination of effective treatment [10].

Currently, diabetes is treated by a synthetic drug such as metformin to prevent diabetes complications, but the treatment does not significantly improve sperm quality parameters [11,12] and somehow it reduces sperm quality parameters [13–15]. As an option, the focus has now shifted to the use of herbs to serve as therapeutic alternatives. *Ficus carica* (FC), also known as the fig, is among the potent medicinal plants in treating diabetes. Irudayaraj et al. [16] suggest that FC is able to enhance the insulin sensitivity in diabetic rats via the repair of one or more defects such as insulin receptor, insulin receptor substrate, or glucose transporter proteins, as demonstrated by the decrease of high blood glucose level, normalization of plasma insulin, and improvement of insulin tolerance level in FC treated group. A previous study also reported the profertility effect of FC on infertile male animal models. Naghdi et al. [17] found that FC improves testis tissue condition, as well as increases sperm number and progressive sperm percentage of formaldehyde-induced mice. FC leaf extract is effective in managing diabetes and also treating male infertility, however, the proteomic data related to its profertility potential on diabetic subjects is not available.

Therefore, this study aims to explore the profertility effect of FC leaf extract on diabetic-induced male rats at particular genetic sequences in protein level.

2. Materials and Methods

2.1. Preparation of FC Aqueous Extract

FC leaves (cultivar B110) were collected from Saf Fa Fig Garden at Kuala Pilah, Negeri Sembilan, Malaysia, under the Department of Chemical and Process Engineering, Faculty of Engineering and Built Environment. FC leaves were deposited in the Universiti Kebangsaan Malaysia herbarium with the voucher number, UKMB40389. FC aqueous extract was prepared as described by Perez et al. [18]. Briefly, FC leaves were rinsed with water, dried, and ground to fine powder. The fine powder was mixed with water in the ratio of 1:9, and boiled at 100 °C for 30 min. The extract was later filtered and freeze dried.

2.2. Experimental Animal

A total of 20 male Sprague-Dawley rats aged eight weeks were divided into four groups. Three groups were normal, positive (diabetic group treated with metformin at 300 mg/kg per body weight) and negative control (diabetic group without treatment), and one group was treated with FC aqueous extract at 400 mg/kg per body weight. All groups were treated by oral gavage every day for 21 consecutive days. The study was approved by the Animal Ethics Committee of Faculty of Medicine, Universiti Kebangsaan Malaysia (FST/2017/MAHANEM/29-MARCH/833-MARCH-2017-FEB.-2019).

2.3. Antihyperglycemic Activity

Diabetes was induced by a single intravenous injection of STZ (50 mg/kg) (Sigma – Aldrich, Saint Louis, Missouri, USA) dissolved in citrate buffer (0.1 M, pH 4.5) (Sigma – Aldrich, Saint Louis, Missouri, USA) after 16 h of fasting. Diabetes was confirmed by measuring fasting blood glucose (FBG) level from the tail tip after five days of STZ-induction with glucometer AccuCheck®Performa (Roche Diagnostic GmbH, Manheim, Germany). The FBG level of 13 mmol/L and above was considered as diabetic. The FBG level after treatment was measured on day 22.

2.4. Sperm Quality Analysis

All rats were sacrificed on day 22, and cauda epididymis was isolated, minced, and suspended in 15 mL of a Biggers-Whitten-Whittingham (BWW) medium. BWW is a medium used effectively to support the fertilizing potential of spermatozoa in vitro, supplemented with balanced salt, bovine serum albumin (BSA), energy substrates such as glucose, pyruvate, and lactate, as well as Ca^{2+} and HCO_3 (Sigma – Aldrich, Saint Louis, Missouri, USA) [19]. The sperm preparation was then incubated in 5% of CO_2 incubator for 30 min at 37 °C to allow sperms to swim up. Sperm count, motility, and viability were assessed using an improved Neubauer haemocytometer in accordance to WHO 2010 laboratory manual. Meanwhile, sperm morphology was determined as described by Seed et al. [20].

Sperm count was assessed using a haemocytometer based on WHO protocol [21]. The sperm sample was pipetted on the middle grid of the haemocytometer, and counted at 10 boxes of that particular grid under light microscope with 100× magnification. Sperm count was recorded in millions (10^6). Then, the same haemocytometer was used in sperm motility determination. Sperm motility was presented in percentage values in accordance to the grade determined by WHO [21]; progressive (P), nonprogressive (NP), and immotile (IM). Next, based on the motility of the sperm, sperm viability was assessed by using the following formula:

$$\frac{\text{Viable sperm (P and NP)}}{200 \text{ sperm}} \times 100\%$$

Lastly, for normal sperm morphology analysis, the sperm sample was pipetted on the glass slide, and smeared over the slide surface. The smeared slide was then dried and stained using Giemsa staining before being observed under a light microscope.

2.5. Sperm Proteomic Analysis

A further study using shotgun proteomics approach was performed to identify and characterize the diabetic rat sperm protein profile in FC group in comparison with the control group. The sperm protein of control groups (normal, negative, and positive) and FC treatment group (dose 400 mg/kg) was used in the proteomic analysis. The protein extraction was conducted based on protocols by Yunianto et al. [22]. The sperm was harvested from the caudal epididymis before being suspended in a Biggers-Whitten-Whittingham (BWW) medium for 30 min at 37 °C in 5% of CO_2 incubator. The sperm samples were centrifuged and lysed with a lysis buffer. The major components of the lysis buffer were urea, 3-cholamidopropyl dimethylammonio 1-propanesulfonate (CHAPS), immobilized pH gradient (IPG) buffer, and phenylmethylsulfonyl fluoride (PMSF). Determination of sperm protein concentration was performed using a Bradford [23] assay to ensure that the extracted sample had sufficient concentration for gel electrophoresis and LCMS/MS analysis. The gel electrophoresis of protein sample was then performed at the voltage of 75 V using sodium dodecyl sulphate (SDS) gel 12.5% and broad range prestained protein markers (Nacalai Tesque, Japan). The protein band in gel was cut and incubated with dithiotreitol and iodoacetamide for reduction and alkylation steps, respectively. This is followed by an overnight incubation with 6 ng/µL trypsin (Promega, Madison, Wisconsin, USA) for protein digestion [24].

Mass spectrometry (MS) analysis was performed using nanoflow reversed phase liquid chromatography (Dionex 3000 Ultimate RSLCnano, Thermo Fisher Scientific, Waltham, MA, USA) coupled to the Orbitrap Fusion mass spectrometer (Thermo Scientific Orbitrap Fusion). The nanoLC system used was the EASY-Spray column Acclaim PepMap C18 100 A°, 50 µm id × 15 cm with particle size of 2 µm. Five microliters digested samples were injected and run in the chromatography system using the gradient mobile phase method with a constant flow of 250 nL/min. This consists of solvent A (0.1% formic acid in water) and solvent B (0.1% formic acid in acetonitrile) with linear gradient running conditions (91 min at 5%–40% B, 2 min at 85% B, 3 min at 85% B, 1 min at 5% B, and 4 min at 5% B). All chemicals used were LC-MS grade purchased from Fisher Scientific (Fair Lawn, NJ, USA).

The MS spectra was acquired with a scan range of 350–1800 m/z, 50 ms injection period, 120,000 resolution, and an accumulation gain control (AGC) target of 4.0 e^5 (400,000). Peptide precursors were selected for MS/MS based on a charge state of 2–7, an assigned monoisotopic m/z value, intensity threshold of 5000 and 20-s dynamic exclusion window. Selected precursors were fragmented using high-energy collision induced dissociation (HCD) (Thermo Fisher Scientific, Waltham, MA, USA) at 28% normalized collision energy. Ion trap MS (ITMS) (Thermo Fisher Scientific, Waltham, MA, USA) was used to analyze the MS/MS spectra with 1.6 m/z isolation window, 250 ms injection time, 60,000 resolving power, rapid scan rate, and $1.0e^2$ (100) AGC target.

Mass spectra data acquired was analyzed by Thermo Scientific Proteome Discoverer Software Version 2.1. The sequence database for rat (*Rattus norvegicus*) was obtained from UniProt database (http://www.uniprot.org) accessed on May 2018. The parameter search included tryptic specific digest with two or less miscleavages and residue modification was set as fixed (cysteine carbamidomethylation) and variable modifications (methionine oxidation and deamidation of aspargine and glutamine). During the main search, parent and fragment ions were permitted with a mass deviation of 10 ppm and 0.6 Da, respectively.

All identified proteins were annotated using Blast2GO software version 5.2 (https://www.blast2go.com) (BioBam Bioinformatics, Valencia, Spain). The blast parameter was set as blastp with an expectation value (E-value) 1×10^{-3} against UniProt database. Then, protein sequences were further examined using InterproScan, Mapping, and Annotation. The gene ontology graph was generated using WEGO (Web Gene Ontology Annotation Plot) program version 2.0 (http://wego.genomics.org.cn) (WEGO 2.0, Beijing Genomics Institute, Shenzen, China).

2.6. Statistical Analysis

The results were presented as mean ± standard error of means (SEM), and analyzed using the One-way analysis of variance (ANOVA). Value $p < 0.05$ was considered as statistically significant.

3. Results

3.1. Antihyperglycemic Activity of FC Extract

The fasting blood glucose (FBG) level of normal rats was in a range between 4.4 ± 0.10 and 4.9 ± 0.20 mmol/L throughout 21 days of experiment, as shown in Figure 1. While the induction of STZ increased, the FBG level of normal rats was from 4.9 ± 0.20 mmol/L to a diabetic stage as resembled by the negative control of 23.7 ± 0.89 mmol/L. Both treatments of 400 mg/kg FC and 300 mg/kg metformin significantly decreased ($p < 0.05$) the FBG level of diabetic rats compared to the negative control (Figure 1). Diabetic rats administered with FC extract showed the reduction of the FBG level from 23.6 ± 2.87 to 12.6 ± 0.66 mmol/L, and it was almost similar to metformin that reduced FBG levels from 24.4 ± 3.27 to 13.22 ± 0.81 mmol/L.

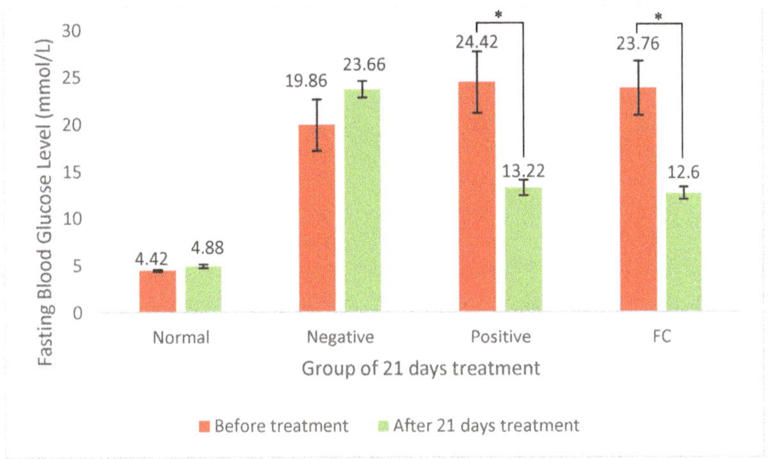

Figure 1. Fasting blood glucose level of rats from normal (normal group with distilled water), negative (diabetic group with distilled water), positive (diabetic group with 300 mg/kg metformin), and *Ficus carica* (FC) (diabetic group with 400 mg/kg *Ficus carica* aqueous extract). Symbol "*" represents significant difference ($p < 0.05$).

3.2. Sperm Quality Analysis

The parameters used for sperm quality assessment were sperm count, motility, the percentage of viability, and percentage of normal morphology. Figure 2 shows the comparison of sperm counts for normal control, negative control (diabetic without treatment), positive control (diabetic treated with metformin), and FC (diabetic with FC treated). Induction of diabetes with STZ significantly reduced ($p < 0.05$) the sperm count of normal control from $(88.6 \pm 6.85) \times 10^6$ sperm to $(2.8 \pm 1.36) \times 10^6$ sperms after 21 days of treatment. However, treatment with FC significantly increased ($p < 0.05$) the number of sperms to $(69.8 \pm 9.81) \times 10^6$ compared to negative control, and it was considerably higher than metformin which was $(14.0 \pm 3.65) \times 10^6$ sperms.

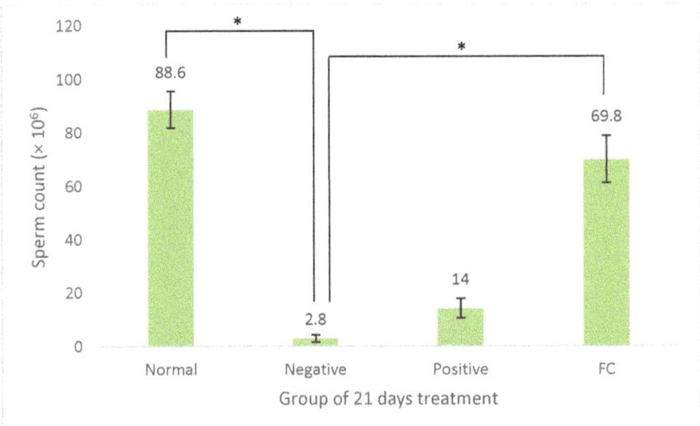

Figure 2. Sperm count ($\times 10^6$) of rats from normal (normal group with distilled water), negative (diabetic group with distilled water), positive (diabetic group with 300 mg/kg metformin), and FC (diabetic group with 400 mg/kg *Ficus carica* aqueous extract). Symbol "*" represents significant difference ($p < 0.05$).

Figure 3 shows the effect of FC treatment on the motility of diabetic rat sperm compared to the negative, positive, and normal control groups. Sperm motility refers to the activeness of sperm movement during microscopic observation which is based on WHO [21], either progressive motility, nonprogressive motility, or immotility. The induction of diabetes resulted in the increase of immotile sperm to the normal control from 25.52% ± 4.95% to 53.32% ± 6.98% as resembled by the negative control. Treatment with FC significantly increased ($p < 0.05$) the percentage of sperm with progressive motility in diabetic rats at 46.48% ± 2.06% compared to the negative control (6.80% ± 4.16%).

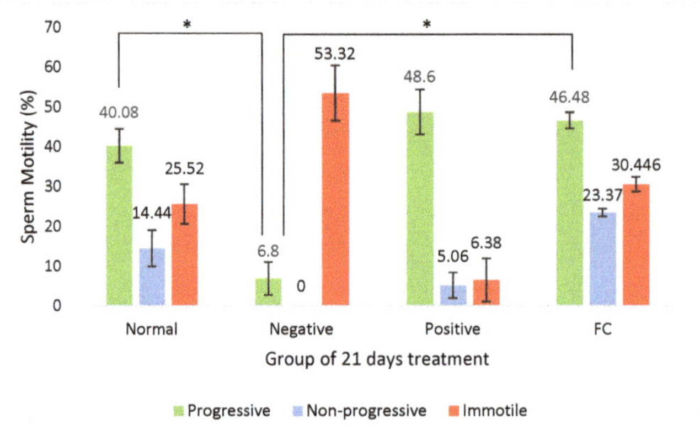

Figure 3. Sperm motility (%) of rats from normal (normal group with distilled water), negative (diabetic group with distilled water), positive (diabetic group with 300 mg/kg metformin), and FC (diabetic group with 400 mg/kg *Ficus carica* aqueous extract). Symbol "*" represents significant difference ($p < 0.05$).

The percentage of viability or percentage of live sperms for each group of rats in this study is shown in Figure 4. The percentage of sperm viability decreased significantly ($p < 0.05$) after the STZ induction, which was 6.68% ± 4.09%, compared to normal control (54.46% ± 4.41%). Diabetic rats treated with FC have shown a significant increase in percentage of sperm viability ($p < 0.05$) which was 69.91% ± 2.04% compared to the negative control (6.68% ± 4.09%), and higher than metformin (53.62% ± 7.4%).

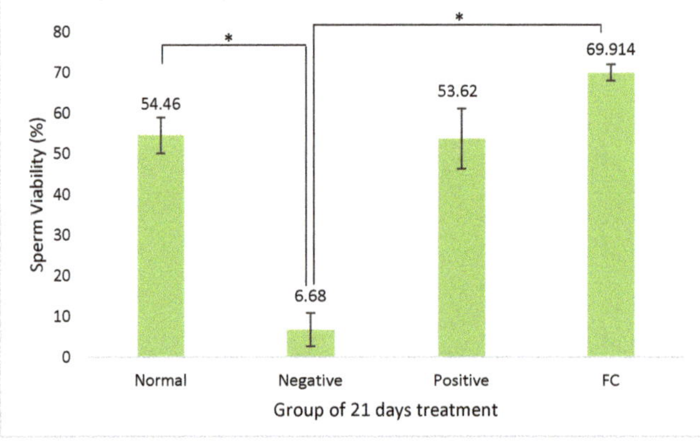

Figure 4. Sperm viability (%) of rats from normal (normal group with distilled water), negative (diabetic group with distilled water), positive (diabetic group with 300 mg/kg metformin), and FC (diabetic group with 400 mg/kg *Ficus carica* aqueous extract). Symbol "*" represents significant difference ($p < 0.05$).

Normal morphology of a sperm was observed on the three main parts of the sperm, which are head, midpiece, and tail. Figure 5 shows the effect of FC on the percentage of normal morphology of a diabetic rat sperm compared to the controls. Diabetes induction decreased the percentage of normal sperm morphology significantly ($p < 0.05$) from 33.74% ± 3.58%, as shown by normal control to 5.05% ± 1.43% as resembled by the negative control. After receiving the FC treatment, the percentage of normal sperm morphology increased significantly ($p < 0.05$) which was 38.04% ± 3.65% compared to negative control, and was observed to surpass the metformin group (21.36% ± 5.18%) and normal control.

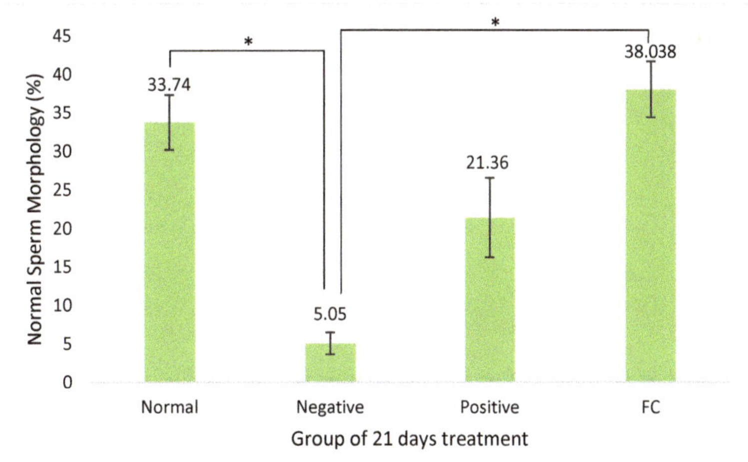

Figure 5. Normal sperm morphology (%) of rats from normal (normal group with distilled water), negative (diabetic group with distilled water), positive (diabetic group with 300 mg/kg metformin), and FC (diabetic group with 400 mg/kg *Ficus carica* aqueous extract). Symbol "*" represents significant difference ($p < 0.05$).

3.3. Sperm Proteomics Analysis

3.3.1. Concentration of Sperm Protein

Table 1 shows the concentration of sperm protein (mg/mL) for each group determined by the Bradford assay [23]. The FC treatment group recorded the highest concentration of protein which was 1.15 mg/mL, followed by normal control 1.13 mg/mL, positive control 1.05 mg/mL, and negative control with the lowest concentration of 0.8 mg/mL.

Table 1. Concentration of sperm protein (mg/mL).

Group	Concentration of Protein (mg/mL)
Normal	1.13
Negative (Diabetic without treatment)	0.8
Positive (Diabetic treated with metformin)	1.05
FC (Diabetic treated with *F. carica* extract)	1.15

3.3.2. Sperm Protein Profile via SDS-PAGE

Based on Figure 6, the negative control group showed a thin protein band similar to positive control, whereas the FC group showed thicker protein band compared to negative control, especially at molecular weight of 9–19 kDa (yellow box) and 38–46 kDa (red box). In addition, the FC treatment group also showed thick protein band similar to normal control, indicating that the treatment might improve the fertility of diabetic male rats.

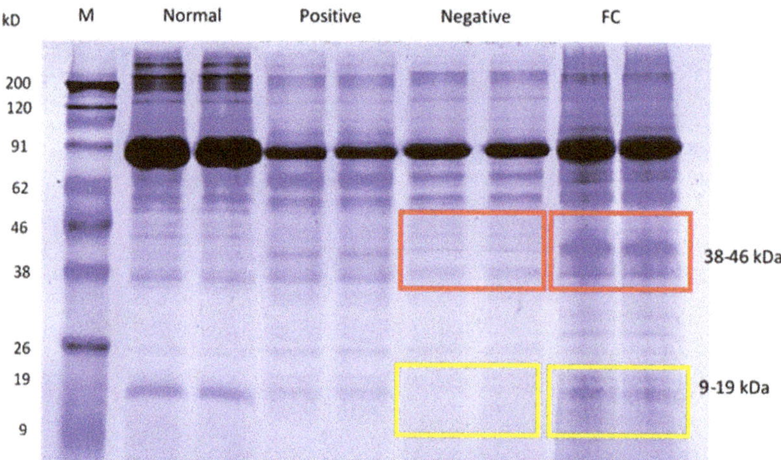

Figure 6. Sperm protein bands from four groups using SDS-PAGE electrophoresis. Column M is a molecular weight marker; normal: Sperm protein from normal group; positive: Diabetic treated with metformin; negative: Diabetic without treatment, and FC: Diabetic treated with *F. carica*.

3.3.3. LC-MS/MS Analysis

LC-MS/MS analysis recorded the total number of proteins present in each group (Tables S1–S4). According to Table 2, the sperms of negative control rat showed lower protein expression which was 55 proteins compared to normal control that showed high protein expression with 149 proteins. The metformin group recorded 90 sperm protein expressions, which was higher than negative control. Meanwhile, the FC group recorded almost a similar number of proteins as normal control which was 155 proteins.

Table 2. The total number of proteins in the sperm sample of FC group and the controls as determined by LC-MS/MS analysis.

Group	Total Number of Protein
Normal	149
Negative (Diabetic without treatment)	55
Positive (Diabetic treated with metformin)	90
FC (Diabetic treated with *F. carica* extract)	155

3.3.4. Gene Ontology Analysis via Blast2GO and WEGO

Based on gene ontology analysis via Blast2GO and WEGO, sperm proteomic analysis was focused on the reproductive process in order to evaluate the protein activity and function involved in diabetic rat fertility, as shown in Figure 7 and Figure S1. Expression of reproductive protein in normal control was 24 proteins, negative control was 11 proteins, positive control was 15 proteins, and FC group with the highest number of 34 proteins.

3.3.5. Sperm Unique Protein upon FC Treatment

Based on the Uniprot database, approximately 32.3% of total sperm proteins in the FC treated group have been identified on fertility-related function and activity. Table 3 shows a total of 14 unique fertility proteins in FC group (negative control did not have these proteins) were identified along with their fertility-related function and molecular weight. The proteins were then classified into two groups according to their functions in the fertility of diabetic rat. The two groups of proteins were protective unique protein (manganese superoxide dismutase, peroxiredoxin-5, endoplasmic reticulum stress regulator ATPase, pyruvate kinase, ATP synthase subunit O and pyruvate dehydrogenase E1 component subunit alpha) and reproductive unique protein (myosin-9, prostaglandin-H2 D-isomerase, histone H2B, dynein light chain 1, ropporin-1, T-complex protein 1 subunit beta, dihydrolipoyl dehydrogenase, and Izumo sperm-egg fusion protein 1). The classification of reproductive unique protein was based on the functions of spermatogenesis and fertilization, as shown in Table 3. The unique protein molecular weight data was then matched back to the FC group protein bands in SDS-PAGE gel to show the distribution of this fertility protein according to molecular weight (Figure 8).

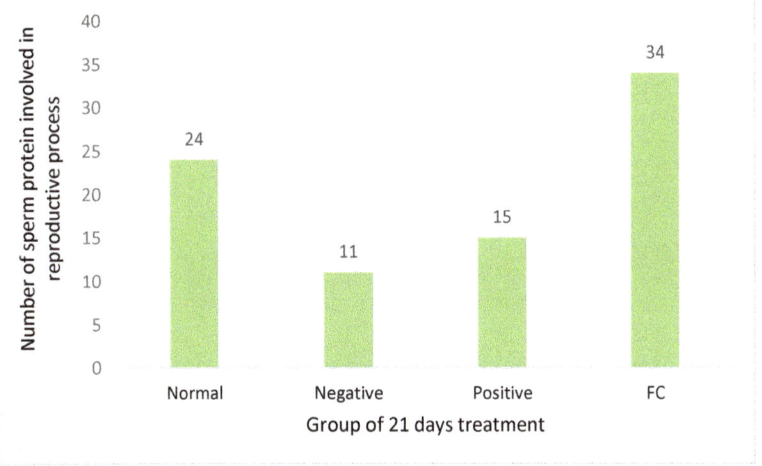

Figure 7. The number of sperm protein involved in reproductive process from normal (normal group with distilled water), negative (diabetic group with distilled water), positive (diabetic group with 300 mg/kg metformin), and FC (diabetic group with 400 mg/kg *Ficus carica* aqueous extract).

Table 3. Classification of 14 unique sperm proteins of the FC treatment group compared to negative controls according to function based on Uniprot database.

Protein ID (Uniprot)	Name	Location	Molecular Weight (kDa)
1. Protective unique protein			
P07895	Manganese superoxide dismutase	mitochondria	24.66
Q9R063	Peroxiredoxin-5, mitochondrial	mitochondria	22.17
P46462	Transitional endoplasmic reticulum ATPase	nucleus	89.29
P11980	Pyruvate kinase PKM	cytoplasm	57.78
Q06647	ATP synthase subunit O	mitochondria	23.38
Q06437	Pyruvate dehydrogenase E1 component subunit alpha	mitochondria	43.37
2. Reproductive unique protein			
i. Spermatogenesis			
Q62812	Myosin-9	cytoplasm	226.2
P22057	Prostaglandin-H2 D-isomerase	nucleus	21.29
Q00729	Histone H2B type 1-A	nucleosome	14.22
P63170	Dynein light chain 1	nucleus	10.36
ii. Fertilization			
Q4KLL5	Ropporin-1	cytoplasm	23.95
Q5XIM9	T-complex protein 1 subunit beta	cytoplasm	57.42
Q6P6R2	Dihydrolipoyl dehydrogenase	mitochondria	54.0
Q6AY06	Izumo sperm-egg fusion protein 1	acrosome	43.55

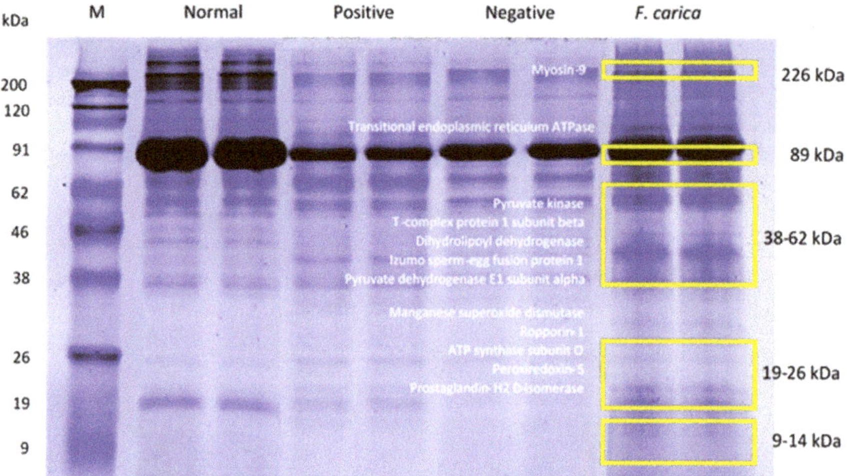

Figure 8. The distribution of unique fertility-related protein of FC treated sperm group based on molecular weight in SDS-PAGE.

4. Discussion

4.1. Antihyperglycemic Activity of FC Extract

STZ [2-deoxy-2-(3-(methyl)-3-nitrosoureido)-D-glucopyranose] is an antimicrobial and alkylated chemotrapeutic agent synthesized by *Streptomycetes achromogenes* [25]. Rakieten et al. [26] reported that STZ was diabetogenic when it caused diabetes via the specific necrosis of β-cells located at *islet of Langerhans* in the pancreas. King [25] describes the mechanism of diabetic induction by STZ, in which it enters β-cells through Glut-2 carrier protein (SLC2A2), and causes the alkylation of DNA. DNA alkylation activates poly (ADP-ribose) polymerase (PARP) that results in depletion of nicotinamide adenine dinucleotide (NAD+) and adenosine triphosphate (ATP), in which both NAD+ and ATP are needed for insulin secretion [27]. Therefore, the depletion of NAD+ and ATP leads to inhibition of insulin production. Pancreatic β-cells produce insulin in order to maintain normal blood glucose levels. Insulin activates glycogen synthase that converts glucose to glycogen, and also activates carrier protein GLUT4 which carries glucose from the blood stream into cells [28]. Thus, when there are problems related to insulin production, the conversion of glucose to glycogen and glucose uptake by the cell will be affected, resulting in the increase of blood glucose level.

Duca et al. [29] reported that the mechanism of blood glucose level reduction by metformin is by decreasing hepatic glucose production via inhibition of mitochondrial complex I, which results in the increase of adenosine monophosphate-activated protein (AMP) and the activation of adenosine monophosphate-activated protein kinase (AMPK). AMPK deactivates the transcription of peroxisome proliferator-activated receptor gamma (PPAR-γ) and sterol regulatory element binding protein 1c (SREBP-1c), in which both of them are regulators in the expression of enzymes involved in gluconeogenesis, glucose transportation, and fatty acid synthesis in the liver [30]. The inhibition of these enzyme expressions eventually causes reduction of blood glucose level.

Meanwhile, FC reduces blood glucose level by facilitating the glucose uptake by the muscle cell, which is the similar action such as insulin in reducing blood glucose level [18]. Nearly 40% of total body weight is represented by the skeletal muscle, in which it is also an important tissue that has been targeted for insulin action and glucose uptake [31]. In vitro experiment done by Perez et al. [18] found that there was a low glucose uptake by skeletal muscle of untreated diabetic rats due to the insufficient amount of insulin, but on the contrary, diabetic rats treated by FC showed a significant increase of glucose uptake by skeletal muscle. Irudayaraj et al. [16] suggest that FC is able to improve the sensitivity of insulin, possibly due to the improvement of defects in insulin function that has been realized after a significant reduction of blood glucose and normalization of plasma insulin levels were found. Mopuri et al. [32] reported that FC also inhibits the activity of digesting enzymes of carbohydrates such as α-amylase and α-glucosidase. The inhibition of these enzymes may slow down the digestion of carbohydrates, thus leading to low glucose levels in the blood stream. Previous studies of some antihyperglycemic plants such as *Gynura procumbens* [14] and *Moringa olifeira* [33] also demonstrated similar mechanism as FC, via the increase of peripheral glucose uptake and inhibition of α-amylase and α-glucosidase. Meanwhile, there were other plants such as *Aloe vera* [34] and *Camillia sinensis* [35] that showed different antihyperglycemic mechanism compared to FC which were an improvement of insulin secretion and pancreatic β-cell function.

Therefore, all studies stated above indicate that there are phytochemicals in FC responsible for its antihyperglycemic property. Zhang et al. [36] conclude that the antihyperglycemic activity of FC may be due to the antioxidant effect of its phytochemicals. Certain antioxidants in FC that may contribute to its antihyperglycemic effect are quercetin [2,37,38], kaempferol [37,39], ficusin [40], ferulic acid [2], and caffeoylmalic acid [39]. It is found that quercetin delays oxidant injury and cell death due to the oxidation by free radicals [37], improves insulin signaling and sensitivity [41], as well as increases glucose uptake by cell [42]. Kaempferol inhibits the breakdown of polysaccharides to glucose by α-glucosidase, as it is a carbohydrate digesting enzyme [43], and activates AMPK to mediate an antidiabetic effect by inhibiting hepatic glucose production [30]. Ficusin restores insulin level in diabetic rats due to regularized β-cell, and improves glucose utilization by increasing the expression level of glucose transporter protein (GLUT) [40]. Next, ferulic acid inhibits gluconeogenesis via the decrease of enzyme of gluconeogenesis, which are phosphoenolpyruvate carboxinase (PEPCK) and glucose-6-phosphatase [44]. Meanwhile, Takahashi et al. [39] found that caffeoylmalic acid is among the abundant polyphenol present in FC that has high antioxidant activity similar to vitamin C.

4.2. Sperm Quality Analysis

Conventional sperm quality analysis such as sperm count, motility, viability, and morphology using haemocytometer is still a relevant, effective, and fundamental investigation method that is able to detect the sign of male infertility as recommended by World Health Organization [45,46]. The sperm count from ejaculation is an early indicator towards the ability of the testis to produce sperm, as normal ejaculation is correlated with testicular volume [21]. A high percentage of progressive motility is important in determining the success of sperms to find and penetrate oocyte during the fertilization process, because the life span of a sperm is short, which is only 6–12 h after ejaculation [47]. Sperm viability is considered as an early indicator for male fertility, as it indicates either the sperms live or die [48]. While normal sperm morphology is crucial to determine the success of fertilization [49], because defects related to head, midpiece, and tail commonly result in nonprogressive motility and immotility [48].

The results showed that STZ-induced diabetic rats were infertile because the number of sperms, the percentage of sperm viability, and normal sperm morphology percentage were at the lowest, and the percentage of immotile sperm recorded was the highest. STZ causes the necrosis of β-cell in the islet of Langerhans in the pancreas which causes insulin production failure [26]. Based on a study by Kim and Moley [50], insulin deficiency causes failure of transcription and activation of GLUT glucose transport proteins. GLUT plays an important role in transporting glucose into sertoli cells to carry out the glycolysis process, which later produces lactate [51]. Lactate is a source of energy that is important in the growth of germ cell and sperm maturation [52]. When lactate fails to be supplied for germ cell nutrition, the spermatogenesis process is affected, and results in the decrease of sperm quality. According to Pavlinkova [4], diabetes decreases sperm concentration, viability, and increases sperm apoptosis due to the changes in protamine 1/protamine 2 ratio, small proteins that bind sperm DNA which are important for DNA stability and sperm maturation. Furthermore, the decrease in sperm quality is closely related to oxidative stress due to hyperglycemia, which causes the excessive production of free radical in the blood such as reactive oxygen species (ROS) [53]. Ding et al. [6] found that high concentration of free radicals induced sperm apoptosis, sperm DNA fragmentation, and changed the expression of fertility proteins, in which they act as antioxidants to protect cells from structural modification by advanced glycation end (AGE) products.

In this study, metformin was found to increase sperm quality parameters, but it was not significant when compared to the negative control. According to Owen et al. [12], metformin increases sperm quality parameters by blocking complex I in the mitochondrial respiratory chain, resulting in the decrease of mitochondrial function and cell respiration, and it also causes the increase of AMPK which eventually leads to anaerobic respiration and the increase of lactate production. Lactate is the main source of energy for germ cells in order to be developed into mature sperms. Bertoldo et al. [11] state that metformin increases the level of mRNA for glucose carrier Glut 1 and lactate dehydrogenase which catalyzes the conversion of pyruvate into lactate. However, some studies have found that metformin lowers male fertility parameters. Research done by Naglaa et al. [15] using alloxan-induced diabetic rats shows a decrease in testicular weight, sperm count, sperm motility, and increase in the number of abnormal sperm when treated with metformin. Furthermore, the study by Hurtado de Llara et al. [13] shows that metformin inhibits sperm mitochondrial membrane potential and motility of boar sperms.

Based on the results of the study, the FC extract was able to improve the quality of sperms in diabetic rats. Since there is no study that had been done for male infertility due to diabetes using FC leaves extract as treatment, we considered choosing the concentration of FC from an antidiabetic study such as Irudayaraj et al. [16], Jayakumar et al. [54], and Mopuri et al. [32] (that used concentration between 400–500 mg/kg) because we believe that the problem is due to diabetes. We also considered the concentration of FC from studying the profertility of FC by Naghdi et al. [17] and Shimaa et al. [55] (that used concentration of FC extract between 200–300 mg/kg), where the concentration was below than what antidiabetic studies used. So based on these studies, we chose 400 mg/kg FC as it was the concentration used in antidiabetic studies and we estimated that the concentration can as well give a good result for profertility. The ability of FC extract to treat diabetes, as well as fertility is closely related to its phytochemical content that acts as an antioxidant in reducing reactive oxygen species production (ROS) [55,56]. Takahashi et al. [39] reported that caffeoylmalic acid is among the most abundant polyphenol found in FC which has the same level of antioxidant activity as vitamin C. The increase in sperm quality parameters is also likely to be due to the positive effect of FC on the hormone involved in male fertility. FC contains saponin which is found to resemble testosterone action, increase testosterone production, and has good effects on follicle stimulating hormone (FSH) [55]. These positive effects accelerate sperm maturation and sperm count. Furthermore, Irudayaraj et al. [16] reported that FC normalizes the plasma insulin levels of diabetic rats. Kim and Moley [50] reported that insulin plays an important role in male reproductive system when the diabetic mice treated with insulin show a significant increase of sperm count and motility compared to diabetic mice without treatment. Mice treated with insulin also show increase of glucose transporter (GLUT) protein expression in leydig and sperm cells. The GLUT protein functions to carry glucose into testicular cells for the continuation of spermatogenesis. Active spermatogenesis activity increases sperm quality parameters such as sperm count, motility, normal morphology, and viability [57].

4.3. Sperm Proteomics Analysis

4.3.1. Sperm Protein Concentration, Protein Expression via SDS-PAGE, LC-MS/MS, and Gene Ontology Analysis

It was consistent that the negative control showed lowest reading for all proteomic parameters; protein concentration, protein band thickness, and total number of proteins. The decrease of the parameters by negative control may be due to the effect of diabetes on endoplasmic reticulum in the sperms. The endoplasmic reticulum is a site of synthesis, folding, and maturation of proteins, which is present in any cell including the sperm [58]. Diabetes and hyperglycemia result in the increase of endoplasmic reticulum stress [59], which causes changes in the homeostasis of the endoplasmic reticulum, and subsequently reduces protein synthesis [60]. Previous studies also supported our findings [61,62]. An et al. [10] reported that no expression of the spermatogenesis regulator protein was recorded in the sperm of diabetic patients compared to normal. Decreased expression of other proteins in the sperm of diabetic patients was also reported by Kim and Moley [50] and Kriegel et al. [61]. Based on gene ontology analysis, diabetes also causes a decrease in the expression of reproductive protein. Gene ontology analysis is a recent development in the molecular biology study that enables genes or proteins to be classified based on their function and involvement in any biological process, hence facilitates scientists to understand the mechanism behind the phenotype [63]. Our study focused on reproductive proteins out of all biological processes since reproduction is an important process that determines fertility. In correlation with sperm quality analysis in our study, it can be speculated at this point that diabetes may reduce sperm quality by depleting the production of reproductive protein.

Interestingly, FC treatment improved all proteomic parameters of diabetic rats. Furthermore, gene ontology analysis revealed that FC treatment increased the number of sperm reproduction protein despite being in a diabetic condition. The increase of the sperm reproduction protein number was in correlation with the increase of sperm quality parameters, as shown in Figures 2–5. It can be inferred that sperm fertility proteins influenced by FC protect male fertility from diabetes complication. The study using another herb by Ghosh et al. [64] found that treatment with *E. jambolana* increases the expression of the protein Apoptosis regulator (BCL2) in the testicular tissue of diabetic mice. BCL2 is a protein that inhibits the process of apoptosis and inflammation of cells [65]. With increased expression of BCL2 protein, damage to testicular cells can be avoided, thus protecting male fertility from diabetes complication.

4.3.2. Sperm Unique Protein upon FC Treatment

Proteomic analysis later was focused on the unique sperm fertility protein of the FC treated group in comparison with the negative controls. This investigation aimed to evaluate how the unique fertility proteins escape diabetes complication and thus maintain male fertility. The heaviest unique protein was myosin-9 with a molecular weight of 226.2 kDa, followed by a transitional endoplasmic reticulum ATPase with a molecular weight of 89.29 kDa. The protein bands between 38 and 62 kDa contained pyruvate kinase protein, T-complex protein 1 subunit beta, dihydrolipoyl dehydrogenase, Izumo sperm-egg fusion protein 1, and pyruvate dehydrogenase E1 alpha subunit. Furthermore, the protein bands between 19 and 26 kDa contained manganese superoxide dismutase, ropporin-1, ATP synthase subunit O, peroxiredoxin-5, and prostaglandin-H2 D-isomerase, while the lowest molecular weight proteins were histone H2B and dynein light chain, both of them in the protein bands between 9 and 14 kDa.

Based on protein molecular weight, many previous studies have concluded that low-molecular weight proteins have high antioxidant activity [66–68]. An in vitro study by Pierro et al. [67] found that the process of hydrolysis of cow's casein by latex FC produced low molecular weight proteins with high antioxidant activity. The LC-MS/MS analysis performed in this study also support the statement, in which we actually found that the rat sperm of FC contained manganese superoxide dismutase (MnSOD) and peroxiredoxin-5 proteins, and both have high antioxidant activity and also low molecular weight proteins (<25 kDa). MnSOD and peroxiredoxin-5 may act to prevent ROS increase in sperm cells, and protect cell organelles, especially mitochondria, which play a key role in ensuring progressive sperm motility as it functions to produce energy in the cell. Manganese superoxide dismutase (MnSOD) is a major antioxidant defense in mitochondria [69]. MnSOD is an enzyme that converts superoxide anion radicals to hydrogen peroxide, thereby reducing the probability of superoxide anions interacting with nitric oxide to form reactive nitrite peroxides [70]. The study by Li et al. [71] found that infants with MnSOD deficiency are genetically exposed to mitochondrial damage in several organs including the heart, causing death within 10 days after birth. This indicates that MnSOD has an important role in the mitochondrial defense system. MnSOD expression in diabetic rats has been shown to suppress the increase of ROS, and inhibit the effect of hyperglycaemia [72–74]. Additionally, Cocchia et al. [75] reported that MnSOD protects spermatozoa from damage by lipid peroxidation, and subsequently increases sperm motility and viability compared to the negative control. Meanwhile, peroxiredocin-5 is also an effective antioxidant enzyme in the inhibition of peroxide in mitochondria. The uniqueness of the peroxiredoxin family compared to other peroxidases is due to the hydrogen peroxide catalysis initiated by the peroxidatic cysteine located at the N-terminus of the protein [76]. Peroxidatic cysteines break down the O-O bond of the hydrogen peroxide which then temporarily produce cysteine sulfenic acid, before returning to its original form of peroxidatic cysteine in order to catalyze other hydrogen peroxide [77]. Kubo et al. [78] reported that the expression of peroxiredoxin-5 protein is found to be responsible for the suppression of apoptosis in pig pericyte cells when the protein is able to reduce oxidative stress induced by diabetes. In addition, O'Flaherty [79] states that peroxiredoxin plays a key role in the protection of spermatozoa function from ROS when it is found that peroxiredoxin deficiency in infertile men is consistent with a decrease in sperm quality.

Another protein, the transitional endoplasmic reticulum ATPase protein or known as the valocin-containing protein (VCP) had the function in regulating endoplasmic reticulum stress induced by diabetes. Excessive endoplasmic reticulum stress induced by diabetes causes alterations in the homeostasis of the endoplasmic reticulum and subsequently reduces protein synthesis [59,60]. In addition, excessive endoplasmic reticulum stress is linked with male infertility [80]. Guzel et al. [81] state that spermatogenesis requires intensive protein synthesis to continue the developmental process from spermatogonia to spermatozoa, however, uncontrolled endoplasmic reticulum stress inhibits spermatogenesis. VCP is a family of ATPases that generally functions in protein folding or unfolding [82]. Additionally, the specific function of VCP is to regulate the degradation of unfolded proteins in response to endoplasmic reticulum stress [83]. Wojcik et al. [84] reported that disruption of VCP RNA results in the reduction of VCP protein expression in cells, thus inducing the changes on over 30 RNA transcripts for other proteins involved in endoplasmic reticulum stress, amino acid starvation, and apoptosis. Hence, the VCP protein presence in FC diabetic rats retains the normal functioning of the endoplasmic reticulum, and thus ensures the continual production of proteins required for sperm development despite being in a diabetic condition.

There are three unique proteins in the sperm of FC group involved in glycolysis and the production of ATP or Nicotinamide adenine dinucleotide (NADH), which are pyruvate kinase, pyruvate dehydrogenase E1 component subunit alpha, and ATP synthase subunit O. Previous studies reported that diabetes disrupts the expression of the proteins in the early pathogenesis and later causes severe complications [85–87]. Disruption of ATP synthase by diabetes stimulate mitochondrial ROS generation and thus promote severe complications on the tissue and organ [85]. The failure of diabetes to disrupt the proteins at the beginning of pathogenesis will terminate the complication of diabetes [86,87]. Qi et al. [86] reported that the activation of pyruvate kinase may protect against diabetes by improving glucose metabolism, inhibiting the production of toxic glucose metabolites and inducing the biogenesis of mitochondria. Another study done by Rahimi et al. [87] concluded that activation of pyruvate dehydrogenase during a diabetic condition reduces available substrate for production of glucose, resulting in low blood glucose level and improved glucose tolerance. It is inferred that treatment with FC expressed these unique proteins thus protecting the sperm from toxicity of diabetes.

The following unique proteins are fertility proteins involved directly in the reproduction process; spermatogenesis (myosin-9, prostaglandin-H2 D-isomerase, histone H2B, and dynein light chain 1) and fertilization (ropporin-1, T-complex protein 1 subunit beta, dihydrolipoyl dehydrogenase, and Izumo sperm-egg fusion protein 1), where the expression of these reproductive unique proteins may be due to the protective effect of the unique proteins discussed before (MnSOD, peroxiredoxin-5, VCP, ATP synthase, pyruvate kinase, and pyruvate dehydrogenase) against diabetes complication.

Myosin-9 functions as a component in the activation of FSH hormone synthesis. The study by Lin et al. [88] reported that myosin-9 is involved in the activation of FSH hormone synthesis by the guanosine triphosphate (Gαh) protein. It was found that myosin-9 inhibition using Blebbistatin, a type of ATPase, inhibited the activation of FSH hormone synthesis by Gαh protein in Sertoli cells. FSH hormone secretion by myosin-9 induces FSH activity in the development of spermatogonia to spermatocytes. The next protein, prostaglandin-H2 D-isomerase was involved in Sertoli cell differentiation. Prostaglandin-H2 D-isomerase or known as prostaglandin D2 synthase (PGDS) is an enzyme that converts the cyclooxygenase product of prostaglandin H2 (PGH2) to prostaglandin D2 (PGD2) [89]. PGD2 is a protein that enhances the differentiation of Sertoli cells, and controls the proliferation of germ cells [90]. The study by Moniot et al. [91] states that expression of the Sox9 gene, which is a transcriptional factor of differentiation of Sertoli cells, is regulated by the PGDS protein. Whereas, Samy et al. [92] reported that increased expression of PGDS was consistent with the increase in germ cell tight junction, indicating that there is a correlation between PGDS expression and junction. The interplay between germ cells, which is part of the blood-testis structure, is important in regulating the molecular entry and development of germ cells [51]. There are two other unique FC proteins identified that help in spermatid development and maturation of spermatozoa which are histone H2B and dynein light chain 1. Histone H2B (HH2B), which is a variant of histone protein, together with DNA binding to nucleosomes in chromatin [93], was known to be involved in spermatid development. Lu et al. [94] reported that the highest HH2B expression was recorded in spermatogonia, followed by spermatocytes and finally spermatids. This implies that HH2B plays a role in every stage of sperm cell in spermatocytogenesis. Dynein light chain 1 (DLC1) is a protein involved in the spermatozoa maturation process. Dynein generally acts to regulate chromosome movement, assembly, and orientation of mitotic spindles, as well as nucleus migration [95]. The study by Wang et al. [96] in rat sperm suggests that DLC1 plays a role in chromatin condensation in the early stages of spermatids, disposal of excess cytoplasm for spermatozoa formation, and release of spermatozoa from the apical compartment into the lumen of the seminal tuberculosis (sperm) when there is intensive DLC1 expression in the spermatid nucleus length and cytoplasm of spermatozoa.

In this proteomic study, a unique sperm protein from FC group was identified to be responsible for fibrous sheath signaling that subsequently influenced sperm motility. The protein is ropporin-1 (ROPN1) which is found only in fibrous sheath of the sperm flagella, and is specifically located at the center and tip of the flagella [97]. ROPN1 binds to the primary components of fibrous sheath, the A-kinase anchoring proteins (AKAP) 3 and 4 to initiate the signaling pathway for sperm motility [98]. Fiedler et al. [98] found that a rat sperm lacked the ROPN1, and ropporin-1 like protein (ROPN1L) had problems with flagella and became immotile. This indicates that ROPN1 is a vital protein in sperm motility and capacity. The discovery of this protein in the sperm of FC group is in line with the results of sperm motility analysis which has shown that treatment with FC helps increase sperm motility in diabetic rats. There is a unique sperm protein in the FC treated group that is known to have activity that helps the binding of the sperm to ZP which is T-complex protein 1 subunit beta. The T-complex protein 1 subunit beta (CCT) is a chaperone-type protein that is commonly known to fold other proteins from its original or nonfoldable state to a folded and functional form [99]. While in fertility, CCTs are the first isolation in mice testes protein that are specifically involved in sperm motility, sperm capacitation, and the ability to bind and penetrate ZP [100]. A proteomic study of CCT by Dun et al. [101] using blue native polyacrylamide gel electrophoresis (BN-PAGE) on spermatozoa found high CCT expression on the surface of the spermatozoa through the capacitation process, while the far Western Blot method found that CCT showed adherence to intact ZP. This clearly indicates that CCT plays an intermediate role in the binding between sperm and ZP. Next, dihydrolipoyl dehydrogenase is a protein identified in the sperm of the FC group which is involved in the acrosome reaction process. Dihydrolipoyl dehydrogenase (DLD) is a post-metabolism pyruvate enzyme, a pyruvate dehydrogenase complex subunit E3 that is involved in the regulation of lactate in spermatozoa [102]. DLD has been reported to play a role during hyperactivation, acrosome reactions [103] and it is also involved in fetal development [104]. A study by Mitra and Shivaji [105] found that the downregulation of DLD activity by inhibitors, specifically 5-methoxyindole-2-carboxylic acid, prevented acrosome reaction and decreased spermatozoa hyperactivation. During adhesion between the sperm membrane and the egg membrane, there are various protein interactions that occur. Sperm proteomic studies have identified that the unique protein of the FC group, Izumo sperm-egg fusion protein 1 is involved in sperm-egg fusion. Izumo sperm-egg fusion protein 1 (IZUMO1) is a protein that is specifically involved in sperm-egg binding and fusion. The protein is identified by Inoue et al. [106] using monoclonal antibodies that inhibit sperm-egg fusion and gene cloning. It is not detectable on the surface of newly released spermatozoa (after ejaculation or in the epididymis), however, the protein can only be detected after acrosome reaction. This is probably because IZUMO1 is hidden inside the plasma membrane and only present on the surface as a result of the acrosome reaction. Inoue et al. [103] also found that mice that lack the IZUMO1 gene have normal sperm morphology and are capable of binding and penetrating ZP, however, they are unable to proceed fusion with the egg. A further study by Bianchi et al. [107] found that IZUMO1 binds to its receptor, namely JUNO protein, located on the surface of the egg, and then these two proteins interact with each other, which subsequently activates sperm-egg fusion.

Therefore, we proposed that FC treatment induces the expression of protective unique proteins that prevent the complication of diabetes on the reproductive unique protein, and thus resulted in the amelioration of male fertility despite being in a diabetic condition.

5. Conclusions

Overall, leaf extract of *Ficus carica* (FC) shows antihyperglycemic and profertility effects on streptozotocin-induced diabetic male rats. Treatment with FC extract shows a significant decrease in blood glucose level closer to normal. These antihyperglycemic effects have been found to prevent the complications of diabetes on male rat fertility in accordance with the recovery of sperm quality parameters after treatment with FC extract. This recovery and protection may be due to the bioactive compounds responsible for the antioxidant activity present in FC extracts such as quercetin, kaempferol, ferulic acid, and caffeoylmalic acid. In addition to acting in order to balance oxidative stress caused by diabetes, these bioactive compounds may also be involved in the secretion of insulin, proteins, and other molecules leading to fertility improvement.

Proteomic studies on the sperm of the FC group show an increase in overall protein expression compared to controls. Furthermore, the comparison between FC group proteins and negative control reveals that the sperm of FC group contains unique proteins that have fertility-related functions. These unique FC group proteins are made up of protective and reproductive unique proteins. The synergy interaction between these unique proteins may prevent the complications of diabetes on the function of testes and sperm cells, and thus restore the fertility of diabetic male rats.

Supplementary Materials: The following materials are available online at http://www.mdpi.com/2227-9717/8/4/395/s1, Table S1: List of sperm proteins in normal control group; Table S2: List of sperm proteins in negative control group; Table S3: List of sperm proteins in positive control group; Table S4: List of sperm proteins in *Ficus carica* treated group; Figure S1: Analysis of sperm proteomic data from (a) normal, (b) negative control, (c) positive control, and (d) *Ficus carica* treated group using the WEGO software.

Author Contributions: Conceptualization, U.A.B., P.S., and M.M.N.; methodology, M.M.N., K.A.K., N.A.K.B., A.K., M.J., W.M.A., and M.S.; software, W.M.A.; validation, M.M.N., W.M.A., and M.J.; formal analysis, U.A.B.; investigation, U.A.B., P.S., N.A.K.B., and A.K.; resources, M.S., M.J., W.M.A., and M.M.N.; data curation, W.M.A. and U.A.B.; writing—original draft preparation, U.A.B.; writing—review and editing, U.A.B., W.M.A., and M.M.N.; visualization, U.A.B. and M.M.N.; supervision, M.M.N., M.J., and W.M.A.; project administration, U.A.B., P.S., N.A.K.B., and A.K.; funding acquisition, M.M.N., M.J., and W.M.A. All authors have read and agreed to the published version of the manuscript.

Funding: This research has been funded by the Research University Grant (GUP-2016-056).

Acknowledgments: The authors wish to thank Universiti Kebangsaan Malaysia for providing the facilities in the Animal House, Faculty of Science and Technology, Universiti Kebangsaan Malaysia. Special thanks to the Mass Spectrometry Technology Section, Malaysia Genome Institute (MGI) for contributing in LC-MS/MS analysis for identification of the sperm protein. Last but not least, thank you to Saf Fa Fig Garden for providing *Ficus carica* leaves as the sample of the research.

Conflicts of Interest: The authors declare no conflict of interest.

References

1. World Health Organization. *Global Report on Diabetes*; WHO: Geneva, Switzerland, 2016.
2. El-Shobaki, F.A.; El-Bahay, A.M.A.; Esmail, R.S.A.; El-Megeid, A.A.A.; Esmail, D.S. Effect of figs fruit (*Ficus carica* L.) and its leaves on hyperglycemia in alloxan diabetic rats. *World J. Dairy Food Sci.* **2010**, *5*, 45–57.
3. Singh, A.K.; Tomarz, S.; Chaudhari, A.R.; Sinqh, R.; Verma, N. Type 2 diabetes mellitus affects male fertility potential. *Indian J. Physiol. Pharmacol.* **2014**, *58*, 403–406. [PubMed]
4. Pavlinkova, G.; Margaryan, H.; Zatecka, E.; Valaskova, E.; Elzeinova, F.; Kubatova, A.; Bohuslavova, R.; Peknicova, J. Transgenerational inheritance of susceptibility to diabetes-induced male subfertility. *Sci. Rep.* **2017**, *7*, 1–14. [CrossRef]
5. Pitetti, J.; Calvel, P.; Zimmermann, C.; Conne, B.; Papaioannou, M.D.; Aubry, F.; Cederroth, C.R.; Urner, F.; Fumel, B.; Crausaz, M.; et al. An essential role for insulin and IGF1 receptors in regulating sertoli cell proliferation, testis size and FSH action in mice. *Mol. Endocrinol.* **2013**, *27*, 814–827. [CrossRef] [PubMed]
6. Ding, G.L.; Liu, Y.; Liu, M.E.; Pan, J.X.; Guo, M.X.; Sheng, J.Z.; Huang, H.F. The effects of diabetes on male fertility and epigenetic regulation during spermatogenesis. *Asian J. Androl.* **2015**, *17*, 948–953. [PubMed]
7. Pennington, S.R.; Wilkins, M.R.; Hochstrasser, D.F.; Dun, M.J. Proteome analysis: From protein characterization to biological function. *Trends Cell Biol.* **1997**, *7*, 168–173. [CrossRef]
8. Rivas, J.D.L.; Luis, A.D. Interactome data and databases: Different types of protein interaction. *Comp. Funct. Genom.* **2004**, *5*, 173–178. [CrossRef]
9. Amaral, A.; Castillo, J.; Ramalho-Santos, J.; Oliva, R. The combined human sperm proteome: Cellular pathways and implications for basic and clinical science. *Hum. Reprod. Update* **2014**, *20*, 40–62. [CrossRef]
10. An, T.; Wang, Y.; Liu, J.; Pan, Y.; Liu, Y.; He, Z.; Mo, F.; Li, H.; Gu, Y.; Lv, B.; et al. Comparative analysis of proteomes between diabetic and normal human sperm: Insights into the effects of diabetes on male reproduction based on regulation of mitochondria-related proteins. *Mol. Reprod. Dev.* **2017**, *85*, 7–16. [CrossRef]
11. Bertoldo, M.J.; Faure, M.; Dupont, J.; Froment, P. Impact of metformin on reproductive tissues: an overview from gametogenesis to gestation. *Ann. Transl. Med.* **2014**, *2*, 1–13.
12. Owen, M.R.; Doran, E.; Halestrap, A.P. Evidence that metformin exerts its anti-diabetic effects through inhibition complex I of the mitochondrial respiratory chain. *Biochem. J.* **2000**, *348*, 607–614. [CrossRef] [PubMed]
13. Hurtado de Llara, A.; Martin-Hidalgo, D.; Garcia-Marin, L.J.; Bragado, M.J. Metformin blocks mitochondrial membrane potential and inhibits sperm motility in fresh and refrigerated boar spermatozoa. *Reprod. Dom. Anim.* **2018**, *2018*, 1–9. [CrossRef] [PubMed]
14. Khaidatul Akmar, K.; Mahanem, M.N. *Gynura procumbens* leaf improves blood glucose level, restores fertility and libido of diabetic-induced male rats. *Sains Malays.* **2017**, *46*, 1471–1477. [CrossRef]
15. Naglaa, Z.H.E.; Hesham, A.M.; Fadil, H.A. Impact of metformin on immunity and male fertility in rabbits with alloxan-induced diabetes. *J. Am. Sci.* **2010**, *6*, 417–426.
16. Irudayaraj, S.S.; Sunil, C.; Stalin, A.; Duraipandiyan, V.; Al-Dhabi, N.A.; Ignacimuthu, S. Protective effects of Ficus carica leaves on glucose and lipids levels, carbohydrate metabolism enzymes and β-cells in type 2 diabetic rats. *Pharm. Biol.* **2017**, *55*, 1074–1081. [CrossRef]
17. Naghdi, M.; Maghbool, M.; Seifalah-Zade, M.; Mahaldashtian, M.; Makoolati, Z.; Kouhpayeh, S.A.; Ghasemi, A.; Ferezdouni, N. Effects of common fig (Ficus carica) leaf extracts on sperm parameters and testis of mice intoxicated with formaldehyde. *Evid.-Based Comp. Alt.* **2016**, *2016*, 2539127. [CrossRef]
18. Perez, C.; Dominguez, E.; Canal, J.R.; Campillo, J.E.; Torres, M.D. Hypoglycaemic activity of aqueous extract from *Ficus carica* (Fig tree) leaves in streptozotocin diabetic rats. *Pharm. Biol.* **2000**, *38*, 181–186. [CrossRef]
19. Lin, M.; Zhang, X.; Murdoch, R.N.; Aitken, R.J. Swim-up of tammar wallaby (Macropus eugenii) spermatozoa in Biggers, Whitter and Whittingham (BWW) medium: maximisation of sperm motility, minimization of impairment of sperm metabolism and induction of sperm hyperactivation. *Reprod. Fert. Develop.* **2015**, *29*, 345–356. [CrossRef]

20. Seed, J.; Chapin, R.E.; Clegg, E.D.; Dostal, L.A.; Foote, R.H.; Hurtt, M.E.; Klinefelter, G.R.; Makris, S.L.; Perreault, S.D.; Schrader, S.; et al. Methods for assessing sperm motility, morphology and counts in the rat, rabbit and dog: A consensus report. *Reprod. Toxicol.* **1996**, *10*, 237–244. [CrossRef]

21. WHO. *Laboratory Manual for the Examination and Processing of Human Semen*, 5th ed.; Cambridge University Press: New York, NY, USA, 2010.

22. Yunianto, I.; Nurul Ain, K.B.; Mahanem, M.N. Antifertility properties of *Centella asiatica* ethanolic extract as a contraceptive agent: Preliminary study of sperm proteomic. *Asian Pac. J. Reprod.* **2018**, *6*, 212–216.

23. Bradford, M.M. A rapid and sensitive method for the quantitation of microgram quantities of protein utilizing the principle of protein-dye binding. *Anal. Biochem.* **1976**, *72*, 248–254. [CrossRef]

24. Shevchenko, A.; Tomas, H.; Havli, J.; Olsen, J.V.; Mann, M. In-gel digestion for mass spectrometric characterization of proteins and proteomes. *Nat. Protoc.* **2006**, *1*, 2856–2860. [CrossRef] [PubMed]

25. King, A.J.F. The use of animal models in diabetes research. *Br. J. Pharmacol.* **2012**, *166*, 877–894. [CrossRef]

26. Rakieten, N.; Rakieten, M.L.; Nadkarni, M.V. Studies on the diabetogenic action of streptozotocin. *Cancer Chemother. Rep.* **1963**, *29*, 91. [PubMed]

27. Jitrapakdee, S. Regulation of insulin secretion: role of mitochondrial signaling. *Diabetologia* **2010**, *53*, 1019–1032. [CrossRef]

28. Wilcox, G. Insulin and insulin resistance. *Clin. Biochem. Rev.* **2005**, *26*, 19–39.

29. Duca, F.A.; Cote, C.D.; Rasmussen, B.A.; Zadeh-Tahmasebi, M.; Rutter, G.A.; Filippi, B.M.; Lam, T.K.T. Metformin activates a duodenal Ampk-dependent pathway to lower hepatic glucose production in rats. *Nat. Med.* **2015**, *21*, 506–511. [CrossRef]

30. Zang, Y.; Zhang, L.; Igarashi, K.; Yu, C. The anti-obesity and anti-diabetic effects of kaempferol glycosides from unripe soybean leaves in high-fat-diet mice. *Food Funct.* **2015**, *6*, 834–841. [CrossRef]

31. Jensen, J.; Rustad, P.I.; Kolnes, A.J.; Lai, Y.C. The role of skeletal muscle glycogen breakdown for regulation of insulin sensitivity by exercise. *Front. Physiol.* **2011**, *2*, 1–11. [CrossRef]

32. Mopuri, R.; Ganjayi, M.; Meriga, B.; Koorbanally, N.A.; Islam, M.S. The effects of *Ficus carica* on the activity of enzymes related to metabolic syndrome. *J. Food Drug Anal.* **2018**, *26*, 201–210. [CrossRef]

33. Amira, K.; Malina, J.; Mahanem, M.N. Ameliorative effect of *Moringa oleifera* fruit extract on reproductive parameters in diabetic-induced male rats. *Pharmacogn. J.* **2018**, *10*, S54–S58.

34. Noor, A.; Gunasekaran, S.; Vijayalakshmi, M.A. Improvement of insulin secretion and pancreatic β-cell function in streptozotocin-induced diabetic rats treated with *Aloe vera* extract. *Pharmacogn. Res.* **2017**, *9*, S99–S104. [CrossRef] [PubMed]

35. Tang, W.; Li, S.; Liu, Y.; Huang, M.; Ho, C. Anti-diabetic activity of chemically profiled green tea and black tea extracts in a type 2 diabetes mice model via different mechanisms. *J. Funct. Foods* **2013**, *5*, 1784–1793. [CrossRef]

36. Zhang, Y.J.; Gan, R.Y.; Li, S.; Zhou, Y.; Li, A.N.; Xu, D.P.; Li, H.B. Antioxidant phytochemicals for the prevention and treatment of chronic diseases. *Molecules* **2015**, *20*, 21138–21156. [CrossRef]

37. Abdel-Nasser, B.S.; Nahla, A.A.; Eman, N.A.; Nada, M.M. Antioxidant and hepatoprotective activities of Egyptian moraceous plants against carbon tetrachloride-induced oxidative stress and liver damage in rats. *Pharm. Biol.* **2010**, *48*, 1255–1264.

38. Boukhalfa, F.; Kadri, N.; Bouchemel, S.; Cheikh, S.A.; Chebout, I.; Madani, K.; Chibane, M. Antioxidant activity and hypolipidemic effect of *Ficus carica* leaf and twig extracts in Triton-WR-1339-induced hyperlipidemic mice. *Med. J. Nutr. Metab.* **2018**, *11*, 37–50. [CrossRef]

39. Takahashi, T.; Okiura, A.; Saito, K.; Kohno, M. Identification of phenylpropanoids in fig (*Ficus carica* L.) leaves. *J. Agric. Food Chem.* **2014**, *62*, 10076–10083. [CrossRef]

40. Irudayaraj, S.S.; Stalin, A.; Sunil, C.; Duraipandiyan, V.; Al-Dhabi, N.A.; Ignacimuthu, S. Antioxidant, antilipidemic and antidiabetic effects of ficusin with their effects on GLUT4 translocation and PPARγ expression in type 2 diabetic rats. *Chem. Biol. Interact.* **2016**, *2016*, 1–38. [CrossRef]

41. Kannappan, S.; Anuradha, C.V. Insulin sensitizing actions of fenugreek seed polyphenols, quercetin and metformin in a rat model. *Indian J. Med. Res.* **2009**, *129*, 401–408.

42. Fang, X.K.; Gao, J.; Zhu, D.N. Kaempferol and quercetin isolated from Euonymus alatus improve glucose uptake of 3T3-L1 cells without adipogenesis activity. *Life Sci.* **2008**, *82*, 615–622. [CrossRef]

43. Ibitoye, O.B.; Uwazie, J.N.; Ajiboye, T.O. Bioactivity-guided isolation of kaempferol as the antidiabetic principle from Cucumis sativus L. fruits. *J. Food Biochem.* **2017**, *2017*, e12479. [CrossRef]

44. Narasimhan, A.; Chinnaiyan, M.; Karundevi, B. Ferulic acid exerts its antidiabetic effect by modulating insulin signaling molecules in the liver of high fat diet and fructose-induced type-2 diabetic adult male rat. *Appl. Physiol. Nutr. Metab.* **2015**, *40*, 769–781. [CrossRef] [PubMed]

45. Talarczyk-Desole, J.; Berger, A.; Taszarek-Hauke, G.; Hauke, J.; Pawelczyk, L.; Jedrzejczak, P. Manual vs. computer-assisted sperm analysis: can CASA replace manual assessment of human semen in clinical practice? *Ginekol. Pol.* **2017**, *88*, 56–60. [CrossRef] [PubMed]

46. Eravuchira, P.J.; Mirsky, S.K.; Barnea, I.; Balberg, M.; Shaked, N.T. Individual sperm selection by microfluidics integrated with interferometric phase microscopy. *Methods* **2018**, *136*, 152–159. [CrossRef] [PubMed]

47. Turner, R.M. Tales from the tail: What do we really know review about sperm motility? *J. Androl.* **2003**, *24*, 790–803. [CrossRef] [PubMed]

48. Vasan, S.S. Semen analysis and sperm function tests: How much to test? *Indian J. Urol.* **2011**, *27*, 41–48. [CrossRef] [PubMed]

49. Sun, Y.; Li, B.; Fan, L.Q.; Zhu, W.B.; Chen, X.J.; Feng, J.H.; Yang, C.L.; Zhang, Y.H. Does sperm morphology affect the outcome of intrauterine insemination in patients with normal sperm concentration and motility? *Andrologia* **2012**, *44*, 299–304. [CrossRef]

50. Kim, S.T.; Moley, K.H. Paternal effect on embryo quality in diabetic mice is related to poor sperm quality and associated with decreased glucose transporter expression. *Reproduction* **2008**, *136*, 313–322. [CrossRef]

51. Alves, M.G.; Martins, A.D.; Cavaco, J.E.; Socorro, S.; Oliveira, P.F. Diabetes, insulin-mediated glucose metabolism and Sertoli/blood-testis barrier function. *Tissue Barriers* **2013**, *1*, e23992. [CrossRef]

52. Mita, M.; Hall, P.F. Metabolism of round spermatids from rats: Lactate as the preferred substrate. *Biol. Reprod.* **1982**, *26*, 445–455. [CrossRef]

53. Mahanem, M.N.; Nani, R.M.R. Anti-hyperglycemic effect of Gynura procumbens methanolic extract on fertility and libido of induced diabetic male rats. *Sains Malays.* **2012**, *41*, 1549–1556.

54. Jayakumar, S.P.; Sen, M.; Jagadeesan, M.; Sundararajan, R. Antidiabetic and antioxidant effects of ethanolic extract of *Anjir* leaves (*Ficus carica*) in alloxan induced diabetic rat. *Asian J. Phytomed. Clin. Res.* **2014**, *2*, 1–10.

55. Shimaa, A.H.; Tamer, S.I.; Omar, A.A. Combination of Ficus carica leaves extract and ubiquinone in a chronic model of lithium induce reproductive toxicity in rats: Hindrance of oxidative stress and apoptotic marker of sperm cell degradation. *IOSR J. Pharm. Biol. Sci.* **2017**, *12*, 64–73.

56. Shahinuzzaman, M.; Yaakob, Z.; Abdullah Sani, N.; Akhtar, P.; Zahidul Islam, M.D.; Afsana Mimi, M.S.T.; Shamsudin, S.A. Optimization of extraction parameters for antioxidant and total phenolic content of *Ficus carica* L. latex from White Genoa cultivar. *Asian J. Chem.* **2019**, *31*, 1859–1865. [CrossRef]

57. Ambiye, V.R.; Langade, D.; Dongre, S.; Aptikar, P.; Kulkarni, M.; Dongre, A. Clinical evaluation of the spermatogenic activity of the root extract of aswagandha (*Withania somnifera*) in oligospermic males: A pilot study. *Evid.-Based Compl. Altern. Med.* **2013**, *2013*, 571420.

58. Ying, X.; Liu, Y.; Guo, Q.; Qu, F.; Guo, W.; Zhu, Y.; Ding, Z. Endoplasmic reticulum protein 29 (ERp29), a protein related to sperm maturation is involved in sperm-oocyte fusion in mouse. *Reprod. Biol. Endocrinol.* **2010**, *8*, 1–9. [CrossRef]

59. Zhong, Y.; Li, J.; Chen, Y.; Wang, J.J.; Ratan, R.; Zhang, S.X. Activation of endoplasmic reticulum stress by hyperglycemia is essential for Muller cell-derived inflammatory cytokine production in diabetes. *Diabetes* **2012**, *61*, 492–504. [CrossRef]

60. Back, S.H.; Kaufman, R.J. Endoplasmic reticulum stress and type 2 diabetes. *Annu. Rev. Biochem.* **2012**, *81*, 767–793. [CrossRef]

61. Kriegel, T.M.; Heldenreich, F.; Kettner, K.; Pursche, T.; Hoflack, B.; Grunewald, S.; Poenicke, K.; Glander, H.; Paasch, U. Identification of diabetes and obesity associated proteomic changes in human spermatozoa by difference gel electrophoresis. *Reprod. BioMed Online* **2009**, *19*, 660–670. [CrossRef]

62. Khaidatul Akmar, K.; Mohd Aizat, W.; Mahanem, M.N. Gynura procumbens improved fertility of diabetic rats: Preliminary study of sperm proteomic. *Evid.-Based Compl. Altern. Med.* **2018**. [CrossRef]

63. Dimmer, E.C.; Huntley, R.P.; Barrell, D.G.; Binns, D.; Draghici, S.; Camon, E.B.; Hubank, M.; Talmud, P.J.; Apweiler, R.; Lovering, R.C. The gene ontology—Providing a functional role in proteomic studies. *Pract. Proteom.* **2008**, *1*, 2–11. [CrossRef]

64. Ghosh, A.; Jana, K.; Ali, K.M.; De, D.; Chatterjee, K.; Ghosh, D. Corrective role of Eugenia jambolana on testicular impairment in streptozotocin-induced diabetic male albino rat: An approach through genomic and proteomic study. *Andrologia* **2013**, *46*, 296–307. [CrossRef] [PubMed]

65. Bruey, J.; Bruey-Sedano, N.; Luciano, F.; Zhai, D.; Balpai, R.; Xu, C.; Kress, C.L.; Bailly-Maitre, B.; Li, X.; Osteman, A.; et al. Bcl-2 and Bcl-XL regulate proinflammatory caspase-1 activation by interaction with NALP1. *Cell* **2007**, *129*, 45–56. [CrossRef] [PubMed]

66. Pena-Ramos, E.A.; Xiong, Y.L. Antioxidative activity of soy protein hydrolysates in a liposomal system. *J. Food Sci.* **2002**, *67*, 2952–2956. [CrossRef]

67. Pierro, G.D.; O'Keeffe, M.B.; Poyarkov, A.; Lomolino, G.; FitzGerald, R.J. Antioxidant activity of bovine casein hydrolysates produced by *Ficus carica* L.—derived proteinase. *Food Chem.* **2014**, *156*, 305–311. [CrossRef] [PubMed]

68. Wu, H.; Chen, H.; Shiau, C. Free amino acids and peptide as related to antioxidant properties in protein hydrolysates of mackerel (Scomber austriasicus). *Food Res. Int.* **2003**, *36*, 949–957. [CrossRef]

69. Guo, W.; Adachi, T.; Matsui, R.; Xu, S.; Jiang, B.; Zou, M.; Kirber, M.; Lieberthal, W.; Cohen, R.A. Quantitative assessment tyrosine nitration of manganese superoxide dismutase in angiotensin II-infused rat kidney. *Am. J. Physiol. Heart Circ. Physiol.* **2003**, *285*, H1396–H1403. [CrossRef]

70. Maritim, A.C.; Sanders, R.A.; Watkins, J.B. Diabetes, oxidative stress an antioxidants: A review. *J. Biochem. Mol. Toxicol.* **2002**, *17*, 24–38. [CrossRef]

71. Li, Y.; Huang, T.; Carlson, E.J.; Melov, S.; Ursell, P.C.; Olson, J.L.; Noble, L.J.; Yoshimura, M.P.; Berger, C.; Chan, P.H.; et al. Dilated cardiomyopathy and neeonatal lethality in mutant mice lacking manganese superoxide dismutase. *Nat. Genet.* **1995**, *11*, 376–381. [CrossRef]

72. Nishikawa, T.; Edelstein, D.; Du, X.L.; Yamagishi, S.; Matsumura, T.; Kaneda, Y.; Yorek, M.A.; Beebe, D.; Oates, P.J.; Hammes, H.; et al. Normalizing mitochondrial superoxide production blocks three pathways of hyperglycemic damage. *Nature* **2000**, *404*, 787–790. [CrossRef]

73. Kowluru, R.A.; Kowluru, V.; Xiong, Y.; Ho, Y. Overexpression of mitochondrial superoxide dismutase in mice protects the retina from diabetes-induced oxidative stress. *Free Radic. Biol. Med.* **2006**, *41*, 1191–1196. [CrossRef] [PubMed]

74. Shen, X.; Zheng, S.; Metreveli, N.S.; Epstein, P.N. Protection of cardiac mitochondria by overexpression of MnSOD reduces diabetic cardiomyopathy. *Diabetes* **2006**, *55*, 798–805. [CrossRef] [PubMed]

75. Cocchia, N.; Pasolini, M.P.; Mancini, R.; Petrazzuolo, O.; Cristofaro, I.; Rosapane, I.; Sica, A.; Tortora, G.; Lorizio, R.; Paraggio, G.; et al. Effect of SOD (superoxide dismutase) protein supplementation in semen extenders on motility, viability, acrosome status and ERK (extracellular signal-regulated kinase) protein phosphorylation of chilled stallion spermatozoa. *Theriogenology* **2011**, *75*, 1201–1210. [CrossRef] [PubMed]

76. Cox, A.G.; Winterbourn, C.; Hampton, M.B. Mitochondrial peroxiredoxin involvement in antioxidant defence and redox signaling. *Biochem. J.* **2010**, *425*, 313–325. [CrossRef]

77. Knoops, B.; Goemaere, J.; Eecken, V.V.D.; Declercq, J. Peroxiredoxin 5: Structure, mechanism and function of the mammalian atypical 2-Cys Peroxiredoxin. *Antioxid. Redox Sign.* **2011**, *15*, 817–829. [CrossRef]

78. Kubo, E.; Singh, D.P.; Fatma, N.; Akagi, Y. TAT-mediated peroxiredoxin 5 and 6 protein transduction protects against high glucose-induced cytotoxicity in retinal pericytes. *Life Sci.* **2009**, *84*, 857–864. [CrossRef]

79. O'Flaherty, C. Peroxiredoxins: Hidden players in the antioxidant defence of human spermatozoa. *Basic Clin. Androl.* **2014**, *24*, 1–10. [CrossRef]

80. Chow, C.Y.; Avila, F.W.; Clark, A.G.; Wolfner, M.F. Induction of excessive endoplasmic reticulum stress in the drosophila male accessory gland results in infertility. *PLoS ONE* **2015**, *10*, e0119386. [CrossRef]

81. Guzel, E.; Arlier, S.; Guzeloglu-Kayisli, O.; Tabak, M.S.; Ekiz, T.; Semerci, N.; Larsen, K.; Schatz, F.; Lockwood, C.J.; Kayisli, U.A. Endoplasmic reticulum stress and homeostasis in reproductive physiology and pathology. *Int. J. Mol. Sci.* **2017**, *18*, 792. [CrossRef]

82. Ye, Y.; Tang, W.K.; Zhang, T.; Xia, D. A mighty "Protein Extractor" of the cell: Structure and function of the p97/CDC48 ATPase. *Front. Mol. Biosci.* **2017**, *4*, 1–20. [CrossRef]

83. Sitia, R.; Braakman, I. Quality control in the endoplasmic reticulum protein factory. *Nature* **2003**, *426*, 891–894. [CrossRef] [PubMed]

84. Wojcik, C.; Rowicka, M.; Kudlicki, A.; Nowis, D.; McConnell, E.; Kujawa, M.; DeMartino, G.N. Endoplasmic reticulum stress and of the degradation of N-end rule and ubiquitin-fusion degradation pathway substrates in mammalian cells. *Mol. Biol. Cell.* **2006**, *17*, 4606–4618. [CrossRef] [PubMed]

85. Ni, R.; Zheng, D.; Xiong, S.; Hill, D.J.; Sun, T.; Gardiner, R.B.; Fan, G.; Lu, Y.; Abel, E.D.; Greer, P.A.; et al. Mitochondrial calpain-1 disrupts ATP synthase and induces superoxide generation in type 1 diabetic hearts: A novel mechanism contributing to diabetic cardiomyopathy. *Diabetes* **2016**, *65*, 255–267. [CrossRef] [PubMed]

86. Qi, W.; Keenan, H.A.; Li, Q.; Ishikado, A.; Kannt, A.; Sadowski, T.; Yorek, M.A.; Wu, I.; Lockhart, S.; Coppey, L.J.; et al. Pyruvate kinase M2 activation may protect against the progression of diabetic glomerular pathology and mitochondrial dysfunction. *Nat. Med.* **2017**, *23*, 753–762. [CrossRef] [PubMed]

87. Rahimi, Y.; Camporez, J.G.; Petersen, M.C.; Pesta, D.; Perry, R.J.; Jurczak, M.J.; Cline, G.W.; Shulman, G.I. Genetic activation of pyruvate dehydrogenase alters oxidative substrate selection to induce skeletal muscle insulin resistance. *Proc. Natl. Acad. Sci. USA* **2014**, *111*, 16508–16513. [CrossRef] [PubMed]

88. Lin, Y.; Yeh, T.; Chen, S.; Tsai, Y.; Chou, C.; Yang, Y.; Huang, H. Nonmuscle Myosin IIA (Myosin Heavy Polypeptide 9): A novel class of signal transducer mediating the activation of Gαh/Phospholipase C-δ1 pathway. *Endocrinology* **2010**, *151*, 876–885. [CrossRef]

89. Adams, I.R.; McLaren, A. Sexually dimorphic development of mouse primordial germ cells: switching from oogenesis to spermatogenesis. *Development* **2002**, *129*, 1155–1164.

90. Moniot, B.; Ujjan, S.; Champagne, J.; Hirai, H.; Aritake, K.; Nagata, K.; Dubois, E.; Nidelet, S.; Nakamura, M.; Urade, Y.; et al. Prostaglandin D2 acts through the Dp2 receptor to influence male germ cell differentiation in the foetal mouse testis. *Development* **2014**, *141*, 3561–3571. [CrossRef]

91. Moniot, B.; Declosmenil, F.; Barrionuevo, F.; Scherer, G.; Aritake, K.; Malki, A.; Marzi, L.; Cohen-Solal, A.; Georg, I.; Klattig, J.; et al. The PGD2 pathway, independently of FGF9, amplifies SOX9 activity in Sertoli cells during male sexual differentiation. *Development* **2009**, *136*, 1813–1821. [CrossRef]

92. Samy, E.T.; Li, J.C.H.; Grima, J.; Lee, W.M.; Silvestrini, B.; Cheng, C.Y. Sertoli cell Prostaglandin D2 Synthetase is a mulitifunctional molecule: Its expression and regulation. *Endocrinology* **2000**, *141*, 710–721. [CrossRef]

93. Gelato, K.A.; Fischle, W. Role of histone modifications in defining chromatin structure and function. *J. Biol. Chem.* **2008**, *389*, 353–363. [CrossRef] [PubMed]

94. Lu, S.; Xie, Y.M.; Li, X.; Luo, J.; Shi, X.Q.; Hong, X.; Pan, Y.H.; Ma, X. Mass spectrometry analysis of dynamic post-translational modifications of TH2B during spermatogenesis. *Mol. Hum. Reprod.* **2009**, *15*, 373–378. [CrossRef] [PubMed]

95. King, S.M. The dynein microtubule motor. *BBA* **2000**, *1496*, 60–75. [CrossRef]

96. Wang, R.; Zhao, M.; Meistrich, M.L.; Kumar, R. Stage-specific expression of Dynein Light Chain-1 and its interacting Kinase, p21-activated Kinase-1, in rodent testes: Implications in spermatogenesis. *J. Histochem. Cytochem.* **2005**, *53*, 1235–1243. [CrossRef] [PubMed]

97. Chen, J.; Wang, Y.; Wei, B.; Lai, Y.; Yan, Q.; Gui, Y.; Cai, Z. Functional expression of Ropporin in human testis and ejaculated spermatozoa. *J. Androl.* **2011**, *32*, 26–32. [CrossRef] [PubMed]

98. Fiedler, S.E.; Sisson, J.H.; Wyatt, T.A.; Pavlik, J.A.; Gambling, T.M.; Carson, J.L.; Carr, D.W. Loss of ASP but not ROPN1 reduces mammalian ciliary motility. *Cytoskeleton* **2012**, *69*, 22–32. [CrossRef] [PubMed]

99. Kubota, H.; Hynes, G.; Willison, K. The chaperonin containing t-complex polypeptide 1 (TCP-1): Multisubunit machinery assisting in protein folding and assembly in the eukaryotic cytosol. *Eur. J. Biochem.* **1995**, *230*, 3–16. [CrossRef]

100. Redgrove, K.A.; Anderson, A.L.; Dun, M.D.; McLaughlin, E.A.; O'Bryan, M.K.; Aitken, R.J.; Nixon, B. Involvement of multimeric protein complexes in mediating the capacitation-dependent binding of human spermatozoa to homologous zonae pellucidae. *Dev. Biol.* **2011**, *356*, 460–474. [CrossRef]

101. Dun, M.D.; Smith, N.D.; Baker, M.A.; Lin, M.; Aitken, R.J.; Nixon, B. The chaperoning containing TCP1 complex (CCT/TRiC) is involved in mediating sperm-oocyte interaction. *J. Biol. Chem.* **2011**, *286*, 36875–36887. [CrossRef]

102. Panneerdoss, S.; Siva, A.B.; Kameshwari, D.B.; Rangaraj, N.; Shivaji, S. Association of lactate, intracellular pH and intracellular calcium during capacitation and acrosome reaction: Contribution of hamster sperm Dihydrolipoamide dehydrogenase, the E3 subunit of Pyruvate dehydrogenase complex. *J. Androl.* **2012**, *33*, 699–710. [CrossRef]

103. Kumar, V.; Kota, V.; Shivaji, S. Hamster sperm capacitation: Role of Pyruvate dehydrogenase A and Dihydrolipoamide dehydrogenase. *Biol. Reprod.* **2008**, *79*, 190–199. [CrossRef] [PubMed]

104. Johnson, M.T.; Yang, H.; Magnuson, T.; Patel, M.S. Targeted disruption of the murine dihydrolipoamide dehydrogenase gene (Dld) results in perigastrulation lethality. *Proc. Natl. Acad. Sci. USA* **1997**, *94*, 14512–14517. [CrossRef] [PubMed]

105. Mitra, K.; Shivaji, S. Novel tyrosine-phosphorylated post-pyruvate metabolic enzyme, Dihydrolipoamide dehydrogenase, involved in capacitation of hamster spermatozoa. *Biol. Reprod.* **2004**, *70*, 887–899. [CrossRef] [PubMed]

106. Inoue, N.; Ikawa, M.; Isotani, A.; Okabe, M. The immunoglobulin superfamily protein Izumo is required for sperm to fuse with eggs. *Nature* **2005**, *434*, 234–238. [CrossRef] [PubMed]

107. Bianchi, E.; Doe, B.; Goulding, D.; Wright, G.J. Juno is the egg Izumo receptor and is essential for mammalian fertilization. *Nature* **2014**, *508*, 483–487. [CrossRef] [PubMed]

 *processes*

Article

Nutritional Quality and Physico-Chemical Characteristics of Selected Date Fruit Varieties of the United Arab Emirates

Krishnamoorthy Rambabu [1,*], Govindan Bharath [1], Abdul Hai [1], Fawzi Banat [1,2,*], Shadi W. Hasan [1,2], Hanifa Taher [1] and Hayyiratul Fatimah Mohd Zaid [3]

[1]   Department of Chemical Engineering, Khalifa University of Science and Technology, P.O. Box 127788 Abu Dhabi, UAE; bharath.govindan@ku.ac.ae (G.B.); abdul.hai@ku.ac.ae (A.H.); shadi.hasan@ku.ac.ae (S.W.H.); hanifa.alblooshi@ku.ac.ae (H.T.)
[2]   Center for Membranes and Advanced Water Technology (CMAT), Khalifa University, P.O. Box 127788 Abu Dhabi, UAE
[3]   Fundamental and Applied Sciences Department, Centre of Innovative Nanostructures & Nanodevices (COINN), Institute of Autonomous System, Universiti Teknologi PETRONAS, Bandar Seri Iskandar 32610, Malaysia; hayyiratul.mzaid@utp.edu.my
*   Correspondence: replyram123@gmail.com or rambabu.krishnamoorthy@ku.ac.ae (K.R.); fawzi.banat@ku.ac.ae (F.B.)

Received: 5 February 2020; Accepted: 19 February 2020; Published: 25 February 2020

Abstract: Production of highly soluble date sugar powder from the nutritive date fruits will be a suitable and superior alternative to commercial refined sugar, providing sustainability in date palm cultivation. A good understanding of the nutritional and phytochemical composition of date fruits is imperative for this purpose. In this work, 11 different date fruit species commonly cultivated in the United Arab Emirates were studied for their chemical composition, physical properties, amino acids, minerals, and anti-nutritional contents. The results revealed that the date fruits contain moisture, protein, lipid, and ash content in the ranges of 14.8%–20.5%, 2.19%–3.12%, 0.25%–0.51%, and 1.37%–1.97%, respectively. Potassium was identified as the major microelement in all the date varieties. Amino acid assay depicted that the date fruits mainly contained glutamine and aspartic acids, along with other essential acids. Monosaccharides (glucose and fructose) were more prevalent in the date fruits than polysaccharides (sucrose), exhibiting the potential of date fruit for non-diabetic sugar production. Phytoconstituents present in date samples, such as flavonoids, oxalates, tannins, saponins, alkaloids, and cyanides, were also evaluated and reported. Results showed that although all date fruit varieties were nutritious, they contain significant variation in their nutritional, physical, elemental, and phytochemical properties.

Keywords: date fruits; proximate analysis; physico-chemical characteristics; date sugar; phytoconstituents; amino acids; biomaterials

---

1. Introduction

The date palm (*Phoenix dactylifera*) fruit is the major staple food and the primary source of agricultural wealth in the United Arab Emirates (UAE) [1]. Although the exact origin of the fruit is still unknown, it is believed that date palm cultivation originated in the southern parts of modern Iraq about 6000 years ago [2]. The good adaptability, storage durability, and high food value of the date fruit has led to the spread of date palm cultivation in other parts of the world, especially in arid and semi-arid conditions. Today, the tree is cultivated as one of the major crops in south central Asia, northern parts of Africa, parts of Europe, and in California and Arizona states in the United States of America [3]. The total production of date fruit around the world was more than 7.5 million tons in

2017 [4], with an equivalent utilization rate. Egypt (1.08 million tons), Iran (0.95 million tons), Saudi Arabia (1.08 million tons), Iraq (1.08 million tons), and the United Arab Emirates (1.08 million tons) are the five major growing countries in the world, which contribute 54% of the total production of date fruit globally [5]. Date palm cultivation in the UAE has experienced tremendous growth in last two decades, with over 250 varieties of crop being farmed in various part of the nation, especially in Al Ain and Abu Dhabi emirates. The total growth rate of date fruit is paralleled by a high consumption rate in the UAE, with a per capita daily intake of 10 to 200 g in Abu Dhabi emirate alone [6].

Research on date fruit has shown that the fruit has a wide range of uses and applications. Primarily, the fruit is seen as a rich source of energy, minerals, and vitamins for human health [7]. Additionally, the fruit is also administrated as a medicinal cure for cancer and various other infectious diseases [8]. The high anti-oxidant and antimutagenic contents present in date fruit makes it an ideal source for the production of a wide range of confectionary, medicinal, and cosmetic products [9]. In recent times, date fruits have been critically analyzed for the production of fruit sugar, which is seen as a very healthy and nutritious substitute for commercial refined sugar [10]. The date seeds obtained from the date fruit have also been widely studied for the production of a variety of potential useful products, such as activated carbon, bio-oil production, cellulose separation, biohydrogen synthesis, and water treatment [11–15].

Although several works have been carried out on date fruit, research studies concerned with the nutritional and physico-chemical characteristics of the fruit are very limited. Especially, key information on various minerals, amino acids, and the anti-nutritional content present in date fruits is not available for various date varieties found in the UAE. This presents a considerable research gap, especially regarding the cultivars, industries, and consumers of the dates, in understanding the complete potential of the fruit. Additionally, a comparative study of the nutritional assay of date varieties for sugar, mineral, and amino acid contents, along with their physical properties, would provide pivotal information for production of soluble fruit sugar and other related food products for the industry. Thus, this paper attempts to report the nutritional and physico-chemical properties of the 11 abundantly grown date fruit varieties in the UAE. Various critical characterizations, such as proximate analysis, total sugar estimation, mineral composition, amino acid quantification, and anti-nutritional assessment, were performed on the fruits. The physical properties of the date varieties, including the fruit weight, were also studied.

2. Materials and Methods

2.1. Sample Collection and Preparation

Eleven date palm fruit varieties, namely Barhe, Bumaan, Dabbas, Fard, Jabri, Khalas, Lulu, Maktoomi, Raziz, Shiakt, and Shishi, were directly procured from the cultivation sites of Abu Dhabi and Al Ain farmlands of the UAE. All the samples were at the "Tamr stage", which is the final phase of the fruit growth. The as-received samples were thoroughly rinsed with distilled water, followed by air drying at ambient temperature. The dried palm fruits were then pitted, segmented, crushed into paste, and stored in air-tight containers for subsequent analysis.

2.2. Physico-Chemical and Proximate Analysis

Physical characteristics, such as mass, volume, density, and fruit content, of the date palm samples were analyzed by randomly selecting fifteen fruits from each variety. The mass (g) of the individual samples was analyzed using digital mass balance, the volume (mL) of each sample was determined by water displacement method, and density (g/mL) was calculated as mass by volume ratio. For fruit content analysis, the mass of thirty date fruits (with and without seeds) from each variety were measured. The fruit content (%) was measured by the following Equation (1).

$$Wt.\ of\ Fruit\ content\ (\%) = \frac{Wt.\ of\ date\ fruit\ with\ seeds - Wt.\ of\ date\ fruit\ with\ seeds)}{Wt.\ of\ date\ fruit\ with\ seeds} \tag{1}$$

Proximate analysis (moisture content, protein, lipid, and ash contents) of date fruits were carried out in accordance to the standard analytical procedure for food analysis, as reported by Shaba et al. [16]. Generally, for moisture content analysis, a pre-weighed crucible was filled with 2 grams of the sample and kept in an oven at 105 °C overnight. The dried samples were weighed again and the difference in sample weight was expressed as percentage moisture content. However, for ash content analysis, the same procedure was repeated by keeping the moisture-free samples in a muffle furnace at 600 °C for 3 h. Additionally, the lipid content of date fruits was determined by extracting the samples in petroleum ether at 60–80 °C for 5 h using a soxhlet extractor. The Kjeldahl method was implied to determine the total protein content in date palm samples. For this purpose, 0.5 g of the sample was digested in a concentrated H_2SO_4 solution, followed by distillation and titration with 40% NaOH and 0.01 M HCl solutions, respectively.

2.3. Total Sugar Content

The total and reducing sugar content in different varieties of date fruits were analyzed according to the standard procedure reported by Ismail et al. [7]. Generally, 20 g of the fruit sample was homogenized in boiling water for 90 s. The resultant suspension was then diluted by deionized water followed by filtration. The sample was then examined and compared with the calibration curves of the two standard solutions (fructose and glucose) by utilizing a Waters High Performance Liquid Chromatography (HPLC) system (Waters 717 Plus Autosampler, Milford, NH, USA). The same method was repeated for all the date varieties examined in this research work.

2.4. Determination of Mineral Contents

The mineral contents, namely potassium (K), phosphorus (P), magnesium (Mg), calcium (Ca), sodium (Na), and iron (Fe), in the date samples were determined as reported by Heckman et al. [17]. A known amount of date fruit sample was heated in a furnace at 550 °C for 4 h. The cooled sample was then boiled with 3N HCl solution for 10 min, followed by cooling and filtration. The mineral contents in the resultant solution were then tested by using inductively coupled plasma mass spectrometry (ICP-MS 7900, Agilent Technologies, Waldbron, Germany).

2.5. Amino Acid Analysis

The relative distribution of essential amino acids present in the date fruits was measured by oxidation followed by hydrolysis using hydrogen peroxide/formic acid/phenol and 6 M hydrogen peroxide solution, respectively, as described by Al-Barnawi [18]. The amino acids were then separated and analyzed by ion-exchange chromatography and photometric detection (440 and 570 nm) using ninhydrin reagent, respectively [18].

2.6. Quantitative Determination of Phytoconstituents

To ascertain the composition of anti-nutritional constituents, an aqueous solution containing 20 wt % date fruit sample was boiled at 95 °C for 30 min. After filtering the resultant suspension, the solution was treated with different solvents, such as isoamyl-alcohol, H_2SO_4, $FeCl_3$, HCl + NaOH, and hydrogen cyanic acid, to measure the various phytoconstituents, namely flavonoids, oxalates, tannins, saponins, and cyanides, respectively. After titration (color development), the resultant solutions were tested by spectrophotometer at 490 nm wavelength. Moreover, the alkaline precipitation gravimetric method was employed to quantitatively analyze the alkaloid content in the date fruits. Shaba et al. [16] and Selmani et al. [19] have presented the detailed analytical procedure for determining the phytoconstituents in the *Phoenix dactylifera L.* pollen and fruits, respectively.

2.7. Statistical Analysis

Experimental results were statistically analyzed through Tukey's multiple test and analysis of variance (ANOVA) modes using Minitab software (Minitab Inc., State College, PA, USA). The significance level was set to $p < 0.05$ for the analysis.

3. Results

3.1. Physico-Chemical and Proximate Analysis of Date Fruits

The physical properties, such as mass (g), volume (mL), density (g/mL), and fruit content (%), of the date fruits are significant factors in describing their diet suitability and storage. Figure 1A depicts the physical properties of various date fruit varieties cultivated in the UAE. Jabri (10.49 g) and Buman (9.78 g) recorded the highest and lowest masses for the studied date selections in the "Tamr" stage. Maximum and minimum fruit volumes were measured as 11.43 mL and 4.35 mL for Maktoomi and Dabbas breeds, respectively. The results also revealed that the density of date fruits varied in the range of 0.8 g/mL (Lulu) to 1.43 g/mL (Raziz). Moreover, the percentage of fruit content was also analyzed for the studied *Phoenix Dactylifera* species. Their variation is illustrated in Figure 1B. It can be observed that the percentages of fruit content differed between 88.51% (Raziz) and 92.13% (Fard). The differences in physical properties for various date species depend on the geographical and cultivation conditions, seed nature, chemical fertilization, and pre- and post-harvesting treatment techniques.

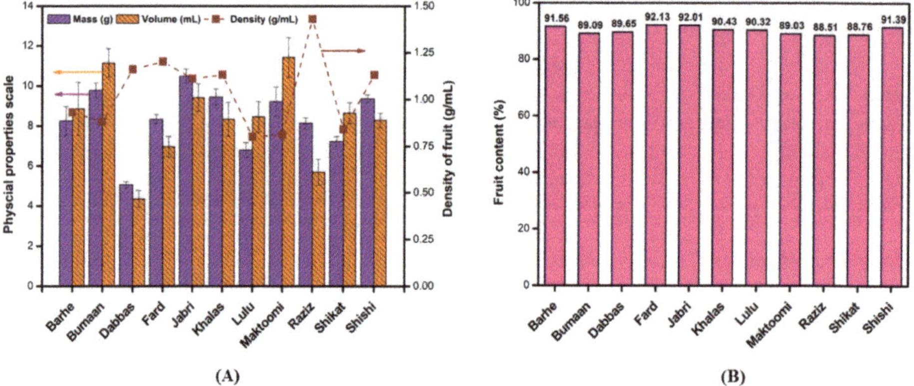

Figure 1. Physical properties of date fruit varieties of the UAE: (**A**) mass, volume, and density of the fruit; (**B**) fruit content of the dates.

The results for proximate analysis of the 11 different varieties of date fruits cultivated in the UAE are presented in Table 1. The results revealed that among various date fruit samples, Maktoomi and Khalas exhibited the lowest (13.6%) and highest (20.5%) moisture content, respectively. Moreover, the protein and lipid contents for the fruit varieties ranged from 2.19% (Jabri) to 3.12% (Khalas) and from 0.25% (Raziz) to 0.51% (Shiakt), respectively. Shishi and Fard reported the minimum (1.37%) and maximum (1.97%) ash content, respectively. No noticeable variation was observed in ash or lipid contents for Barhe, Bumaan, Fard, Khalas, Maktoomi, and Raziz dates. In summary, for all the studied date varieties, slight deviations in the protein, lipid, and ash contents were observed compared to moisture content. The variation in the proximate analysis of various date fruit samples might be because of differences in geographic locations and agro-climatic and environmental conditions.

Table 1. Proximate analysis for the date fruit varieties of the UAE.

Date Variety	Moisture *	Protein *	Lipid *	Ash *
Barhe	16.4 ± 0.1 [a]	2.73 ± 0.04 [a]	0.29 ± 0.005 [a]	1.96 ± 0.02 [a]
Bumaan	17.7 ± 0.2 [b]	2.36 ± 0.02 [b]	0.38 ± 0.002 [b]	1.77 ± 0.01 [b]
Dabbas	19.5±0.3 [c]	2.54±0.01 [c,d]	0.41±0.004 [c]	1.64 ± 0.02 [c]
Fard	18.1 ± 0.2 [b]	2.89 ± 0.03 [e]	0.33 ± 0.001 [d]	1.97 ± 0.03 [a]
Jabri	15.4 ± 0.4 [d]	2.19 ± 0.02 [f]	0.48 ± 0.002 [e]	1.53 ± 0.02 [d]
Khalas	20.5 ± 0.3 [e]	3.12 ± 0.05 [g]	0.31 ± 0.003 [f]	1.72 ± 0.01 [b,e]
Lulu	16.9 ± 0.1 [a]	2.63 ± 0.02 [h,i]	0.44 ± 0.005 [g]	1.68 ± 0.05 [c,e]
Maktoomi	13.6 ± 0.2 [f]	2.58 ± 0.03 [c,i]	0.37 ± 0.003 [h]	1.85 ± 0.03 [f]
Raziz	15.3 ± 0.2 [d]	2.67 ± 0.02 [a,h]	0.25 ± 0.001 [i]	1.87 ± 0.04 [f]
Shikat	13.8 ± 0.1 [f]	2.46 ± 0.01 [d]	0.51 ± 0.000 [j]	1.64 ± 0.01 [c]
Shishi	14.8 ± 0.3 [d]	2.53 ± 0.04 [c,d]	0.35 ± 0.002 [k]	1.37 ± 0.02 [g]

* Results are presented as g/100 g of fruit flesh's weight; mean values superscripted with different alphabets within the same column are significantly different ($p < 0.05$), as established by Tukey's test.

3.2. Sugar Content in Date Fruits

Sugar is the main component of date fruits and is present mainly in the form of monosaccharides (glucose and fructose) and polysaccharides (sucrose). Figure 2 illustrates the amount of total and reducing sugars analyzed in 11 different date varieties of the UAE. It can be inferred that a maximum of 90.5% total sugar content was present in Maktoomi dates, while Fard dates contained the minimum sugar content of 71.8%, as compared to other varieties. The compositions of sucrose, glucose, and fructose, along with the glucose to fructose ratio (G/F), are summarized in Table 2. It can be observed that the major components of date sugar were glucose and fructose, whereas a relatively small amount of sucrose was also found in all of the examined dates. Interestingly, Barhe had the lowest sucrose content and higher fructose content than glucose, as evident from its low G/F ratio of 0.88. In contrast, some dates, such as Dabbas, Khalas, Lulu, and Maktoomi, showed a higher concentration of glucose than fructose, with G/F values > 1. Overall, the G/F ratios of all analyzed date types varied in the range of 0.88 (Barhe) to 1.33 (Maktoomi).

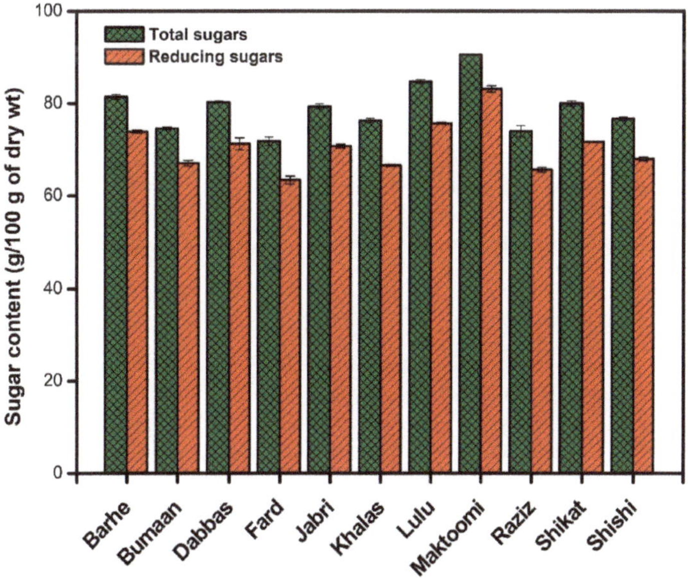

Figure 2. Sugar content of date fruit varieties of the United Arab Emirates (UAE).

Table 2. Sucrose, glucose, and fructose contents for the date fruit varieties of the UAE.

Date Variety	Sucrose *	Glucose(G) *	Fructose (F) *	G/F Ratio
Barhe	4.2 ± 0.1 [a]	34.6 ± 0.1 [a,b]	39.3 ± 0.2 [a]	0.88
Bumaan	4.4 ± 0.3 [a]	32.3 ± 0.1 [c]	34.8 ± 0.1 [b,c]	0.93
Dabbas	6.4 ± 0.7 [b,c,d]	36.3 ± 0.5 [d]	35 ± 0.2 [b,d]	1.04
Fard	5.2 ± 0.2 [a,e,f,g]	31.3 ± 0.2 [c]	32.1 ± 0.2 [e]	0.98
Jabri	5.8 ± 0.1 [c,d,e,f]	35.2 ± 0.2 [a]	35.6 ± 0.3 [f]	0.99
Khalas	6.1 ± 0.5 [b,c,d,e]	33.4 ± 0.3 [e]	33.2 ± 0.1 [g]	1.01
Lulu	6.7 ± 0.3 [b,c]	40.2 ± 0.5 [f]	35.5 ± 0.1 [d,f]	1.13
Maktoomi	6.9 ± 0.1 [b]	47.4 ± 0.4 [f]	35.7 ± 0.2 [f]	1.33
Raziz	4.5 ± 0.1 [a,g]	31.8 ± 0.7 [c]	33.9 ± 0.1 [h]	0.94
Shikat	4.9 ± 0.2 [a,f,g]	35.1 ± 0.4 [a]	36.6 ± 0.3 [i]	0.96
Shishi	5.5 ± 0.6 [d,e,f,g]	33.7 ± 0.2b [e]	34.3 ± 0.2 [c]	0.98

* Results are presented as g/100 g of fruit flesh weight; mean values superscripted with different alphabets within the same column are significantly different ($p < 0.05$), as established by Tukey's test.

It is worth mentioning the fact that the diverse variety of dates or the same variety cultivated in different regions of the world may have different G/F ratios based on their geographical and environmental aspects in addition to cultivation conditions. Convincingly, Barhe dates of UAE contained significantly higher levels of total sugar (81.4%) with maximum fructose (39.3%) and minimum G/F ratio of among other stated varieties. Hence, this date variety is one of the ideal date fruit breed for solid date sugar production.

3.3. Mineral Composition of Date Flesh

Commonly, date fruits possess substantial amount of essential micronutrients necessary for human health. Concentration of various essential micronutrients such as K, P, Mg, Ca, Na and Fe in the examined date variants are displayed in Table 3. Results showed that potassium was the major mineral element present in all date varieties. The date selections contained potassium, phosphorus, magnesium, calcium and sodium in the range of 281.74 (Bumaan)–478.29 (Lulu), 48.36 (Dabbas)–77.94 (Raziz), 42.17 (Dabbas)–70.38 (Barhe), 15.46 (Bumaan)–42.39 (Barhe) and 6.25 (Bumaan)–17.52 (Maktoomi) mg/100 g of fruit flesh (FF), respectively. Iron, an essential trace element, was present in good levels in all date breeds with a maximum (1.51 mg/100 g of FF) in Raziz and minimum (0.78 mg/100 g of FF) in Dabbas. Interestingly, Barhe dates comprised noticeably virtuous amount of all the vital minerals such as K (444.69 mg/100 g of DF), P (65.84 mg/100 g of DF), Mg (70.38 mg/100 g of DF) and Ca (42.39 mg/100 g of DF) compared to other date species. In general, potassium and phosphorous elements impart strength for human cells regeneration, while magnesium and calcium are vital for healthy bones development [8,20]. Iron is very critical for blood production and tissue respiration [6]. Hence, due to the presence of these essential elements, date fruits are considered to be healthy diet containing a rich source of minerals and play a pivotal role in development of the immune system for humans.

3.4. Amino Acid Composition of Date Flesh

Date fruit comprises various amino acids essential for human metabolism, especially for cell growth and regeneration [16]. In the present study, the date varieties of UAE were assessed for the composition of essential amino acids (glutamine, aspartic acid, glycine, proline, histidine, and valine) present in them. Figure 3 presents the results of the amino acids assessment for different varieties of UAE dates on a relative percentage scale for comparative quantification. The results revealed that glutamine was the major amino acid found in all date varieties, except for Bumaan and Shikat. Additionally, it was observed that all date selections contained high levels of glutamine and aspartic acid (>50% composition) as compared to other essential amino acids. Histidine was present at the lowest concentration in all date varieties. These essential amino acids present in the date fruits are vital in muscle building and generation of red and white blood cells in humans [21]. The consumption of date fruits overcomes the problems caused by malnutrition, and serves as a building block for protein and energy regulation in humans [22,23].

Table 3. Composition of various mineral content in the date fruits of UAE.

Date Variety	Ca *	P *	Na *	K *	Mg *	Fe *
Barhe	42.39 ± 0.84 [a]	65.84 ± 0.14 [a]	14.73 ± 0.09 [a]	444.69 ± 3.63 [a]	70.38 ± 0.59 [a]	1.37 ± 0.05 [a]
Bumaan	15.46 ± 0.57 [b]	59.21 ± 0.65 [b]	6.25 ± 0.14 [b]	281.74 ± 2.18 [b]	55.92 ± 0.31 [b]	1.05 ± 0.03 [b]
Dabbas	35.71 ± 1.27 [c]	48.36 ± 0.65 [c]	14.38 ± 0.32 [a,c]	419.05 ± 6.11 [c]	42.17 ± 1.44 [c]	0.78 ± 0.02 [c]
Fard	34.23 ± 0.79 [c,d]	55.04 ± 1.27 [d]	11.56 ± 0.17 [d]	466.28 ± 4.50 [d,e]	61.56 ± 0.62 [d]	0.84 ± 0.01 [c,d]
Jabri	29.48 ± 1.44 [e]	71.45 ± 0.38 [e]	12.12 ± 0.58 [d]	398.51 ± 7.29 [f]	50.97 ± 0.47 [e]	0.93 ± 0.03 [d,e,f]
Khalas	40.25 ± 0.39 [a]	52.19 ± 0.47 [f]	8.41 ± 0.07 [e]	431.17 ± 1.63 [c]	46.28 ± 0.83 [f]	1.42 ± 0.04 [a,g]
Lulu	36.19 ± 0.65 [c]	66.82 ± 0.77 [a]	13.78 ± 0.26 [c]	478.29 ± 5.34 [d]	50.56 ± 0.36 [e]	1.26 ± 0.02 [h]
Maktoomi	30.14 ± 1.08 [e]	74.17 ± 1.36 [g]	17.52 ± 0.43 [f]	424.81 ± 4.93 [c]	67.53 ± 0.19 [g]	0.89 ± 0.01 [d,f]
Raziz	53.82 ± 0.36 [f]	77.94 ± 0.82 [h]	9.26 ± 0.11 [g]	462.42 ± 2.72 [e,g]	42.35 ± 0.85 [c]	1.51 ± 0.06 [g]
Shikat	32.67 ± 0.51 [d]	65.56 ± 0.96 [a]	11.41 ± 0.05 [d]	450.96 ± 3.26 [a,g]	62.41 ± 0.09 [d,h]	1.00 ± 0.04 [b,e]
Shishi	30.11 ± 0.86 [e]	67.28 ± 0.30 [a]	14.87 ± 0.37 [a]	375.33 ± 0.98 [h]	64.01 ± 0.24 [h]	0.97 ± 0.03 [b,e,f]

* Results are presented as mg/100 g of fruit flesh's weight; Mean values superscripted with different alphabets within the same column are significantly different ($p < 0.05$) as established by Tukey's test.

3.5. Anti-Oxidant and Anti-Nutritional Assessment of Date Fruits

The nutritional aspects of date fruits could also be expressed in terms of phytoconstituents [24]. Measurement of flavonoids indicates the anti-oxidant capacity, while concentrations of oxalates, tannins, saponins, alkaloids, and cyanides highlight the anti-nutritional nature of the fruit. The results for the anti-oxidant and anti-nutritional contents of various varieties of *Phoenix dactylifera* fruits cultivated in UAE are presented in Table 4. The oxalate content in the date samples ranged from 0.26% to 1.18% for Maktoomi and Raziz, respectively. Among the different fruit varieties examined, Barhe, Fard, Lulu, Shikat, and Shishi had imperceptible oxalate content, while the remaining samples had negligible oxalate content, indicating that even the excessive utilization of these date samples has no deleterious impact on the human body. The flavonoid and alkaloid contents in the dates ranged between 24.33% (Shikat) and 54.26% (Barhe), and between 0.09% (Dabbas) and 1.28% (Lulu), respectively. The high levels of flavonoids in Barhe variety exhibited their better anti-oxidant capacity as compared to other date varieties. The cyanide, saponin, and tannin contents in the date samples were in the ranges of 0.0%1–0.04%, 0.02%–1.06%, and 0.19%–0.92%, respectively. Overall, the composition of the anti-nutrients was very low in all fruit types, ensuring their safety and health benefits for consumption. Remarkably, the Barhe fruit possessed high levels of flavonoids and extremely low levels of anti-nutrients, which makes it the best date sample amongst all other studied varieties.

Table 4. Anti-nutritional assessment of date fruit varieties of the UAE.

Date Variety	Flavanoid *	Oxalate *	Tannin *	Saponin *	Alkaloid *	Cyanide *
Barhe	54.26 ± 2.17 [a]	-	-	0.05 ± 0.00 [a,b]	0.93 ± 0.03 [a]	-
Bumaan	36.37 ± 1.63 [b]	1.09 ± 0.03 [a,b]	-	0.22 ± 0.04 [a,c]	0.64 ± 0.05 [b]	0.02 ± 0.00 [a,b]
Dabbas	46.35 ± 0.99 [c]	0.87 ± 0.19 [b,c]	0.52 ± 0.03 [a]	-	0.09 ± 0.01 [c]	-
Fard	29.56 ± 0.87 [d,e]	-	0.78 ± 0.12 [b,c]	0.39 ± 0.06 [c,d]	0.57 ± 0.04 [b]	-
Jabri	37.21 ± 0.32 [b,f]	0.43 ± 0.06 [d,e]	0.19 ± 0.02 [d]	1.06 ± 0.14 [e]	-	0.04 ± 0.01 [c]
Khalas	45.18 ± 0.72 [c]	0.59 ± 0.05 [c,d]	-	0.16 ± 0.01 [a,b]	-	0.04 ± 0.01 [c]
Lulu	32.49 ± 0.55 [d,g]	-	0.64 ± 0.05 [a,c]	-	1.28 ± 0.14 [d]	0.01 ± 0.00 [b]
Maktoomi	29.14 ± 0.61 [e]	0.26 ± 0.02 [e]	0.92 ± 0.04 [b,c]	0.02 ± 0.00 [b]	0.72 ± 0.09 [b]	-
Raziz	39.72 ± 0.82 [f]	1.18 ± 0.14 [a]	0.23 ± 0.02 [d]	-	0.11 ± 0.02 [c]	0.01 ± 0.00 [b]
Shikat	24.33 ± 0.63 [h]	-	0.63 ± 0.01 [a,c]	0.41 ± 0.05 [d]	-	0.03 ± 0.01 [a,c]
Shishi	34.28 ± 1.01 [b]	-	-	0.77 ± 0.09 [f]	0.26 ± 0.04 [c]	-

* Results are presented as % concentration; mean values superscripted with different alphabets within the same column are significantly different ($p < 0.05$), as established by Tukey's test.

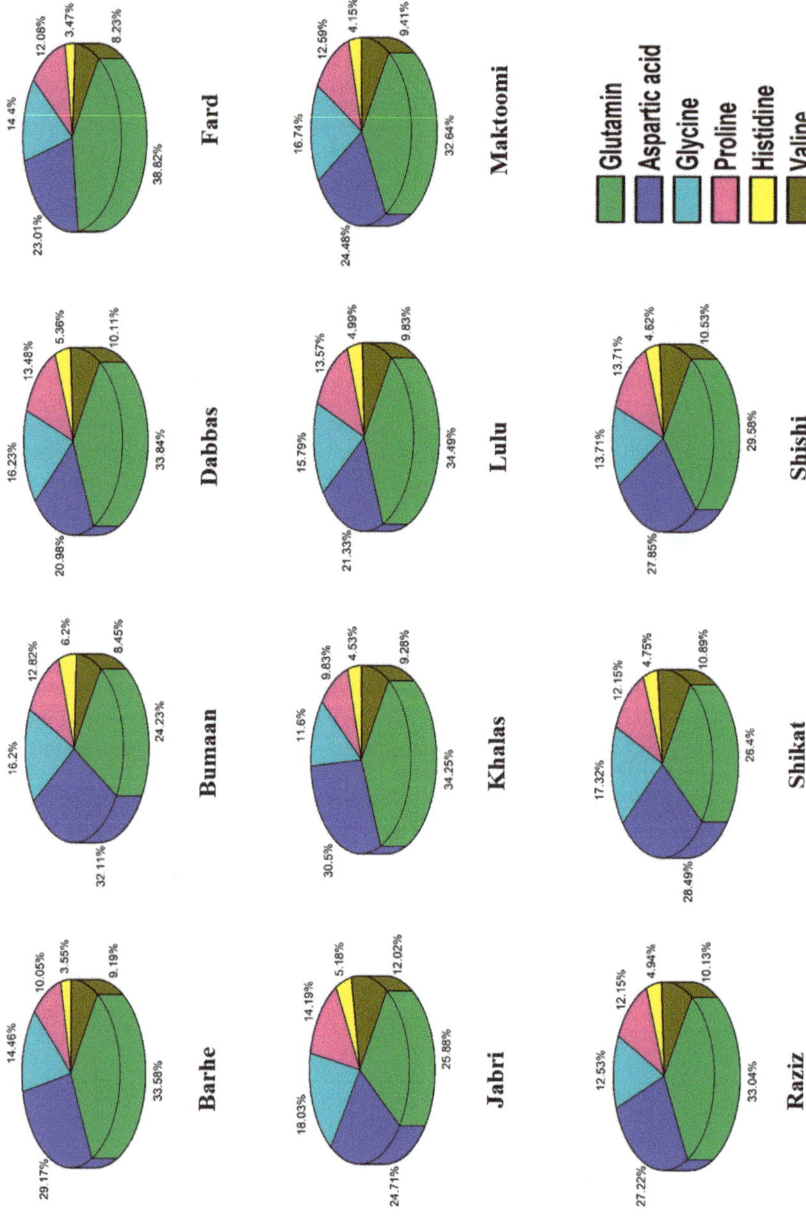

Figure 3. Relative distribution of essential amino acids present in the date fruit varities of the UAE.

4. Discussions

Date fruits have been predominately utilized as an integral component of the UAE food system to provide a wide range of nutrients and supplements for human health. Very recently, the scope of the date fruit has been extensively researched in UAE for production of numerous products that are expected to have versatile applications. These applications includes medicinal products, cosmetics, functional foods and beverages, specialty chemicals, biofuels, bioplastics, water treatment, animal fodder, and as mechanical sealants. Thus, a complete understanding of the nutritional and physico-chemical properties of the popular date palm fruits in the UAE is very essential for the intensified growth of the UAE industries producing date-fruit-related products. The presented study, which caters exactly to the above-specified requirement, provides vital information to these industries to help them effectively process the date fruits to achieve these products in an efficient manner. The reported physical properties for the date samples would serve both agricultural and industrial sectors in choosing the appropriate date type for cultivation and end-product application for a given purpose.

Details on the proximate analysis of the local date fruit variants provides the key information for processing, preservation, storage, and grading of the fruits [10]. The sugar assay for the specified varieties of UAE date fruits would aid the pharmaceutical and food and beverages industries for effective synthesis of commercial functional foods and medicinal items. The quantitative determination of the sugar components in date fruits is highly crucial for production of novel, soluble, solid date fruit sugar (fructose and glucose), which is foreseen to be a suitable and superior alternative to commercial refined sugar (sucrose). Additionally, the sugar analysis in conjunction with the phytochemical analysis of the date fruits provides the preliminary and pivotal data for synthesis of new types of healthy and non-alcoholic beverages through fermentation and thermo-chemical methods.

The abundant amount of minerals and salts present in the UAE date fruits at their respective computed levels are reported in this work. These nutrients would serve the pharmaceutical and cosmetics industries in identifying the right candidate for the production of versatile products, specifically in anti-cancer and anti-hyperlipidemic drugs, supplement capsules, skin creams, and lotions [18]. As date fruits are generally prescribed as natural supplements for minerals, the presented statistics would help the consumers in the UAE to opt for the suitable fruit breed for a particular mineral deficiency. The presented work is also an initial attempt to assess the amino acids in the eleven date palm fruit types cultivated in the UAE. High levels of amino acids, especially glutamine and aspartic acid, are detected in UAE-grown date fruits as compared to other topographical date fruit types. The amino acid assay for the UAE date breeds reported through this work could plausibly open a new field of bioengineering research. Studies on the selective extraction of amino acids from the date fruits using an efficient methodology is a promising research field with good commercialization potential. Additionally, novel efforts to bioengineer the native date palm to enhance the levels of existing amino acids (as well as to introduce new types of amino acids) in the fruit could be attempted. Additionally, since the local date fruit variants are rich in glutamine, an extended study on the influence of the ratio of glutamic acid (the precursor of glutamine) to glutamine on different stages of the date fruit growth and the shelf life of the final fruit could also be carried out [25]. The amino acids data would also help the pharmaceutical and food industries to pick the right candidates for synthesized drugs and potions rich in protein, especially for human blood, immune, and reproductive systems.

The anti-oxidant and anti-nutritional evaluation indicated the presence of significant levels of flavonoids in the abundantly cultivated local date fruit breeds, while the anti-nutritional components were almost negligible and non-detectable in most cases. This highlights the rich phytochemical nature of the fruit, which paves the way for research towards isolation, identification, and characterization of the various classes of flavonoids present in the fruits, with an emphasis on their functionality. Additionally, the preliminary results regarding the phytochemical nature of the fruit assures promising research potential in the development of active packaging biopolymer films and anti-oxidant bioplastic sealants from the date fruit waste, which is abundant in the UAE. Additionally, the overall data presented in the work would benefit the industries, consumers, and regulatory bodies of the UAE and related agencies towards effective and efficient utilization of the UAE date fruit variants for new functionalities.

Although the presented work covers a range of characteristic features of the locally grown UAE date fruits types, there are still certain properties of the fruits (mostly secondary in nature) that are yet to be analyzed. For instance, the mechanical properties and the thermal responses of the fruits are important to understand the complete nature of these fruits for their intensified industrial processing. Additionally, the results presented for anti-oxidants and anti-nutritional assessment are elementary, and a detailed analysis for the same properties could be carried out as a possible extension of this work.

5. Conclusions

In summary, a detailed and comprehensive analysis of 11 popular varieties of *Phoenix dactylifera* in the UAE was presented to assess their nutritional and anti-nutritional aspects. The amounts of nutrients present in the fruits showed considerable variation based on the type of date fruit grown, either in the same region or elsewhere. Proximate analysis showed that the date fruits possessed good levels of proteins. Glucose and fructose were identified as major sugar constituents in all of the date types. All of the samples contained a reasonably significant amount of micronutrients (K, Mg, Ca, and P). Amino acid quantification and anti-nutritional assessment for the UAE date fruit varieties were reported for the first time. Studies showed that all date types predominately contained glutamine and aspartic amino acids. Other essential amino acids, such as glycine, proline, and valine, were also present to significant levels, with histidine showing the lowest concentration in all examined date breeds. Anti-nutritional assay of the date fruits confirmed their food safety and health benefits. Hence, these fruits could be effectively used as a nutritional source in food industries and as a precursor to the synthesis of a wide range of functional products. A close analysis of the results showed that among the examined date types, the Barhe variety possessed better nutritional and physio-chemical characteristics, such as high fructose level, substantial amounts of microelements, better anti-oxidants potential, and very low anti-nutrient contents. These features make the Barhe type an ideal candidate for soluble solid date sugar production through suitable processing technology.

Author Contributions: Conceptualization, K.R., and F.B.; methodology, K.R., G.B., and A.H.; software, K.R., and A.H.; data curation, S.W.H., and H.T.; writing—review and editing, K.R., A.H., F.B., and H.F.M.Z.; project administration, K.R., and F.B.; funding acquisition, F.B., and H.F.M.Z.; All authors have read and agreed to the published version of the manuscript.

Funding: The authors would like to thank Khalifa University of Science and Technology (KUST) for providing the primary financial support for this work under CIRA-2019-028 project. The authors also acknowledge the funding support from Yayasan Universiti Teknologi PETRONAS under grant number YUTP 015LC0-047.

Acknowledgments: The authors would like to thank Khalifa University of Science and Technology (KUST) for providing all the necessary technical and administrative support to perform the reported research work.

Conflicts of Interest: The authors declare no conflict of interest.

References

1. Oladipupo Kareem, M.; Edathil, A.A.; Rambabu, K.; Bharath, G.; Banat, F.; Nirmala, G.S.; Sathiyanarayanan, K. Extraction, characterization and optimization of high quality bio-oil derived from waste date seeds. *Chem. Eng. Commun.* **2019**, 1–11. [CrossRef]

2. Chao, C.T.; Krueger, R.R. The date palm (Phoenix dactylifera L.): Overview of biology, uses, and cultivation. *HortScience* **2007**, *42*, 1077–1082. [CrossRef]

3. Ortiz-Uribe, N.; Salomón-Torres, R.; Krueger, R. Date Palm Status and Perspective in Mexico. *Agriculture* **2019**, *9*, 46. [CrossRef]

4. Mohd Jaih, A.A.; Rahman, R.A.; Razis, A.F.; Ariffin, A.A.; Al-Awaadh, A.A.; Suleiman, N. Fatty acid, triacylglycerol composition and antioxidant properties of date seed oil. *Int. Food Res. J.* **2019**, *26*, 5120527.

5. Food and Agricultural Organinzation of the United Nations. Available online: http://www.fao.org/faostat/en/#rankings/countries_by_commodity (accessed on 15 January 2020).

6. Habib, H.M.; Ibrahim, W.H. Nutritional quality of 18 date fruit varieties. *Int. J. Food Sci. Nutr.* **2011**, *62*, 544–551. [CrossRef]

7. Ismail, B.; Haffar, I.; Baalbaki, R.; Mechref, Y.; Henry, J. Physico-chemical characteristics and total quality of five date varieties grown in the United Arab Emirates. *Int. J. Food Sci. Technol.* **2006**, *41*, 919–926. [CrossRef]

8. Assirey, E.A.R. Nutritional composition of fruit of 10 date palm (Phoenix dactylifera L.) cultivars grown in Saudi Arabia. *J. Taibah Univ. Sci.* **2015**, *9*, 75–79. [CrossRef]

9. Miller, C.J.; Dunn, E.V.; Hashim, I.B. The glycaemic index of dates and date/yoghurt mixed meals. Are dates 'the candy that grows on trees'? *Eur. J. Clin. Nutr.* **2003**, *57*, 427–430. [CrossRef]

10. Ashraf, Z.; Hamidi-Esfahani, Z. Date and Date Processing: A Review. *Food Rev. Int.* **2011**, *27*, 101–133. [CrossRef]

11. Krishnamoorthy, R.; Govindan, B.; Banat, F.; Sagadevan, V.; Purushothaman, M.; Show, P.L. Date pits activated carbon for divalent lead ions removal. *J. Biosci. Bioeng.* **2019**, *128*, 88–97. [CrossRef]

12. Rambabu, K.; Banat, F.; Nirmala, G.S.; Velu, S.; Monash, P.; Arthanareeswaran, G. Activated carbon from date seeds for chromium removal in aqueous solution. *Desalin. Water Treat.* **2019**, *156*, 267–277. [CrossRef]

13. Swathy, R.; Rambabu, K.; Banat, F.; Ho, S.H.; Chu, D.T.; Show, P.L. Production and optimization of high grade cellulase from waste date seeds by Cellulomonas uda NCIM 2353 for biohydrogen production. *Int. J. Hydrogen Energy* **2019**. [CrossRef]

14. Rambabu, K.; Show, P.L.; Bharath, G.; Banat, F.; Naushad, M.; Chang, J.S. Enhanced biohydrogen production from date seeds by Clostridium thermocellum ATCC 27405. *Int. J. Hydrogen Energy* **2019**. [CrossRef]

15. Hai, A.; Bharath, G.; Babu, K.R.; Taher, H.; Naushad, M.; Banat, F. Date seeds biomass-derived activated carbon for efficient removal of NaCl from saline solution. *Process Saf. Environ. Prot.* **2019**, *129*, 103–111. [CrossRef]

16. Shaba, E.Y.; Ndamitso, M.M.; Mathew, J.T.; Etsunyakpa, M.B.; Tsado, A.N.; Muhammad, S.S. Nutritional and anti-nutritional composition of date palm (Phoenix dactylifera L.) fruits sold in major markets of Minna Niger State, Nigeria. *Afr. J. Pure Appl. Chem.* **2015**, *9*, 167–174.

17. Heckman, M. Collaborative study of a copper in feeds by atomic absorption spectrophotometry. *J. Assoc. Off. Anal. Chem.* **1971**, *54*, 666–668. [CrossRef]

18. Al-barnawi, H.M. Nutritional Composition of Protein Extract from Date Palm Fruit (Phoenix dactylifera L.) Cultivar Grown in Saudi Arabia. *SciFed J. Protein Sci.* **2018**, *1*, 1.

19. Selmani, C.; Chabane, D.; Bouguedoura, N. Ethnobotanical survey of phoenix dactylifera l. Pollen used for treatment of infertility problems in algerian oases. *Afr. J. Tradit. Complement. Altern. Med.* **2017**, *14*, 175–186. [CrossRef]

20. Gasim, A.A.A. Changes in sugar quality and mineral elements dur-ing fruit development in five date palm cultivars in Al-MadinahAl-Munawwarah. *J. King Abdul Aziz Univ.* **1994**, *6*, 29–36.

21. Xu, Y.T.; Ma, X.K.; Wang, C.L.; Yuan, M.F.; Piao, X.S. Effects of dietary valine:lysine ratio on the performance, amino acid composition of tissues and mRNA expression of genes involved in branched-chain amino acid metabolism of weaned piglets. *Asian-Australas. J. Anim. Sci.* **2018**, *31*, 106–115. [CrossRef]

22. Vayalil, P.K. Date Fruits (Phoenix dactylifera Linn): An Emerging Medicinal Food. *Crit. Rev. Food Sci. Nutr.* **2012**, *52*, 249–271. [CrossRef] [PubMed]

23. Zhang, C.R.; Aldosari, S.A.; Vidyasagar, P.S.P.V.; Shukla, P.; Nair, M.G. Health-benefits of date fruits produced in Saudi Arabia based on in vitro antioxidant, anti-inflammatory and human tumor cell proliferation inhibitory assays. *J. Saudi Soc. Agric. Sci.* **2017**, *16*, 287–293. [CrossRef]

24. Aletor, O.; Adebayo, A.O. Comparative Evaluation of the Nutritive and Physico-Chemical Characteristics of the Leaves and Leaf Protein Concentrates from Two Edible Vegetables. *J. Food Technol.* **2007**, *5*, 152–156.

25. Pratta, G.; Zorzoli, R.; Boggio, S.B.; Picardi, L.A.; Valle, E.M. Glutamine and glutamate levels and related metabolizing enzymes in tomato fruits with different shelf-life. *Sci. Hortic.* **2004**, *100*, 341–347. [CrossRef]

 *processes*

Article

Application of a Liquid Biphasic Flotation (LBF) System for Protein Extraction from *Persiscaria Tenulla* Leaf

Hui Shi Saw [1], Revathy Sankaran [2,3], Kuan Shiong Khoo [4], Kit Wayne Chew [5,*], Win Nee Phong [6], Malcolm S.Y. Tang [2], Siew Shee Lim [4,*], Hayyiratul Fatimah Mohd Zaid [7], Mu. Naushad [8] and Pau Loke Show [4,*]

[1] School of Biosciences, Faculty of Science and Engineering, University of Nottingham Malaysia, Jalan Broga, Semenyih, 43500 Selangor Darul Ehsan, Malaysia; huishi.saw@gmail.com
[2] Institute of Biological Sciences, Faculty of Science, University of Malaya, 50603 Kuala Lumpur, Malaysia; revathy@um.edu.my (R.S.); Malct32@gmail.com (M.S.Y.T.)
[3] Research and Training Department, Graduate School, University of Nottingham Malaysia, Jalan Broga, Semenyih, 43500 Selangor Darul Ehsan, Malaysia
[4] Department of Chemical and Environmental Engineering, Faculty of Science and Engineering, University of Nottingham Malaysia, Jalan Broga, Semenyih, 43500 Selangor Darul Ehsan, Malaysia; kuanshiong.khoo@hotmail.com
[5] School of Mathematical Sciences, Faculty of Science and Engineering, University of Nottingham Malaysia, Jalan Broga, Semenyih, 43500 Selangor Darul Ehsan, Malaysia
[6] School of Molecular and Life Sciences, Curtin University, Bentley, GPO Box U1987, Perth, WA 6845, Australia; pwinnee@gmail.com
[7] Fundamental and Applied Sciences Department, Centre of Innovative Nanostructures and Nanodevices, Institute of Autonomous Systems, Universiti Teknologi PETRONAS, 32610 Bandar Seri Iskandar, Malaysia; hayyiratul.mzaid@utp.edu.my
[8] Department of Chemistry, College of Science, King Saud University, Riyadh 11451, Saudi Arabia; shad81@rediffmail.com
* Correspondence: KitWayne.Chew@nottingham.edu.my (K.W.C.); SiewShee.Lim@nottingham.edu.my (S.S.L.); PauLoke.Show@nottingham.edu.my (P.L.S.); Tel.: +603-8924-8719 (K.W.C.); +603-8924-8180 (S.S.L.); +603-8924-8605 (P.L.S.)

Received: 10 January 2020; Accepted: 17 February 2020; Published: 21 February 2020

Abstract: *Persiscaria tenulla*, commonly known as *Polygonum*, is a plant belonging to the family Polygonaceae, which originated from and is widely found in Southeast Asia countries, such as Indonesia, Malaysia, Thailand, and Vietnam. The leaf of the plant is believed to have active ingredients that are responsible for therapeutic effects. In order to take full advantage of a natural medicinal plant for the application in the pharmaceutical and food industries, extraction and separation techniques are essential. In this study, an emerging and rapid extraction approach known as liquid biphasic flotation (LBF) is proposed for the extraction of protein from *Persiscaria tenulla* leaves. The scope of this study is to establish an efficient, environmentally friendly, and cost-effective technology for the extraction of protein from therapeutic leaves. Based on the ideal conditions of the small LBF system, a 98.36% protein recovery yield and a 79.12% separation efficiency were achieved. The upscaling study of this system exhibited the reliability of this technology for large-scale applications with a protein recovery yield of 99.44% and a separation efficiency of 93.28%. This technology demonstrated a simple approach with an effective protein recovery yield and separation that can be applied for the extraction of bioactive compounds from various medicinal-value plants.

Keywords: extraction; leaf; liquid biphasic flotation; polygonum; protein

---

1. Introduction

The usage of traditional herbs for preventive health care is widespread, and plants are the source of numerous natural antioxidants that could be utilized for the development of novel medicines. Natural antioxidants and bioactive compounds that originate from traditional herbal medicines have received escalating attention for their ability in treating specific human diseases. For example, traditional herbal medicine has been used widely in treating cancer patients [1] and to treat neurodegenerative disorders [2]. Plants comprising high medicinal value are currently screened for a variety of pharmacological properties.

Persiscaria tenulla (formerly known as *Polygonum* (*P. minus*)) or frequently recognized as "kesum" in Malaysia, have been used as a flavoring ingredient and food additive in Malaysia. Polygonum plant has been reported to contain a wide range of pharmacological properties and many studies have been conducted to evaluate the phytochemical and pharmacological aspects of the plant, which include anti-inflammatory activity [3], antiproliferative effects [4], anti-microbial activity [5], cytotoxic activity [6], gastric cytoprotective activity [7], and antiviral activity [8]. It has been proven that *P. minus* comprises many high-value components that include protein, flavonoids, and antioxidant vitamins, such as carotenes, retinol equivalents, and vitamin C. However, there are limited studies concerning the effective extraction techniques of the biomolecules from the plant extract.

The major challenge in the extraction of the high-value components from herbal plants is the downstream processing. Up to date, there is a lack of effective and efficient techniques for high yield and cost-effective biomolecule extraction. In this study, a novel method known as liquid biphasic flotation (LBF) system is introduced to extract protein from *P. tenulla* leaves. The LBF process comprises the incorporation of two processes, which are an aqueous two-phase system (ATPS) and a solvent sublation (SS). The conventional flotation system that is commonly known as SS was first introduced by Sebba [3,9]. The SS process is a type of non-foaming adsorptive bubble separation technique in which the surface-active or hydrophobic compounds in aqueous phase are adsorbed on the bubble surfaces of an ascending gas stream and then collected in an immiscible apolar organic solution layer placed on top of the aqueous phase. The mass transfer of SS involves the air bubbles that are produced from the sublation column. The air bubbles drag a sheath of water into the top organic solvent, which eventually drains as water droplets, depleted of solute, and descend back into the bottom aqueous phase via the gravitational force [5,10]. As for the LBF process, the mass transfer comprises the integration of ATPS and SS, which are the utilization of aqueous two-phase systems as a liquid medium for facilitating the mass transfer of biomolecules and the involvement of mass transport from SS. The LBF system is known to be a newly developing separation process that has many benefits over conventional processes. LBF has several advantages, such as a high separation efficiency, high yield, simple separation, and is an economical technique [11]. Recently, the LBF process has gained much attention and many studies have been conducted using this technique for various biomolecules extraction, which include lipase from bacteria, penicillin G, puerarin, a-lactalbumin, lincomycin, C-phycocyanin, polyphenols, betacyanins, and protein from microalgae [12].

In this current study, the LBF method is utilized to extract protein from *P. tenulla* and the optimization of the technique is performed to obtain optimal operating conditions for protein recovery. Parameters that were evaluated in this research were the effect of types of alcohol, types of salt, the concentration of alcohol and salt, amount of kesum biomass, pH, flotation time, and scale-up capability. This study aimed to assess the feasibility of LBF in protein extraction from kesum leaf and to demonstrate that the LBF system is an effective method that has a high possibility to be employed in large scale productions.

2. Materials and Methods

2.1. Materials

Ammonium sulphate, dipotassium hydrogen phosphate, magnesium sulphate, disodium hydrogen phosphate, sodium hydroxide, hydrochloric acid, and Bradford reagent were purchased from R&M Chemicals, (Selangor, Malaysia). Food grade 99.8% ethanol and 2-propanol were acquired from R&M Chemicals,(Selangor, Malaysia). Fresh Kesum leaves were obtained from Aeon supermarket (Selangor, Malaysia, distributed by Edsam Trading Sdn. Bhd.).

2.2. Equipment

The LBF equipment was made up of a glass column, where the small LBF column had an internal diameter of 2.4 cm, height of 15 cm, and could accommodate a solution with a maximum volume of 50 mL. For the large-scale LBF system, the internal diameter was 7 cm, 15 cm in height, and the capacity of the system was 500 mL. The base of the column was built using a G4 sintered disk with a pore size of 10 μm manufactured by DONEWELL RESOURCES SDN. BHD, (Selangor Malaysia) and the base was connected to an air pump for air bubble generation. A Dwyer flowmeter (model RMA-26-SSV) with the range of 25–250 cc/min was used to measure the flow rate of air supplied to the column. Figure 1 displays the experimental framework of the LBF method that was used in this study.

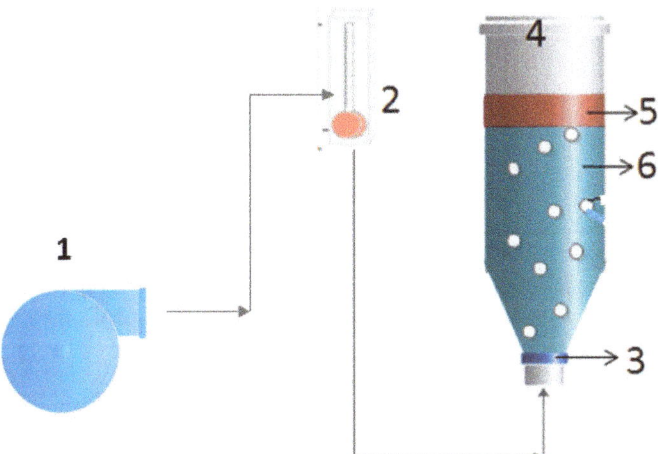

Figure 1. Schematic diagram illustrating the apparatus set-up of liquid biphasic flotation (LBF) system for protein extraction. 1: Air pump; 2: flowmeter; 3: sintered disk; 4: LBF column; 5: top alcohol phase; 6: bottom salt phase.

2.3. Methodology

2.3.1. Preparation of Kesum Leaf Powder

Fresh *Persiscaris tenulla* obtained from a supermarket were used in this study. The leaves with petiole attached were removed from the stem and were cleaned. The leaves were cut into smaller pieces and were desiccated in a silica gel box overnight. The dried leaves were then ground with a mortar and pestle into powder form.

2.3.2. LBF Extraction

This study was performed in batches and was repeated thrice. A one variable at a time (OVAT) method was used in this study to assess the effect of different parameters on the protein recovery.

The initial condition of 250 g/L of ammonium sulphate was dissolved in 15 mL of distilled water. Fifteen milliliters of salt solution, which served as the bottom phase, was pipetted into the flotation system. Fifteen milliliters of 100% ethanol was added to 300 mg of ground leaves and vortexed. The top phase was then poured into the LBF tube gently from the edge. The system was capped with a lid and then immediately timed using a stopwatch. The amount of bubbles that formed was maintained by adjusting the pressure of the flotation system. Adjustment of the pH was made via the addition of hydrochloric acid (1 M) or sodium hydroxide (1 M). After the system had settled for 10 min, the top phase was pipetted into a tube and the remaining bottom phase was poured into another tube. The volumes of the top and bottom phases were measured.

2.3.3. Protein Assay

The protein concentration was obtained by applying the Bradford method. An extracted protein sample with a volume of 0.25 mL was mixed with 2.5 mL of Bradford reagent in a cuvette. After 10 min of reaction time, the absorbance was measured using a UV-Vis spectrophotometer at the wavelength of 595 nm. The absorbance reading obtained was converted to a protein concentration by using a standard calibration curve that was established using a standard protein, namely BSA. The results were expressed as a mean of triplicate readings.

2.3.4. Calculation of the Separation Efficiency (E) and Recovery Yield (R)

The separation efficiency (E) describes the concentration of protein being extracted in the alcohol phase (top phase). The efficiency is obtained by calculating the concentration of protein present in the bottom phase before and after the flotation process and it was evaluated by employing Equation (1):

$$E = \left(1 - \frac{c_B}{c_{Bi}}\right) \times 100\%, \tag{1}$$

where c_B represents protein concentration in bottom phase after flotation, while c_{Bi} signifies protein concentration in bottom phase before flotation. The E value determines the concentration of protein being successfully recovered in the alcohol-rich top phase.

The total recovery yield (R) of protein was assessed by applying Equation (2). The C_T describes the protein concentration that is recovered in the top phase, while V_T defines the volume of the top phase. Based on the protein concentration obtained from the top phase, the amount is compared with the theoretical protein content in mg to obtain the recovery yield. The amount of protein present in the leaf is based on theoretical value obtained from Revathy Sankaran et al. [11].

$$R\ (\%) = (C_T \times V_T)/(\text{Amount of protein content based on theory}) \times 100\% \tag{2}$$

3. Results and Discussion

3.1. Effect of Alcohol Types on the Protein Recovery and Separation Efficiency

In this study, water-miscible pure alcohols (100%), namely ethanol and 2-propanol, demonstrated the ability to form LBF with 250 g/L of ammonium sulphate $(NH_4)_2SO_4$. Based on the results, LBF formed using $(NH_4)_2SO_4$/ethanol showed a 74.93% separation efficiency and a 32.35% protein recovery. In contrast, the separation efficiency and recovery yield achieved using LBF containing $(NH_4)_2SO_4$/2-propanol were 57.95% and 28.2%, respectively. The results clearly show that LBF formed using $(NH_4)_2SO_4$/ethanol was more efficient in recovering protein from kesum leaves than LBF containing $(NH_4)_2SO_4$/2-propanol. Ethanol exhibited a better performance possibly due to its property of high solute solubility that can assist in the desorption of the solute from the substrate [12]. Additionally, by comparing both solvents, ethanol is more environmentally friendly compared to propanol [12]. From the industrial point of view, ethanol is a better selection for large-scale production as it can be easily reused [13]. The findings suggest that the efficiency of LBF in protein separation is dependent on the type of alcohol used in the system. In this case, LBF formed using 250 g/L of $(NH_4)_2SO_4$ and 100% of ethanol was chosen for the subsequent studies.

3.2. Effect of Types of Salt on Protein Recovery and Separation Efficiency

Selecting a proper phase-separation salt with a high salting-out ability is a key step in developing an efficient LBF system for maximum protein recovery from kesum leaves. The type of salt selected is considered to be an important variable to take into account when designing an LBF for protein separation owing to their strong effects on the salting-out effect and the partition coefficient of protein [14]. While keeping the other variables constant, such as alcohol type, alcohol concentration, salt concentration, flotation time, and the amount of starting material constant, the relative salting-out effectiveness of salt types was investigated in this study. The salts used were ammonium sulphate, di-potassium hydrogen phosphate, magnesium sulphate, and disodium hydrogen phosphate.

In this study, the protein separation efficiency was found to be varied with the types of salts used. There was no value for the separation efficiency in the LBF formed using magnesium sulphate/ethanol and disodium hydrogen phosphate/ethanol due to the precipitation that occurred in these two systems (Figure 2b). A similar conclusion was also reached by Phong et al. [14], who reported that the protein separation efficiency as a result of the salting-out effect is greatly influenced by the types and complexation of the cations and anions of salt in the LBF system.

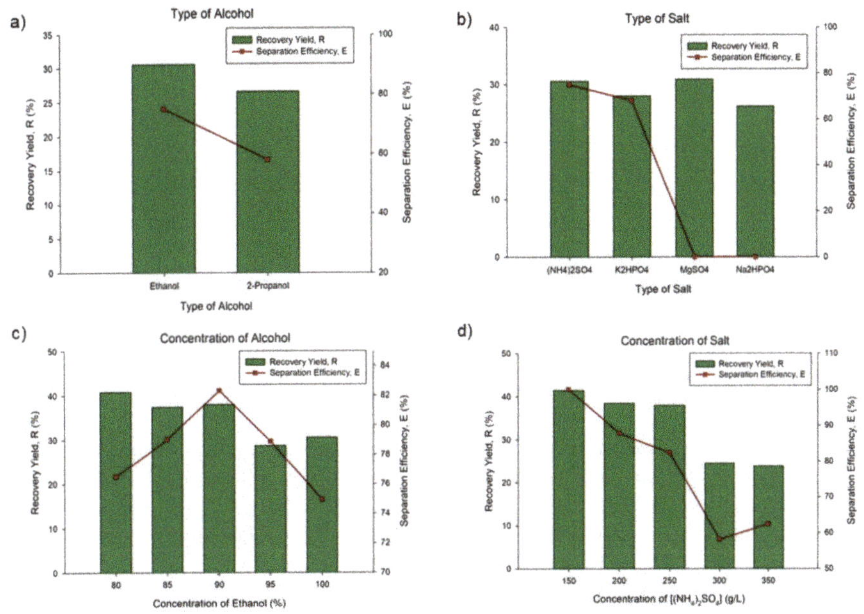

Figure 2. Effect of different conditions on the protein recovery and separation efficiency: (**a**) Effect of the alcohol type, (**b**) Effect of the types of salt, (**c**) Effect of the ethanol concentration, and (**d**) Effect of the ammonium salt concentration.

Among all the salts, LBF formed using ammonium sulphate achieved the highest separation efficiency and recovery yield, with the values of 74.93% and 32.35%, respectively, as shown in Figure 2b. This trend supports previous findings in the literature. The relative effectiveness of salt types was found to follow the well-known Hofmeister series [14], in which ammonium sulphate forms two ions at the ends of their respective Hofmeister series [15]. Apart from this, ammonium sulphate has been reported as the most commonly used salt for salting out of proteins due to its effectiveness, high solubility, cheapness, availability of pure material, lack of toxicity, and their ions possess the stabilizing effect on protein structure and bioactivity [15]. All these characteristics have made ammonium sulphate a popular choice for use in protein precipitation [16]. As such, LBF formed using 250 g/L of $(NH_4)2SO_4$ and 100% ethanol was selected for the next experiment.

3.3. Effect of Concentration of Alcohol on Protein Recovery and Separation Efficiency

While keeping the other variables constant, the relative effectiveness of salting-out at different concentrations of ethanol was investigated in this study. It is stated that a high extraction yield can be obtained by using ethanol-water compared to pure ethanol [17]. In this research, the influence of ethanol-water with several different concentrations and pure ethanol on the protein extraction was examined. The result obtained is similar to the study done by the Machado group in which they discovered greater extraction yields of blackberry residues attained by applying pressurized liquid extraction when ethanol-water was utilized compared to pure ethanol [18].

Based on Figure 2c, the results show that the protein separation efficiency increased from 80% to 90% of ethanol concentration and reached an optimum level of 82.33% at 90% ethanol concentration. However, the separation efficiency started to show a decreasing trend with ethanol concentration higher than 90%. In the case of the recovery yield, the highest yield obtained was 43.19% in LBF containing 80% ethanol, followed by a 40.2% recovery yield at 90% ethanol. The findings indicate that there was no correlation relationship between the two variables of separation efficiency and recovery yield in the same system. The addition of water to the organic solvent in this case ethanol possibly created a relatively polar medium that facilitated the extraction of protein [18]. In this study, LBF formed using 250 g/L of $(NH_4)_2SO_4$ and 90% ethanol was identified as being the most proficient at protein separation and was thus chosen for further optimization.

3.4. Effect of Salt Concentration on Protein Recovery and Separation Efficiency

The concentration of salt in the liquid biphasic system is another important factor to consider because different salts interact differently with the protein, water, and other chemicals. The effects of the concentration of salt on protein recovery and separation are well documented. The presence of salt in the solution will impact the surface tension of water, which will then increase the hydrophobic interaction between the protein and water [19]. Following this change, the targeted protein will shift to or from the aqueous phase depending on the nature of the protein [20]. For this work, the ammonium sulphate salt concentration was varied from 150 g/L to 350 g/L with 50 g/L increments. The alcohol content was set at 90%, while the mass of leaves used was 300 mg. The flotation time for this experiment was set at 10 min.

One important observation for this experiment is that the volume of top and bottom phases changed as the concentration of salt increased. The volume of the two phases reached a plateau when the concentration of salt reached 300 g/L and above. Another important observation to note is that at a low salt concentration (150 g/L), the two phases did not form. This suggests that the lower boundary for salt concentration that allows for two-phase formation is higher than 150 g/L. This is because an increase in salt or alcohol concentration in an alcohol/salt system could increase the tie-line-length (TLL), which could facilitate phase separation [21].

A plot of the effect of salt concentration on the recovery yield and separation efficiency is shown in Figure 2d. From Figure 2d, it is seen that the recovery yield and separation efficiency decreased in tandem with the increase in salt concentration. At a concentration of 200 g/L, the highest recovery yield of 40.69% and separation efficiency of 87.81% was recorded. As the concentration of salt increased, however, the recovery yield and separation efficiency gradually decreased. As the concentration of salt increased, the solubility of protein decreased. This is commonly known as the salting-out effect. Since different proteins salt-out at different salt concentrations, the effect can help us to determine the upper boundary of salt concentration for the LBF system. A higher salt concentration results in a higher salting-out effect. This could then lead to a higher protein partition coefficient Ke [21]. A high Ke would result in a low yield.

At concentrations of 300 g/L and 350 g/L, however, there was a slight increase in separation efficiency compared to a decrease in recovery yield. There was a sharp decrease in both separation efficiency and recovery yield at a concentration of 300 g/L. This could be the concentration at which the salting-out effect occurred. As the concentration of salt increased, it caused more water to enter the bottom phase. This way, the protein was then impelled to the top phase [21], which caused the slight increase in the recovery yield at a concentration of 350 g/L. Therefore, we can conclude that the optimum salt concentration for the extraction of protein from kesum leaf was between 150 g/L and 300 g/L. The concentration of 200 g/L was then used in the experiments with other parameters in this work.

3.5. Effect of the Kesum Leaf Biomass Amount on the Protein Recovery and Separation Efficiency

The influence of kesum leaf biomass, or the mass of protein source, is another important factor to consider. Generally, increasing the concentration of protein sources can have a profound effect on the performance of phase separation due to the specific partition behavior of the target protein [22]. For this work, the concentration of salt was set to 200 g/L, ethanol concentration was set to 90%, and the flotation time was set to 10 min. The mass of kesum leaves was varied between 100 mg and 400 mg with 100 mg increments.

From Figure 3, it is seen that as the mass of leaves increased, the yield decreased. There was a significant drop in yield (over 40%) when the mass of leaves increased from 100 mg to 200 mg, and the drop continued gradually as the amount of leaves continued to increase. For the separation efficiency, the highest efficiency occurred at 300 mg, which then dropped to its lowest point at 400 mg. It was interesting to see that the yield decreased as the mass of leaves increased. In general, increasing the biomass concentration of the LBF also increases the number of contaminants and impurities in the system, thereby reducing the performance of the LBF separation [23]. In addition, a higher biomass content also increases the precipitation at the interface of the two phases, which could adversely affect the yield [23].

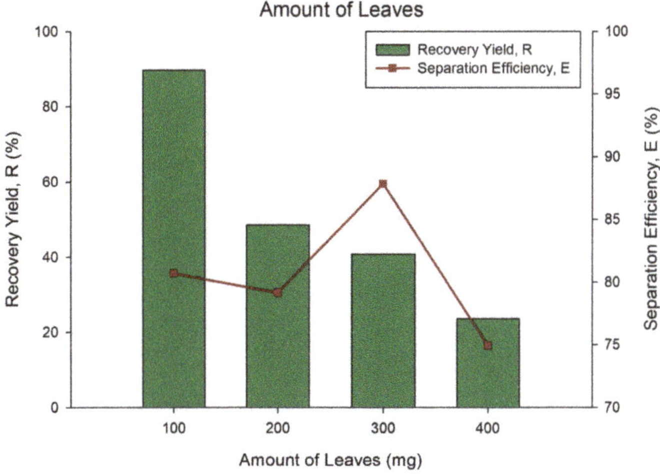

Figure 3. Effect of the kesum biomass amount on the protein recovery yield and separation efficiency.

Based on the definition highlighted in the materials section, separation efficiency is highly dependent on the protein activity in the bottom phase after the LBF process. It determines the concentration of protein extracted and this represents the efficiency of the system in extracting the protein. As for the recovery yield, it represents the total amount of protein being recovered in the top phase. At 300 mg, the high separation efficiency was obtained with 87.81%; however, the protein recovery yield was low with only 40.69%. The low recovery yield was possibly due to high impurities present in the top phase. Several possibilities contributed to the low recovery yield: (1) protein retrieved in the upper phase was low (CT), or (2) the low phase volume of the top phase (VT) at 300 mg. These could be caused by the decreasing LBF performance as the level of impurities increased due to the increase in kesum leaf mass. Due to the high recovery yield value (89.58%) and separation efficiency of 80.68%, a 100 mg kesum leaves mass was used for the next step of this experiment.

3.6. Influence of pH on the Recovery Yield and Separation Efficiency

The pH value of an LBF system affects the separation outcome by altering the surface properties of the target protein, including the surface net charge, molecular shape, surface hydrophobicity, and the presence of specific binding sites [24]. A simple example is the case of a biomolecule with both polar and non-polar groups that experiences changes in its net charge and surface properties with changing pH values [25]. For this work, the pH of the system was varied from pH 4 to pH 8.

From Figure 4, it is seen that despite the high absorbance at pH 5, the highest recovery yield and separation efficiency occurred at pH 6. The lowest yield and E, on the other hand, occurred at pH 7. This suggests that for the extraction of biomolecules from kesum leaves, the LBF system should be kept in an acidic condition. Both the recovery yield and separation efficiency did not fluctuate much as the pH of the system increased. An interesting observation is that there was no formation of two-phases at pH 8. This experiment shows that as the pH approached basic pH, the hydrophobicity of the system was impacted to the point where it induced a salt-out. In order to prevent a salt-out, the pH of the system should be kept below the neutral level. The optimized pH condition that gave the maximum separation efficiency and recovery yield was at pH 6 with 87.19% and 96.37%, respectively.

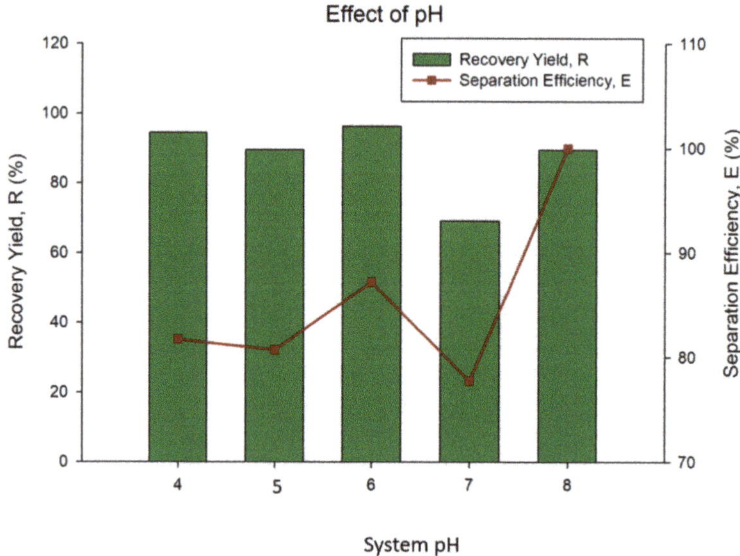

Figure 4. Effect of pH on the protein recovery yield and separation efficiency.

3.7. The Influence of the Flotation Time on the Recovery Yield and Separation Efficiency

The effect of flotation time is one of the most important factors that could affect the LBF recovery yield and separation efficiency. The flotation time affects the outcome of the process by influencing the area of the air–water interface per unit volume of aqueous solution over time [26]. For this part, the concentration of salt was set to 200 g/L, the pH of the system was maintained at pH 5.0, the ethanol concentration was maintained at 90%, and the mass of kesum leaves consumed was 100 mg. The flotation time was varied from 5 min to 15 min with 2.5 min increments. The results of the experiment are provided in Figure 5.

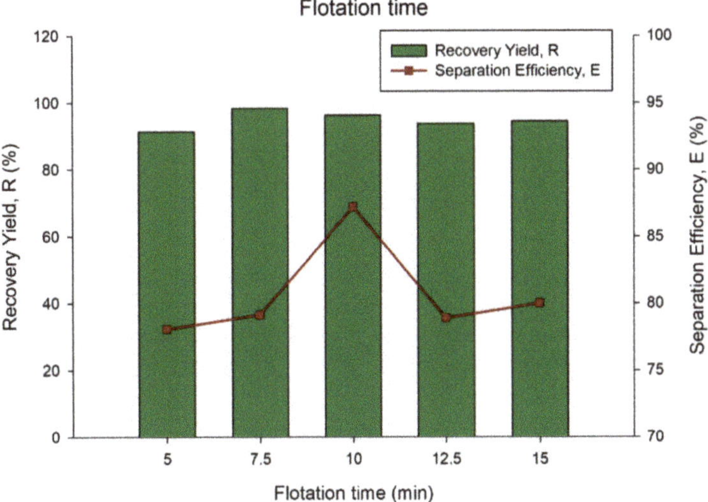

Figure 5. Effect of the the flotation time on the protein recovery yield and separation efficiency.

The resulting yield and separation efficiency, as shown in Figure 5, shows that the flotation time of 7.5 min gave the highest recovery yield with 98.36%, while the flotation time of 5 min gave the lowest recovery yield of 91.42%. For the separation efficiency, the highest occurred at a 10 min flotation time with 87.19% but at 7.5 min, 79.12% was recorded. One possible explanation for this phenomenon is that longer flotation time allowed for the accumulation and build-up of biomolecules in the LBF phases with the movement of gas bubbles [25]. As the flotation time increased, the concentration of targeted biomolecules in the top and bottom phases changed based on the kinetic processes. This explains the general trend of the separation efficiency, which showed a positive gradient, i.e., increasing with increasing flotation time. However, as the flotation time continued to increase, the level of impurity carried upward by the gas bubbles also increased, which then affected the separation performance of the LBF system.

According to Iqbal et al., the flotation force is highly dependent on the flow properties of the phases [27]. As the flotation time increases, the concentration of targeted biomolecules in the top and bottom phases changes based on the kinetic processes [25]. Based on Figure 5, as the flotation time increased, the yield was reduced. From this study, although 10 min gave the highest separation efficiency, 7.5 min was selected as an optimized condition because the focus of this study was to obtain a high protein recovery yield. Additionally, a long flotation time requires high energy consumption, which is costly, non-environmentally friendly, and it is not suitable for large-scale processes.

3.8. Large-Scale Protein Extraction Using the LBF System

In this section, a scale-up study of the protein extraction using the optimized conditions were assessed. By using the operating conditions that were optimized previously, a large-scale study was conducted. This assessment was performed to corroborate the consistency and efficiency of the LBF technique on a large scale. In the large-scale study, it was discovered that the amounts of both top organic phase and bottom phases increased ten-fold compared with the small-scale experiment. A total of 300 mL of working volume with 150 mL of bottom phase and 150 mL of top phase were employed. Following the results achieved from Table 1, it can be seen that a comparatively higher recovery yield and separation efficiency of 99.44% and 93.28%, respectively, were obtained. Many reported studies have shown that the separation efficiencies of LBF for the recovery of different kind of biomolecules in the comparative study were between 85%–98.5%, which were much higher than SS, which achieved separation efficiencies of 48%–70%. LBF is preferable compared for this case as the separation efficiency achieved was more than 90%.

Table 1. Comparison between small- and large-scale LBF systems for the protein extraction of *P. tenulla* leaf.

LBF System	Recovery Yield (%)	Separation Efficiency (%)
Small scale	98.36	79.12
Large scale	99.44	93.28

Based on the results achieved, the large-scale version of the LBF system was validated for its reliability, which is beneficial for the extraction of other biomolecules on an industrial scale. Other studies that demonstrated that LBF can be an alternative technology that can be utilized in industries for the extraction of various medicinal components include ortho-phenylphenol [28], puerarin [29], antioxidant peptides from trypsin hydrolysates of whey protein [30], baicalin [31], lipase enzyme [32], C-phycocyanin [33], and betacyanins [34]. All these studies have proven that the separation efficiency and recovery of the biocomponents were enhanced with the utilization of LBF system as their extraction method.

4. Conclusions

The findings from this study revealed that a high protein recovery and separation efficiency can be obtained using this LBF approach. Based on the experiment conducted, the optimized conditions for highest protein recovery and separation efficiency were 90% ethanol, 200g/L of ammonium sulphate, 100 mg of kesum leaf biomass, pH 6, and a flotation time of 7.5 min. The highest protein recovery yield achieved was 98.36% and the separation efficiency was 79.12% for the small-scale system. The study on the large-scale LBF system demonstrated the reliability and consistency of the system in which a recovery yield and separation efficiency of 99.44% and 93.28%, respectively, were attained. The application of the LBF system for protein extraction from herbal leaves involves a simple procedure, short processing time, and cost-effective process with a high recovery yield. This study shows the significance of LBF in downstream processing, especially for the extraction of value-added biomolecules. LBF can be an alternative technology that can be utilized in industries for the extraction of various medicinal leaf extracts. This LBF system is beneficial in the pharmaceutical and nutraceutical industries for the improvement of overall production and biotechnology fields.

Author Contributions: Conceptualization, K.W.C.; data curation, H.S.S. and K.S.K.; methodology, software, and resources: H.S.S., K.S.K., and M.N.; original draft writing: R.S., W.N.P., and M.S.Y.T.; review and editing: R.S., K.W.C., and P.L.S.; supervision: S.S.L., P.L.S., and R.S.; project administration: K.W.C., S.S.L., P.L.S., and H.F.M.Z.; funding acquisition: P.L.S. and H.F.M.Z. All authors have read and agreed to the published version of the manuscript.

Funding: This research was funded by the Fundamental Research Grant Scheme, Malaysia (FRGS/1/2019/STG05/UNIM/02/2). The APC was funded by Yayasan Universiti Teknologi PETRONAS (YUTP 015LC0-047).

Acknowledgments: One of the authors (M. Naushad) is grateful to the Researchers Supporting Project number (RSP-2019/8), King Saud University, Riyadh, Saudi Arabia, for the support.

Conflicts of Interest: The authors declare no conflict of interest.

References

1. Poonthananiwatkul, B.; Lim, R.H.M.; Howard, R.L.; Pibanpaknitee, P.; Williamson, E.M. Traditional medicine use by cancer patients in Thailand. *J. Ethnopharmacol.* **2015**, *168*, 100–107. [CrossRef] [PubMed]
2. Li, X.; Zhang, S.; Liu, S.; Lu, F. Recent advances in herbal medicines treating Parkinson's disease. *Fitoterapia* **2013**, *84*, 273–285. [CrossRef] [PubMed]
3. George, A.; Chinnappan, S.; Chintamaneni, M.; Kotak, C.V.; Choudhary, Y.; Kueper, T.; Radhakrishnan, A.K. Anti-inflammatory effects of Polygonum minus (Huds) extract (LineminusTM) in in-vitro enzyme assays and carrageenan induced paw edema. *BMC Complement. Altern. Med.* **2014**, *14*, 355. [CrossRef] [PubMed]
4. Hassim, N.; Markom, M.; Anuar, N.; Dewi, K.H.; Baharum, S.N.; Mohd Noor, N. Antioxidant and Antibacterial Assays on Polygonum minus Extracts: Different Extraction Methods. *Int. J. Chem. Eng.* **2015**, *2015*, 1–10. [CrossRef]
5. Johnny, L.; Yusuf, U.K.; Nulit, R. The effect of herbal plant extracts on the growth and sporulation of Colletotrichum gloeosporioides. *J. Appl. Biosci.* **2010**, *34*, 2218–2224.
6. H, F.; Ahmadi, B.; Lajisi, N. Antiviral and Cytotoxic Activities of Some Plants Used in Malaysian Indigenous Medicine. *Pertanika J. Trop. Agric. Sci.* **1996**, *19*, 129–136.
7. Qader, S.W.; Abdulla, M.A.; Chua, L.S.; Sirat, H.M.; Hamdan, S. Pharmacological Mechanisms Underlying Gastroprotective Activities of the Fractions Obtained from Polygonum minus in Sprague Dawley Rats. *Int. J. Mol. Sci.* **2012**, *13*, 1481–1496. [CrossRef]
8. Qader, S.W.; Abdulla, M.A.; Chua, L.S.; Najim, N.; Zain, M.M.; Hamdan, S. Antioxidant, Total Phenolic Content and Cytotoxicity Evaluation of Selected Malaysian Plants. *Molecules* **2011**, *16*, 3433–3443. [CrossRef]
9. Bi, P.Y.; Li, D.Q.; Dong, H.R. A novel technique for the separation and concentration of penicillin G from fermentation broth: Aqueous two-phase flotation. *Sep. Purif. Technol.* **2009**, *69*, 205–209. [CrossRef]
10. Lee, S.Y.; Khoiroh, I.; Ling, T.C.; Show, P.L. Aqueous Two-Phase Flotation for the Recovery of Biomolecules. *Sep. Purif. Rev.* **2016**, *45*, 81–92. [CrossRef]
11. Sankaran, R.; Show, P.L.; Cheng, Y.S.; Tao, Y.; Ao, X.; Nguyen, T.D.P.; Van Quyen, D. Integration Process for Protein Extraction from Microalgae Using Liquid Biphasic Electric Flotation (LBEF) System. *Mol. Biotechnol.* **2018**, *60*, 749–761. [CrossRef] [PubMed]
12. Sankaran, R.; Parra Cruz, R.A.; Show, P.L.; Haw, C.Y.; Lai, S.H.; Ng, E.-P.; Ling, T.C. Recent advances of aqueous two-phase flotation system for the recovery of biomolecules. *Fluid Phase Equilib.* **2019**, *501*, 112271. [CrossRef]
13. He, Z.; Tan, J.S.; Abbasiliasi, S.; Lai, O.M.; Tam, Y.J.; Halim, M.; Ariff, A.B. Primary recovery of miraculin from miracle fruit, Synsepalum dulcificum by AOT reverse micellar system. *LWT Food Sci. Technol.* **2015**, *64*, 1243–1250. [CrossRef]
14. Mustafa, A.; Turner, C. Pressurized liquid extraction as a green approach in food and herbal plants extraction: A review. *Anal. Chim. Acta* **2011**, *703*, 8–18. [CrossRef] [PubMed]
15. Dias, A.L.B.; Arroio Sergio, C.S.; Santos, P.; Barbero, G.F.; Rezende, C.A.; Martínez, J. Ultrasound-assisted extraction of bioactive compounds from dedo de moça pepper (*Capsicum baccatum* L.): Effects on the vegetable matrix and mathematical modeling. *J. Food Eng.* **2017**, *198*, 36–44. [CrossRef]
16. Phong, W.N.; Show, P.L.; Teh, W.H.; Teh, T.X.; Lim, H.M.Y.; Nazri, N.S.; Tan, C.H.; Chang, J.-S.; Ling, T.C. Proteins recovery from wet microalgae using liquid biphasic flotation (LBF). *Bioresour. Technol.* **2017**, *244*, 1329–1336. [CrossRef] [PubMed]
17. Duong-Ly, K.C.; Gabelli, S.B. Salting out of Proteins Using Ammonium Sulfate Precipitation. *Methods Enzymol.* **2014**, *541*, 85–94.
18. Wingfield, P.T. Protein precipitation using ammonium sulfate. *Curr. Protoc. Protein Sci.* **2016**. [CrossRef]
19. Santos, P.H.; Baggio Ribeiro, D.H.; Micke, G.A.; Vitali, L.; Hense, H. Extraction of bioactive compounds from feijoa (Acca sellowiana (O. Berg) Burret) peel by low and high-pressure techniques. *J. Supercrit. Fluids* **2019**, *145*, 219–227. [CrossRef]
20. Machado, A.P.D.F.; Pasquel-Reátegui, J.L.; Barbero, G.F.; Martínez, J. Pressurized liquid extraction of bioactive compounds from blackberry (*Rubus fruticosus* L.) residues: A comparison with conventional methods. *Food Res. Int.* **2015**, *77*, 675–683. [CrossRef]
21. Wingfield, P. Protein Precipitation Using Ammonium Sulfate. *Curr. Protoc. Protein Sci.* **1998**, *53*, A–3F.

22. Chew, K.W.; Chia, S.R.; Lee, S.Y.; Zhu, L.; Show, P.L. Enhanced microalgal protein extraction and purification using sustainable microwave-assisted multiphase partitioning technique. *Chem. Eng. J.* **2019**, *367*, 1–8. [CrossRef]
23. Wang, Y.; Liu, Y.; Han, J.; Hu, S. Application of Water-Miscible Alcohol-Based Aqueous Two-Phase Systems for Extraction of Dyes. *Sep. Sci. Technol.* **2011**, *46*, 1283–1288. [CrossRef]
24. Lin, Y.K.; Show, P.L.; Yap, Y.J.; Tan, C.P.; Ng, E.; Ariff, A.B.; Mohamad Annuar, M.S.B.; Ling, T.C. Direct recovery of cyclodextringlycosyltransferase from Bacillus cereus using aqueous two-phase flotation. *J. Biosci. Bioeng.* **2015**, *120*, 684–689. [CrossRef]
25. Ooi, C.W.; Tey, B.T.; Hii, S.L.; Kamal, S.M.M.; Lan, J.C.W.; Ariff, A.; Ling, T.C. Purification of lipase derived from Burkholderia pseudomallei with alcohol/salt-based aqueous two-phase systems. *Process Biochem.* **2009**, *44*, 1083–1087. [CrossRef]
26. Naimi-Jamal, M.R.; Hamzeali, H.; Mokhtari, J.; Boy, J.; Kaupp, G. Sustainable Synthesis of Aldehydes, Ketones or Acids from Neat Alcohols Using Nitrogen Dioxide Gas, and Related Reactions. *ChemSusChem* **2009**, *2*, 83–88. [CrossRef]
27. Iqbal, M.; Tao, Y.; Xie, S.; Zhu, Y.; Chen, D.; Wang, X.; Huang, L.; Peng, D.; Sattar, A.; Shabbir, M.A.B.; et al. Aqueous two-phase system (ATPS): An overview and advances in its applications. *Biol. Proced. Online* **2016**, *18*, 18. [CrossRef]
28. De Araújo Padilha, C.E.; Dantas, P.V.; Júnior, F.C.; Júnior, S.D.; da Costa Nogueira, C.; de Santana Souza, D.F.; de Oliveira, J.A.; de Macedo, G.R.; dos Santos, E.S. Recovery and concentration of ortho-phenylphenol from biodesulfurization of 4-methyl dibenzothiophene by aqueous two-phase flotation. *Sep. Purif. Technol.* **2017**, *176*, 306–312. [CrossRef]
29. Bi, P.Y.; Dong, H.R.; Yuan, Y.C. Application of aqueous two-phase flotation in the separation and concentration of puerarin from Puerariae extract. *Sep. Purif. Technol.* **2010**, *75*, 402–406. [CrossRef]
30. Jiang, B.; Na, J.; Wang, L.; Li, D.; Liu, C.; Feng, Z. Separation and Enrichment of Antioxidant Peptides from Whey Protein Isolate Hydrolysate by Aqueous Two-Phase Extraction and Aqueous Two-Phase Flotation. *Foods* **2019**, *8*, 34. [CrossRef]
31. Bi, P.Y.; Chang, L.; Mu, Y.L.; Liu, J.Y.; Wu, Y.; Geng, X.; Wei, Y. Separation and concentration of baicalin from Scutellaria Baicalensis Georgi extract by aqueous two-phase flotation. *Sep. Purif. Technol.* **2013**, *116*, 454–457. [CrossRef]
32. Sankaran, R.; Show, P.L.; Lee, S.Y.; Yap, Y.J.; Ling, T.C. Integration process of fermentation and liquid biphasic flotation for lipase separation from Burkholderia cepacia. *Bioresour. Technol.* **2018**, *250*, 306–316. [CrossRef] [PubMed]
33. Chew, K.W.; Chia, S.R.; Krishnamoorthy, R.; Tao, Y.; Chu, D.T.; Show, P.L. Liquid biphasic flotation for the purification of C-phycocyanin from Spirulina platensis microalga. *Bioresour. Technol.* **2019**, *288*, 121519. [CrossRef] [PubMed]
34. Leong, H.Y.; Chang, Y.-K.; Ooi, C.W.; Law, C.L.; Julkifle, A.L.; Show, P.L. Liquid Biphasic Electric Partitioning System as a Novel Integration Process for Betacyanins Extraction from Red-Purple Pitaya and Antioxidant Properties Assessment. *Front. Chem.* **2019**, *7*, 1–11. [CrossRef] [PubMed]

Review

Potential of *Jatropha curcas* L. as Biodiesel Feedstock in Malaysia: A Concise Review

Nurul Husna Che Hamzah [1], Nozieana Khairuddin [1,*], Bazlul Mobin Siddique [2] and Mohd Ali Hassan [3]

[1] Department of Basic Science and Engineering, Faculty of Agriculture and Food Sciences, Universiti Putra Malaysia Bintulu Sarawak Campus, Bintulu 97008, Sarawak, Malaysia; gs56633@student.upm.edu.my
[2] School of Chemical Engineering and Science, Swinburne University of Technology Sarawak Campus, Kuching 93350, Sarawak, Malaysia; msiddique@swinburne.edu.my
[3] Department of Bioprocess Technology, Faculty of Biotechnology and Biomolecular Sciences, Serdang 43400, Selangor, Malaysia; alihas@upm.edu.my
* Correspondence: nozieana@upm.edu.my; Tel.:+60-86-855-823

Received: 24 April 2020; Accepted: 2 June 2020; Published: 6 July 2020

Abstract: Fluctuation in fossil fuel prices and the increasing awareness of environmental degradation have prompted the search for alternatives from renewable energy sources. Biodiesel is the most efficient alternative to fossil fuel substitution because it can be properly modified for current diesel engines. It is a vegetable oil-based fuel with similar properties to petroleum diesel. Generally, biodiesel is a non-toxic, biodegradable, and highly efficient alternative for fossil fuel substitution. In Malaysia, oil palm is considered as the most valuable commodity crop and gives a high economic return to the country. However, the ethical challenge of food or fuel makes palm oil not an ideal feedstock for biodiesel production. Therefore, attention is shifted to non-edible feedstock like *Jatropha curcas Linnaeus* (*Jatropha curcas* L.). It is an inedible oil-bearing crop that can be processed into biodiesel. It has a high-seed yield that could be continually produced for up to 50 years. Furthermore, its utilization will have zero impact on food sources since the oil is poisonous for human and animal consumption. However, Jatropha biodiesel is still in its preliminary phase compared to palm oil-based biodiesel in Malaysia due to a lack of research and development. Therefore, this paper emphasizes the potential of *Jatropha curcas* as an eco-friendly biodiesel feedstock to promote socio-economic development and meet significantly growing energy demands even though the challenges for its implementation as a national biodiesel program might be longer.

Keywords: non-edible; oil; biodiesel production; fuel

---

1. Introduction

The depletion of crude oil reserves coupled with the awareness of environmental issues and escalating petroleum prices have stimulated the search for alternatives to reduce overdependence on conventional fossil fuels [1,2]. Historically, researchers have substituted conventional fuels with renewable energy resources (e.g., biofuels) since the invention of diesel engines [3]. These technologies have since advanced to this day. Typically, conventional diesel and petroleum fuels release harmful gases into the atmosphere, thereby causing global warming and climate change. Furthermore, fossil fuels contribute to pollution through the emission of major greenhouse gases (GHG). In principle, a biofuel is cleaner than any fossil fuel, since it can reduce carbon dioxide (CO_2) emissions by 78% and carbon monoxide (CO) emissions by 50% [4].

The global human population is predicted to increase by 34% by 2050 [5]. This increase in human population has become a contributing factor to the high demand for energy consumption. Moreover, the continuous exploitation and rapid depletion of the Earth's natural and mineral resources will significantly increase the energy demand for transportation, industrial, and other purposes. According to the International Energy Agency (IEA), it is estimated that global energy consumption will soar by 53% by the year 2030 [6]. Another study also reported that global petroleum resources will be completely depleted within 40 years [7]. This situation could result in annual oil price escalation in the near future. Therefore, renewable energy resources such as biofuels urgently need to be adopted as substitutes for conventional fuels.

Biodiesel derived from *Jatropha curcas*, which is suitably planted in tropical or subtropical countries such as Malaysia and Indonesia, could potentially reduce the use of conventional fuels. However, there are many challenges faced during the cultivation, harvesting, and processing of the crop yield. Hence, finding the root cause is important for resolving the issues and ensuring a higher quality of harvested Jatropha seeds. Consequently, Jatropha's enormous potential in the financial, agricultural, environmental, sustainable energy production, and industrial fronts could attract the attention of researchers and policymakers.

2. Distribution and Physicochemical Properties of Biodiesel Feedstock

Bioethanol, biodiesel, and biogas are the main biofuel components of various agricultural biomasses produced from the different biochemical routes [8]. The first generation feedstock was produced from edible oils such as corn, sugarcane, sugar-beet, and others. Unlike the first generation of biofuels, second-generation biofuels targeted non-food biomass and agricultural residues [9]. Biodiesel originated from the first and second-generation of biofuels, as it is processed from agricultural crops and residues. Typically, the sources of biodiesel differ in many regions or countries. European nations use rapeseed due to the surpluses from edible oil production. On the other hand, soybean is commonly utilized for biodiesel production in the United States, which is becoming the main biodiesel-producing country [10]. Singh, V. et al., had summarized all classification of biofuels production starting from the first until the fourth generation as shown in Figure 1.

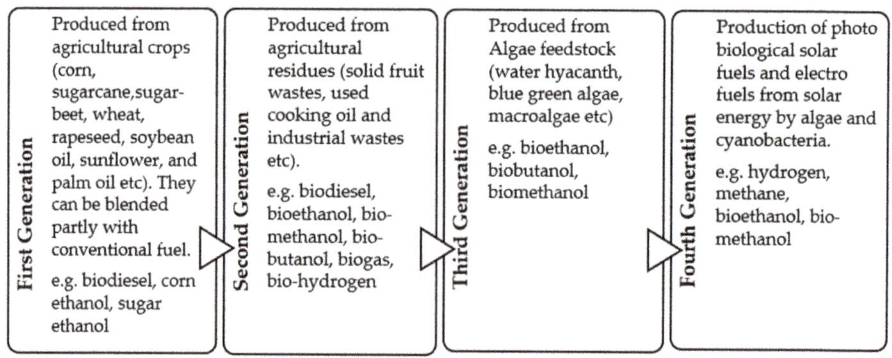

Figure 1. Classification of biofuels (adapted from [11]).

On the contrary, the excess palm oil and coconut oil in Malaysia, Indonesia, and Thailand could be utilized for the synthesis of biodiesel. However, the food versus fuel competition could be overcome by exploring non-edible seed oils such as *Jatropha curcas* and *Karanja* (*Pongamia* second-generation) as raw materials for biodiesel production [12]. Other than vegetable-based biodiesel, waste cooking oil (WCO) is becoming more popular for biodiesel production in Malaysia due to the low cost and the high volume of waste generation in each household. Kabir et al. reported that the average WCO generated in Malaysia per household is 2.34 kg/month [13].

Biodiesel processed from animal fats and vegetable oils is defined as fatty acid alkyl esters or fatty acid methyl esters (FAME) [14]. It is typically synthesized from the reaction of triglycerides with alcohol and a catalyst in a process known as transesterification (Figure 2). The transesterification will produce FAME and simultaneously cause saponficiation or soap formation to occur. Common catalysts usually employed during the reaction are homogeneous alkaline catalysts—for instance, potassium hydroxide (KOH), sodium hydroxide (NaOH), potassium methoxide (CH_3OK), and sodium methoxide (CH_3ONa). However, it is necessary to control the amount of alkali catalyst, because excess alkali enhances the saponification reaction which reduces the yield of product [15]. Nevertheless, the saponification value is important as an indicator of the oil as normal triglycerides, making the oil useful in the soap and shampoo industries [16]. It is reported that the nature of the catalyst employed is crucial to converting triglycerides into biodiesel. For instance, using a homogeneous catalyst will produce glycerol or soap as a by-product, which could risk consuming or even deactivating the catalyst. As a result, the biodiesel purification process would be hampered by the loss of catalytic process [17]. As for heterogeneous catalysts, a higher quality of FAME can be generated after transesterification.

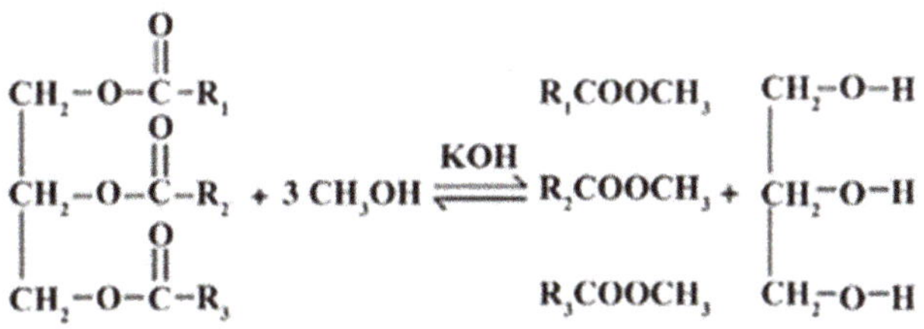

$$
\begin{array}{ccc}
CH_2-O-\overset{\overset{\displaystyle O}{\|}}{C}-R_1 & & \\
CH_2-O-\overset{\overset{\displaystyle O}{\|}}{C}-R_2 \;+\; 3\,CH_3OH \; & \underset{\displaystyle}{\overset{\displaystyle KOH}{\rightleftharpoons}} & R_1COOCH_3 \;\; CH_2-O-H \\
CH_2-O-\overset{\overset{\displaystyle O}{\|}}{C}-R_3 & & R_2COOCH_3 \;+\; CH_2-O-H \\
& & R_3COOCH_3 \;\; CH_2-O-H
\end{array}
$$

Figure 2. Transesterification of triglycerides with alcohol and catalyst [18].

The physicochemical properties of biodiesel depend on numerous factors—for example, the composition of the fatty acid in the raw feedstock, the chain length of the fatty acid, the saturation degree, and branching. Other factors include the production technique and operating conditions for the biodiesel synthesis [19]. The quality of biodiesel may also differ due to the impurities from unreacted feedstock glycerides, the fraction of non-fatty acids, or runaway reactions during the process of transesterification [20]. Typically, a longer fatty acid chain will enhance the synthesis of biodiesel products with a higher cetane number, which results in lower emissions of toxic NOx [21]. The composition of the fatty acid determines the level of saturation with higher compositions, resulting in a higher degree of saturation and viscosity [10].

3. Potential of *Jatropha curcas* as Biodiesel Feedstock

Energy crops are specifically grown for fuel and energy production. Currently, these crops only contribute to a comparably small percentage of the total biomass energy produced each year. However, this percentage is expected to increase over the next few decades. Nevertheless, energy crops compete for land earmarked for food production, environmental protection, and forestry or nature conservation. Generally, the characteristics of ideal energy crops are high yield (maximum dry matter per hectare production), low cost, low energy input, low nutrient or fertilizer requirements, pest resistance, and the composition with the least contaminants generated [22].

In Malaysia, palm oil is considered the most valuable energy crop due to its abundance and productivity. The Malaysian Palm Oil Board (MPOB) reports that the total area of oil palm planted in 2019 was 5.9 million ha [23]. Recently, the Ministry of Primary Industries and the MPOB launched the B20 Biodiesel program (20% palm methyl esters blended with 80% petroleum diesel) to offset the palm vegetable oil demand and stabilize the market price of these products [24]. However, this crop is still perceived as a major source of vegetable oil all over the world rather than as a fuel, and the demand for it as a food ingredient is increasing. Therefore, the idea of utilizing inedible food crops such as *Jatropha curcas* could help overcome the major problems faced by the first generation of biodiesel feedstock.—for example, the food vs. fuel dilemma, the issues of scaling, and the inability to grow on peripheral areas of land, etc. Its average productive life span is also longer than that of oil palm

(50 years and 30 years, respectively). Besides, its oil content is reportedly between 63.16% and 66.4%, which is higher than that of soybean (18.35%), linseed (33.33%), and palm kernel (44.6%) [25].

The tree is a drought-resistant perennial and grows well in marginal land that has little or no agricultural or industrial value due to poor soil and other undesirable characteristics. Correspondingly, unlike some other conventional edible feedstock, the planting of this crop is no threat to existing arable land land and the food chain. In 2012, it was reported that Forest Research Institute Malaysia (FRIM) has successfully planted 6000 *Jatropha curcas* plants in Terengganu, since the state has about 71,000 ha of problematic land along the coast that is left without commercialized agricultural activities [26].

Besides FRIM, other government agencies such as the Malaysian Rubber Board (MRB) estimate that approximately 50 hectares (ha) of *Jatropha curcas* were planted in Sungai Buloh, Selangor, and Kota Tinggi as of June 2012. In the early phase, about 1712 ha of land in total was earmarked for the principal cultivation of this crop in the country. Likewise, a small number of local private companies have indicated their willingness to cultivate *Jatropha curcas* on the scale of 400 ha to 1000 ha. Some stakeholders are planning to expand the cultivation to 57,601 ha in total by the year 2015. The Plantation, Industries, and Commodities Ministry (MPIC) has also initiated an experimental project on Jatropha, for which 300 ha has been allocated [4].

Biodiesel fuel has been widely adopted in most countries because it is biodegradable, non-toxic, and environmentally friendly with lower greenhouse gas (GHG) emissions. Furthermore, biodiesel adoption is considered an excellent method for reducing noise and potentially scaling down air pollutants such as carbon monoxide (CO), sulfur, polycyclic aromatic hydrocarbon (PAH), smoke, and particulate matter (PM) [27]. The most significant fuel characteristics considered for biodiesel application in diesel engines are density, viscosity, cetane number, and flash point [28].

The Cetane number is the principal indicator of fuel quality, particularly ignition and combustion in diesel engines. Typically, a high Cetane number indicates a lower ignition delay time—i.e., the time interval from the injection of the fuel to initialization of ignition in the combustion chamber. Typically, the parameter ensures a good quality fuel combustion, cold start, and engine performance, along with low white smoke formation and emissions [29]. The Cetane number of Jatropha is reported to be as high as 55, which is similar to that of diesel (Table 1). Hence, any biodiesel to be effectively substituted for diesel should retain a higher cetane number.

Table 1. Comparison of vegetable oil with biodiesel specification [30–33].

Properties	Diesel	Palm Biodiesel	Jatropha Biodiesel	ASTM D6751	EN 14214
Cetane number	45–55	52	57	Min. 47	Min. 51
Flash point, °C	50–98	181	135	Min. 130	Min. 120
Viscosity, mm^2/s	2.5–5.7	4.9	4.8	1.9–6.0	3.5–5.0
Density, kg/m^3	816–840	879.3	862	860–900	860–900

In practice, the blend of any vegetable-based biodiesel with petroleum diesel need to comply with the two most referred biodiesel standards, namely, the American Standard Specifications for Biodiesel Fuel (B100) Blend Stock for Distillate Fuels (ASTM 6751) and European Standard for Biodiesel (EN 14214). According to the both standards, biodiesel must meet the minimum flash point—i.e., above 120 °C. The flashpoint is the temperature at which a fuel begins to burn after interaction with fire. Typically, any fuel with a high flash point could result in the deposition of carbon in the combustion compartment. Since Jatropha oil has a lower flash point compared to palm oil (162 °C and 181 °C, respectively), it has a higher potential compared to palm oil as a biodiesel.

Based on Table 1, Jatropha oil has a medium viscosity between palm oil and diesel, which is good for biodiesel utilization. Typically, most vegetable oils have higher viscosities due to their high fatty acid compositions relative to petroleum diesel. The higher viscosity indicates a better lubrication of the fuel, which reduces wear on the moving mechanical parts of the engine. Ultimately, the reduced wear prevents leakage and reduces issues related to power losses and the durability of engines. Viscosity plays an important roles in the atomization efficiency of fuel injection inside the combustion chamber, fuel droplet size distribution, and the mixture uniformity. If viscosity is too high it may lead to pump damage, filter clogging, poor combustion, and increased emissions. A higher viscosity will also lead to greater surface tension and will influence the dissolution of a liquid jet into smaller fuel droplets, which will impose a bad effect on the spray characteristic of the fuel spray injector in a diesel engine. As a result, larger size fuel droplets are injected from an injector nozzle instead of a spray of fine droplets, leading to inadequate air–fuel mixing [34,35].

Furthermore, the price of biodiesel feedstock derived from *Jatropha curcas* is considered to be the most affordable compared to other biodiesel feedstock as shown in Table 2. Its low end price will attract consumers to be using biodiesel on the road. Eventually, it would increase the market demand for biodiesel used in the transportation vehicles and will enhance the economy from people living in rural areas. The data on the price of B100 biodiesel for different feedstock in Table 2 was reported by Lim, S. and Teong, L.K. [36].

Table 2. A comparison of biodiesel prices from different feedstock (adapted from [36]).

Feedstock	Price of B100 Biodiesel (USD/Tonne)
Jatropha	400–500
Palm oil	720–750
Soybean	800–805
Rapeseed	940–965

4. Biodiesel Processing from *Jatropha curcas*

The general processing of biodiesel from *Jatropha curcas* oil involves three major steps, namely seed drying, oil extraction, and transesterification (the processing of pure vegetable oil into biodiesel), as shown in Figure 3. There are also other minor steps that are considered significant, such as the cleaning of the seeds, dehulling, and post-harvest storage. The conventional technique for recovering oils from Jatropha seeds is through the use of a mechanical screw press machine. However, a large proportion of the oil is retained in the kernel, which requires more effective ways to extract the residual oil. The most notable extraction techniques include ultrasound-assisted systems, enzyme extraction, and the utilization of catalytic materials [37]. The catalyst materials are chemicals that enhance the process of transesterification. The extraction method is closely related to the cost of mass biodiesel production in a biorefinery plant.

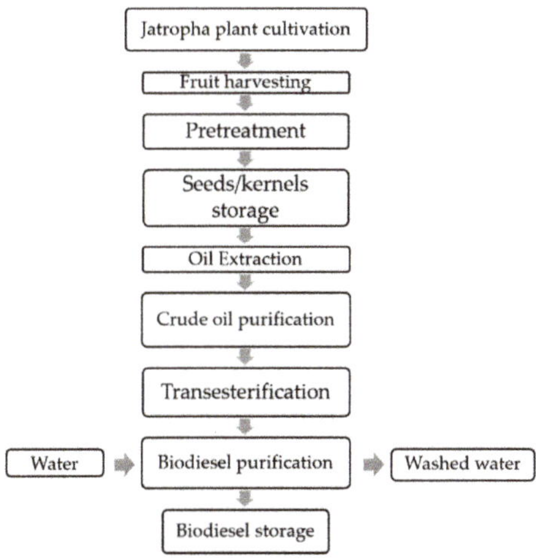

Figure 3. Biodiesel processing of *Jatropha curcas* [38–41].

The extracted oil subsequently undergoes a purification and transesterification process for the production of crude biodiesel. However, the crude biodiesel cannot be directly used as a transportation fuel due to limitations such as the standard requirements for biodiesel in the industry. Therefore, the crude biodiesel is usually blended with pure diesel in certain percentages before utilization in diesel engines. Before blending, the crude biodiesel is purified to remove unwanted moisture and the chemical waste produced during the transesterification process. The most popular method of purification is water washing since it is cheap and easy, although this time-consuming [42] process needs to be run several times until no more glycerol is produced.

The composition of the fatty acid significantly affects the fuel properties of the biodiesel [18]. Typically, inedible oils such as Jatropha comprise high compositions of detrimental free fatty acids (FFA) (>1% *w/w*), which reduces the biodiesel yields. Likewise, the high amount of fatty acid hampers the direct conversion of the oil into biodiesel since the high FFAs promote soap formation, which can hamper the separation of products during or after transesterification. Jatropha oil comprises nearly 14% FFA, which exceeds by far the standard limit of 1% FFA. Therefore, the pretreatment stage is required to lower the feedstock FFAs for an enhanced yield of biodiesel [30]. The typical unwanted saponification reaction that forms soap and water when NaOH catalyst is utilized is presented in Equation (1).

$$R1\text{-COOH (FFA)} + NaOH \text{ (sodium hydroxide)} \rightarrow R1COONa \text{ (soap)} + H_2O \text{ (water)}. \qquad (1)$$

Therefore, the two-step transesterification process is an efficient method used extensively to process crude oil from *Jatropha curcas* that contains significant FFAs. Furthermore, the pretreatment or esterification process using the acid- base catalyst is performed to reduce the FFA content of *Jatropha curcas* oil. Hence, transesterification subsequently results in an optimal yield of 90% methyl ester after two hours [43]. In addition, the acid catalyst reduces the FFA content to <1% through conversion into esters by esterification. The second step involves the transesterification of the triglycerides in *J. curcas* oil into biodiesel in the presence of an alkaline catalyst. The unsaturation of fatty acids in oil is an important factor that determines the biodiesel quality. In this aspect, Malaysia, however, sits on the favourable side, as polyunsaturated fatty acids are lower in Malaysia-grown *Jatropha* oil than in the varities found in neighbouring countries [44]. Interestingly though, the Triacylglycerol (TAG) profile among these different varities of *Jatropha* oil from Malaysia and neighbouring countries did not show significant differences [45]. The kinematic viscosity of *Jatropha* oil is higher than that of regular diesel fuel, which indeed imparts a problem for all use in a diesel engine without blending. On the other side, it is much safer to handle and for storage than regular diesel fuel at higher temperatures [39]. Considering all these promising factors, its a highly promising seed oil to be taken seriously when it comes to boosting the socio-economic conditions in Malaysia.

5. *Jatropha curcas* Planting Challenges in Malaysia

Jatropha curcas is a highly promising crop for biodiesel production, although supportive innovative technologies are required for planting, harvesting, and oil extraction. Furthermore, mechanized crop operations are limited, and hence Malaysia needs to import the knowledge and machines from other countries such as India. Other notable challenges of Jatropha production are the poor seed yield, low impute crop, and pest and disease vulnerability. Although it is a promising crop for biodiesel production, the unavailability of a high-yielding cultivar is a major failure factor [46]. Besides this, high-yield fluctuation among trees and the lack of disease resistance could also hamper the commercialization of Jatropha biodiesel. In addition, it also needs appropriate nutrients and irrigation for growth and maturity, although it could flourish under limited conditions. Recent studies have reported that *Jatropha curcas* is susceptible to virus-related contaminations such as the Cucumber mosaic virus, powdery mildew, leaf spots, and soil fungous diseases. Other notable challenges are insect and rodent attacks, which result in the extensive defoliation of the plant [47,48].

In some environs, Jatropha creates complications such as weeds, which could require higher labour costs during cultivation [49]. Lastly, technologies for harvesting and post-harvesting, such as oil extraction, are still lacking. Furthermore, biodiesel is vulnerable to oxidation when exposed to air, selected storage conditions, and high levels of unsaturated fatty acids [50]. As a result, the oil content deteriorates due to inappropriate handling and storage. In addition, the main ester components of biodiesel could rapidly undergo hydrolysis to form carboxylic acids in the presence of water. Hence, both materials along with the chemical structure of biodiesel affect the swelling characteristics of the elastomer, which in turn depends on its composition and the preparation of the compound [51].

6. Approaches to Enhance the Jatropha Seed Oil

One of the most critical solutions is the cultivation of high oil yield *Jatropha curcas*, although such commercial varieties are lacking. The existing Jatropha breeding scheme is restricted to the traditional approach, which involves the collection of wild plant germplasm capital of Jatropha [52]. Furthermore, the review of modern applications of biotechnology for improving Jatropha is minimal [53]. In particular, research is largely absent on the expression, cloning, and annotation of biotic roles for Jatropha genes, which are responsible for its economic characteristics [46]. The main purpose of cultivation should be to advance the unit seed yield of Jatropha for commercial uses. Therefore, Jatropha cultivation techniques must require the application of numerous field practices such as planting, site planning, tree density, irrigation management, and cropping treatments.

Other notable practices involve fertilization and canopy protection, along with the control of pests and diseases. However, there is limited research that precisely and systematically validates the effect of field activity on the seed yield of *Jatropha curcas*. Selected methodological studies on planting base and management restrict the commercial cultivation of *J. curcas* [54]. Furthermore, there are limited comprehensive field or empirical reports on seed yield under different agronomic or treatment methods. For instance, data on the cultivated Jatropha tree density, the strength and interval of pruning its canopies, the insecticide impact, fertilization, and irrigation efficiency are mostly lacking in the literature [46].

7. Economic and Business Perspectives of Biodiesel from Jatropha Oil

Malaysia is amongst the world's premier biodiesel manufacturers. The immense profit of biodiesel in terms of rising fossil fuel values and the intention of decreasing the emission of greenhouse gasses (GHG) are factors that contribute to the development of the biodiesel market in Malaysia. Additionally, the community also is personally involved in production in order to improve income and eliminate poverty. In Malaysia, the growth of the biodiesel industry has been maintained at the top of its agenda by granting a significant amount of subsidies and farmer support programs. In fact, the government is encouraging private companies to launch more treating plants and improve biodiesel for vehicles and electricity generation. This is parallel with the post estimation of diesel vehicles, which accounts for approximately 5% of the motor vehicle population in Malaysia. The number of vehicles used is as indicated by the registered vehicles from 1996 to 2009. Based on the post estimation, diesel vehicles may possibly provide a larger share of the total in the forthcoming prior to the commencement of B5 and the campaign of the government incentives.

To date, the majority of countries have declared the standards and policies of their biodiesel. All countries have regulated their mandate or aim for biodiesel consumption success and proclaimed the exploitation of biodiesel energy fusion in their policies. As recapped in Figure 4, the national biodiesel policy of Malaysia stated on 21 March 2006 [24] objectives are as follows:

a) The employment of environmentally friendly, sustainable, and viable sources of energy to reduce the dependency on depleting fossil fuels;

b) The enhanced prosperity and well being of all the stakeholders in the agriculture and commodity-based industries through stable and remunerative prices;

c) Reducing the country's dependence on depleting reserves of fossil fuels, promoting the demand for palm oil, and stabilizing its prices.

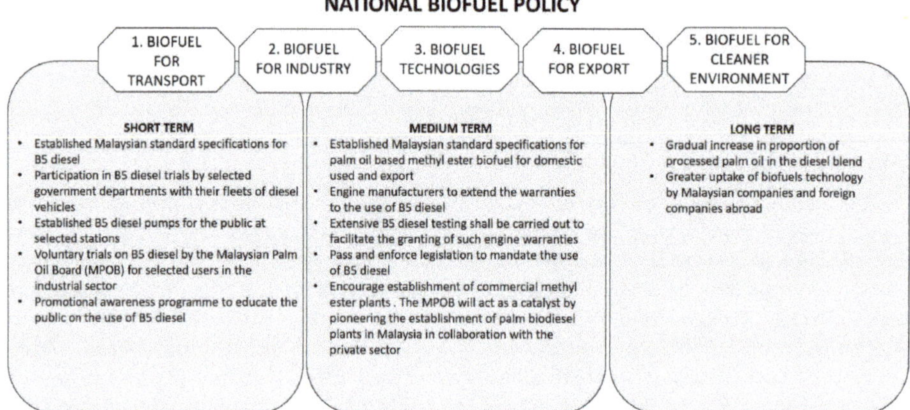

Figure 4. Strategic five thrusts of Malaysia's national biofuel policy and implementation (Adapted from [24]).

8. Conclusions

In conclusion, *Jatropha curcas* has a bright future as the next important biodiesel feedstock, considering the problems currently faced by the oil palm industry in Malaysia. It is necessary to create a higher value of its by-products in order to make Jatropha a viable biofuel in the market. Therefore, other parts of this crop, such as wood, fruit shells, seed husks and kernels, could be used to produce renewable energy [55]. The waste generated after the oil extraction process, such as the pressed cake, could also be utilized as organic fertilizer. Thus, this crop has the same characteristics as the oil palm crop, which can be used as a whole package. However, more research and development needs to be undertaken by researchers to find solutions to the existing challenges outlined in this review.

Biodiesel from *Jatropha curcas* has great potential to be implemented because it has lower carbon and emissions of GHG. It also has a lower cost compared to palm biodiesel. However, our dependency on the foreign workers in the plantation is unavoidable. The mechanization and automation specifically for maintaining the good health of *Jatropha curcas* must be improved and tested in the field beforehand. The workers must be careful while harvesting because the oil yield is dependent on the right timing of harvesting. As we know, this fruit's ripening is uneven, making harvesting a strenuous and time-consuming process. Until 2015, it has been stated that Malaysia has a total of 259,906 hectares of Jatropha crops plantation, and the current planted crops are capable of producing 4.27 tons of dry seeds every year [56,57].

As the world has been affected by global warming and the alarming threat of food security for the growing human population, much attention should be focused on non-edible oil bearing crops as biodiesel feedstock. This renewable green energy will protect the environment from the emission of harmful gases due to the combustion of fossil fuels and become an effective substitution for the depletion of the mineral resources of Earth. While many challenges await as this crop is introduced as a new biodiesel feedstock, it will never be impossible to cope with them when there is ample research and development undertaken and more expertise involved in joint ventures for the research project on *Jatropha curcas*.

Author Contributions: Investigation, N.H.C.H.; writing—original draft preparation, N.H.C.H.; Funding acquisition, N.K.; Supervision, N.K.; M.A.H., and B.M.S.; writing—review and editing, N.K., M.A.H. and B.M.S. All authors have read and agreed to the published version of the manuscript.

Funding: This research was funded by Geran Putra—Inisiatif Putra Berkumpulan grant number (9671301) to support the research and development activities in Universiti Putra Malaysia Bintulu Sarawak Campus, Malaysia.

Acknowledgments: The authors acknowledge the funding received from the Geran Putra—Inisiatif Putra Berkumpulan by Universiti Putra Malaysia, Malaysia.

Conflicts of Interest: The authors declare no conflict of interest.

References

1. Sharma, B.; Ingalls, R.G.; Jones, C.L.; Khanchi, A. Biomass supply chain design and analysis: Basis, overview, modelling, challenges, and future. *Renew. Sustain. Energy Rev.* **2013**, *24*, 608–627. [CrossRef]
2. Tobib, H.M.; Rostam, H.; Mossa, M.A.; Aziz Hairuddin, A.; Noor, M.M. The performance of an HCCI-DI engine fuelled with palm oil-based biodiesel. *IOP Conf. Ser. Mater. Sci. Eng.* **2019**, *469*, 012079. [CrossRef]
3. Demirbas, A. Biodiesel fuels from vegetable oils via catalytic and non-catalytic supercritical alcohol transesterifications and other methods: A survey. *Energy Convers. Manag.* **2003**, *44*, 093–109. [CrossRef]
4. Sulaiman, Z.; Ramlan, M.F. Research and development on Jatropha (Jatropha curcas) by Malaysian Rubber Board. Presented at the Malaysia-Indonesia Scientific Meeting on Jatropha Royal Chulan, Kuala Lumpur, Malaysia, 6 February 2013. [CrossRef]
5. Sadhukhan, J.; Martinez-Hernandez, E.; Murphy, R.J.; Ng, D.K.S.; Hassim, H.; Siew, K.; Kin, W.Y.; Jaye, M.F.I.; Hang, P.L.; Andiappan, V. Role of bioenergy, biorefinery and bioeconomy in sustainable development: Strategic pathways for Malaysia. *Renew. Sustain. Energy Rev.* **2018**, *81*, 1966–1987. [CrossRef]
6. Ong, H.C.; Mahlia, T.M.I.; Masjuki, H.H. A review of energy scenario and sustainable energy in Malaysia. *Renew. Sustain. Energy Rev.* **2011**, *15*, 639–647. [CrossRef]
7. Peng, D. The effect on diesel injector wear, and exhaust emissions by using ultralow sulphur diesel blending with biofuels. *Mater. Trans.* **2015**, *56*, 642–647. [CrossRef]
8. Paudel, S.R.; Banjara, S.P.; Choi, O.K.; Park, K.Y.; Kim, Y.M.; Lee, J.W. Pretreatment of agricultural biomass for anaerobic digestion: Current state and challenges. *Bioresour. Technol.* **2017**. [CrossRef]
9. Food and Agriculture Organization of the United Nations (FAO). *Tackling Climate Change through Livestock*; Food and Agriculture Organization of the United Nations: Rome, Italy, 2013. Available online: http://www.fao.org/3/i3437e.pdf (accessed on 27 April 2020).
10. Ayetor, G.K.; Sunnu, A.; Parbey, J. Effect of biodiesel production parameters on viscosity and yield of methyl esters: *Jatropha curcas*, *Elaeis guineensis* and *Cocos nucifera*. *Alex. Eng. J.* **2015**, *54*, 1285–1290. [CrossRef]
11. Singh, V.; Zhao, M.; Fennell, P.S.; Shah, N.; Anthony, E.J. Progress in biofuel production from gasification. *Prog. Energy Combust. Sci.* **2017**, *61*, 189–248. [CrossRef]
12. Silitonga, A.S.; Masjuki, H.H.; Mahlia, T.M.I.; Ong, H.C.; Atabani, A.E.; Chong, W.T. A global comparative review of biodiesel production from jatropha curcas using different homogeneous acid and alkaline catalysts: Study of physical and chemical properties. *Renew. Sustain. Energy Rev.* **2013**, *24*, 514–533. [CrossRef]
13. Kabir, I.; Yacob, M.; Radam, A. Households' awareness, attitudes and practices regarding waste cooking oil recycling in Petaling, Malaysia. *IOSR-JESTFT* **2014**, *8*, 45–51. [CrossRef]
14. Muanruksa, P.; Kaewkannetra, P. Combination of fatty acids extraction and enzymatic esterification for biodiesel production using sludge palm oil as a low-cost substrate. *Renew. Energy* **2020**, *146*, 901–906. [CrossRef]
15. Babajide, O.; Petrik, L.; Amigun, B.; Ameer, F. Low-cost feedstock conversion to biodiesel via ultrasound technology. *Energies* **2010**, *3*, 1691–1703. [CrossRef]
16. Akbar, E.; Yaakob, Z.; Kamarudin, S.K.; Ismail, M.; Salimon, J. Characteristic and composition of Jatropha Curcas oil seed from Malaysia and its potential as biodiesel feedstock feedstock. *Eur. J. Sci. Res.* **2009**, *29*, 396–403.
17. Ismail, S.A.A.; Ali, R.F.M. Physico-chemical properties of biodiesel manufactured from waste frying oil using domestic adsorbents. *Sci. Technol. Adv. Mater.* **2015**. [CrossRef]
18. Saraf, S.; Thomas, B. Influence of feedstock and process chemistry on biodiesel quality. *Process. Saf. Environ. Prot.* **2007**, *85*, 360–364. [CrossRef]
19. Reddy, A.N.R.; Saleh, A.A.; Islam, M.S.; Hamdan, S.; Rahman, M.R.; Masjuki, H.H. Experimental evaluation of fatty acid composition influence on Jatropha biodiesel physicochemical properties. *J. Renew. Sustain. Energy* **2018**, *10*, 013103. [CrossRef]

20. Knothe, G.; Steidley, K.S. Kinematic viscosity of biodiesel fuel component and related compounds: Influence of compound structure and comparison to petrodiesel fuel components. *Fuel* **2005**, 1059–1065. [CrossRef]

21. Knothe, G.; Matheaus, A.C.; Ryan, T.W., III. Cetane numbers of branched and straight-chain fatty esters determined in an ignition quality tester. *Fuel* **2003**, *82*, 971–975. [CrossRef]

22. Sims, R.E.H.; Hastings, A.; Schlamadinger, B.; Taylor, G.; Smith, P. Energy crops: Current status and future prospects. *Glob. Chang. Biol.* **2006**, *12*, 2054–2076. [CrossRef]

23. Malaysian Palm Oil Board Homepage. Available online: http://bepi.mpob.gov.my (accessed on 15 April 2020).

24. The Malaymail. B20 Biodiesel Programme to Be Expanded Nationwide in June 2021, Says Teresa Kok. Available online: https://www.malaymail.com (accessed on 27 March 2020).

25. Mofijur, M.; Masjuki, H.H.; Kalam, M.A.; Hazrat, M.A.; Liaquat, A.M.; Shahabuddin, M.; Varman, M. Prospects of biodiesel from Jatropha in Malaysia. *Renew. Sustain. Energy Rev.* **2012**, *16*, 5007–5020. [CrossRef]

26. The Star. FRIM Ready to Produce Biodiesel. Available online: https://www.thestar.com.my (accessed on 27 March 2020).

27. Ong, H.C.; Mahlia, T.M.I.; Masjuki, H.H.; Norhasyima, R.S. Comparison of palm oil, *Jatropha curcas* and *Calophyllum inophyllum* for biodiesel: A review. *Renew. Sustain. Energy Rev.* **2011**, *15*, 3501–3515. [CrossRef]

28. Patel, C.; Chandra, K.; Hwang, J.; Agarwal, R.A.; Gupta, N. Comparative compression ignition engine performance, combustion, and emission characteristics, and trace metals in particulates from Waste cooking oil, Jatropha and Karanja oil derived biodiesels. *Fuel* **2019**, 1366–1376. [CrossRef]

29. Ramos, M.J.; Fernández, C.M.; Casas, A.; Rodríguez, L.; Pérez, Á. Influence of fatty acid composition of raw materials on biodiesel properties. *Bioresour. Technol.* **2009**, *100*, 261–268. [CrossRef] [PubMed]

30. Atadashi, I.M.; Aroua, M.K.; Aziz, A.A. High-quality biodiesel and its diesel engine application: A review. *Renew. Sustain. Energy Rev.* **2010**, *14*, 1999–2008. [CrossRef]

31. Chongkhong, S.; Tongurai, C.; Chetpattananondh, P.; Bunyakan, C. Biodiesel production by esterification of palm fatty acid distillate. *Biomass Bioenergy* **2007**, *31*, 563–568. [CrossRef]

32. Sahoo, P.K.; Das, L.M. Process optimization for biodiesel production from Jatropha *Karanja* and Polanga oils. *Fuel* **2009**, *88*, 1588–1594. [CrossRef]

33. Tiwari, A.K.; Kumar, A.; Raheman, H. Biodiesel production from Jatropha (*Jatropha curcas*) with high free fatty acids: An optimized process. *Biomass Bioenergy* **2007**, *31*, 569–575. [CrossRef]

34. Ejim, C.E.; Fleck, B.A.; Amirfazli, A. Analytical study for atomization of biodiesels and their blends in a typical injector: Surface tension and viscosity effects. *Fuel* **2007**, *86*, 1534–1544. [CrossRef]

35. Abedin, M.J.; Masjuki, H.H.; Kalam, M.A.; Sanjid, A.; Rahman, S.M.A.; Fattah, I.M.R. Performance, emissions, and heat losses of palm and jatropha biodiesel blends in a diesel engine. *Ind. Crops Prod.* **2014**, *59*, 96–104. [CrossRef]

36. Lim, S.; Teong, L.K. Recent trends, opportunities and challenges of biodiesel in Malaysia: An overview. *Renew. Sustain. Energy Rev.* **2010**, *14*, 938–954. [CrossRef]

37. Koh, M.Y.; Idaty, T.; Ghazi, M. A review of biodiesel production from *Jatropha curcas* L. oil. *Renew. Sustain. Energy Rev.* **2011**, *15*, 2240–2251. [CrossRef]

38. Jongh, J.A.; van der Putten, E. Contributors. In *The Jatropha Handbook. From Cultivation to Application*; FACT Foundation: Omaha, NE, USA, 2010; ISBN1 9081521918, ISBN2 9789081521918.

39. Kalam, M.A.; Ahamed, J.U.; Masjuki, H.H. Land availability of Jatropha production in Malaysia. *Renew. Sustain. Energy Rev.* **2012**, *16*, 3999–4007. [CrossRef]

40. Mehla, S.K. *Biodiesel Production Technologies*; Joshi, D.C., Sutar, R.F., Parmar, M.R., Singh, S.N., Eds.; Pointer Publishers: Jaipur, India, 2007; Chapter 11, ISBN1 10: 8171325173, ISBN2 13: 9788171325177.

41. Vairavan, K.; Thukkaiyannan, P.; Paramathma, M.; Venkatachalam, P.; Sampathrajan, A. *Biofuel Crops Cultivation and Management: Jatropha, Sweet Sorghum and Sugarbeet*; Agrobios: Jodhpur, India, 2007; ISBN1 8177543164, ISBN2 9788177543162.

42. Ali, R.M.; Farag, H.A.; Amin, N.A.; Farag, I.H. Abu-Tartour phosphate rock catalyst for biodiesel production from waste frying oil. *JOKULL* **2015**, *65*, 233–244.

43. Berchmans, H.J.; Hirata, S. Biodiesel production from crude *Jatropha curcas* L. seed oil with a high content of free fatty acids. *Bioresour. Technol.* **2008**, *99*, 1716–1721. [CrossRef]

44. Augustus, G.D.P.S.; Jayabalan, M.; Seiler, G.J. Evaluation and bioinduction of energy components of Jatropha curcas. *Biomass Bioenergy* **2002**, *23*, 161–164. [CrossRef]

45. Emil, A.; Yaakob, Z.; Kumar, M.N.S.; Jahim, J.M.; Salimon, J. Comparative evaluation of physicochemical properties of jatropha seed oil from Malaysia, Indonesia and Thailand. *JAOCS J. Am. Oil Chem. Soc.* **2010**, *87*, 689–695. [CrossRef]

46. Moniruzzaman, M.; Yaakob, Z.; Shanizuzzaman, M.; Khatun, R.; Islam, A.K.M.A. *Jatropha Biofuel Industry: The Challenges*; INTECH OPEN: London, UK, 2017. [CrossRef]

47. Singh, B.; Singh, K.; Rao, G.R.; Chikara, J.; Kumar, D.; Mishra, D.K.; Saikia, S.P.; Pathre, U.V.; Raghuvanshi, N.; Rahi, T.S.; et al. Agrotechnology of Jatropha curcas for diverse environmental conditions in India. *Biomass Bioenergy* **2013**, *48*, 191–202. [CrossRef]

48. Everson, C.S.; Mengistu, M.G.; Gush, M.B. A field assessment of the agronomic performance and water use of Jatropha curcas in South Africa. *Biomass Bioenergy* **2013**, *59*, 59–69. [CrossRef]

49. Goswami, K.; Choudhury, H.K. Economic benefits and costs of *Jatropha plantation* in North-East India. *Agric. Econ. Res. Rev.* **2011**, *24*, 99–108.

50. Haseeb, A.S.M.A.; Fazal, M.A.; Jahirul, M.I.; Masjuki, H.H. Compatibility of automotive materials in biodiesel: A review. *Fuel* **2011**, *90*, 922–931. [CrossRef]

51. Thomas, E.W.; Fuller, R.E.; Terauchi, K. Fluoroelastomer compatibility with biodiesel. *Fuels* **2007**. [CrossRef]

52. Zhang, G.W. Existing problems and countermeasures for *Jatropha curcas* industrialization in China. *J. Anhui Agric. Sci.* **2009**, *8*, 182.

53. Moniruzzaman, M.; Yaakob, Z.; Khatun, R. Biotechnology for Jatropha improvement: A worthy exploration. *Renew. Sustain. Energy Rev.* **2016**, *54*, 1262–1277. [CrossRef]

54. Yu, B.; Tang, X.Z.; Zhu, Z.Z.; Yang, J.Y.; Zou, X.; Pang, D.B. The current situation and countermeasures of *Jatropha curcas* L. in Sichuan Province. *Sichuan For. Explor. Des.* **2007**, *3*, 16–18.

55. Singh, R.; Singh, R.N.; Vyas, D.K.; Srivastava, N.S.L.; Narra, M. SPRERI experience on holistic approach to utilize all parts ofJatropha curcasfruit for energy. *Renew. Energy* **2008**, *33*, 1868–1873. [CrossRef]

56. Then, K. The Potential of *Jatropha Carcus* planting as renewable energy crop under Malaysia weather condition. In Proceedings of the 16th TSAE National Conference & the 8th TSAE International Conference, Bangkok, Thailand, 17–19 March 2015.

57. Ministry of Plantation Industries & Commodities (MPIC) Homepage. Available online: https://www.mpic.gov.my/mpi (accessed on 24 April 2020).

Article

Enzymatic Saccharification with Sequential-Substrate Feeding and Sequential-Enzymes Loading to Enhance Fermentable Sugar Production from Sago Hampas

Nurul Haziqah Alias, Suraini Abd-Aziz, Lai Yee Phang and Mohamad Faizal Ibrahim *

Department of Bioprocess Technology, Faculty of Biotechnology and Biomolecular Sciences, Universiti Putra Malaysia, Serdang 43400, Selangor, Malaysia; nhaziqahalias95@gmail.com (N.H.A.); suraini@upm.edu.my (S.A.-A.); phanglaiyee@upm.edu.my (Y.L.P.)
* Correspondence: faizal_ibrahim@upm.edu.my; Tel.: +603-9769-1936

Abstract: Sago hampas composed of a high percentage of polysaccharides (starch, cellulose and hemicellulose) that make it a suitable substrate for fermentation. However, the saccharification of sago hampas through the batch process is always hampered by its low sugar concentration due to the limitation of the substrate that can be loaded into the system. Increased substrate concentration in the system reduces the ability of enzyme action toward the substrate due to substrate saturation, which increases viscosity and causes inefficient mixing. Therefore, sequential-substrate feeding has been attempted in this study to increase the amount of substrate in the system by feeding the substrate at the selected intervals. At the same time, sequential-enzymes loading has been also evaluated to maximize the amount of enzymes loaded into the system. Results showed that this saccharification with sequential-substrate feeding and sequential-enzymes loading has elevated the solid loading up to 20% (w/v) and reduced the amount of enzymes used per substrate input by 20% for amylase and 50% for cellulase. The strategies implemented have enhanced the fermentable sugar production from 80.33 g/L in the batch system to 119.90 g/L in this current process. It can be concluded that sequential-substrate feeding and sequential-enzymes loading are capable of increasing the total amount of substrate, the amount of fermentable sugar produced, and at the same time maximize the amount of enzymes used in the system. Hence, it would be a promising solution for both the economic and waste management of the sago hampas industry to produce value-added products via biotechnological means.

Keywords: sago hampas; amylase; cellulase; substrate feeding; saccharification; biomass

Citation: Alias, N.H.; Abd-Aziz, S.; Phang, L.Y.; Ibrahim, M.F. Enzymatic Saccharification with Sequential-Substrate Feeding and Sequential-Enzymes Loading to Enhance Fermentable Sugar Production from Sago Hampas. *Processes* **2021**, *9*, 535. https://doi.org/10.3390/pr9030535

Academic Editor: Pietro Bartocci

Received: 31 December 2020
Accepted: 27 January 2021
Published: 18 March 2021

Publisher's Note: MDPI stays neutral with regard to jurisdictional claims in published maps and institutional affiliations.

1. Introduction

Sago palm, scientifically known as *Metroxylon sagu*, can be found in tropical Southeast Asia. This plant grows healthily in the environment with an average temperature of 25 °C and an approximate humidity of 70% [1]. Its ability to thrive in a swampy area and grow naturally without the need for pesticide or herbicide has made sago palm cultivation increase in recent decades [2]. Approximately 90% of commercially grown sago palm in Malaysia is in Sarawak, a state located in the east of Malaysia. Sago palm became an important economic species and resource for this region as the production of sago starch was reported to be approximately 15–25 tons/ha. The starch composition in sago palm is the highest (25 tons/ha) as compared with other types of the plant such as rice (6 tons/ha), corn (5.5 tons/ha), wheat (5 tons/ha) and potato (2.5 tons/ha) [3]. The commercial production of sago starch was established in Malaysia in the 1970s and became one of the most important industries in terms of its contribution to the export revenue [4]. Due to the upward trend in sago starch production, the amount of waste generated from this industry has significantly increased due to the numerous of sago processing mills. The industry of sago palm has generated an extensive amount of waste including sago bark, sago hampas and sago wastewater [5]. The polluting effects caused by these agro-wastes have become the main concern and started to generate attention among the researchers attempting to find a solution with a sustainable approach.

In Malaysia, the mass production of sago starch from 600 logs of sago palm per day was 15.6 tons of woody bark, 237.6 tons of wastewater and 7.1 tons of starch fibrous sago pith residue [6]. Starch fibrous sago pith residue or commonly known as sago hampas composed of starchy and lignocellulosic components, which are 54.6% of starch, 31.7% of cellulose and hemicellulose, and 3.3% of lignin [7]. The high polysaccharide content and low lignin composition in sago hampas make this agricultural residue a promising feedstock for fermentation operation. More importantly, there is no pretreatment required before the saccharification process due to the low lignin content in sago hampas [8]. The pretreatment process is one of the crucial and costly processes in the bioconversion of agricultural residue into fermentable sugar before fermentation [9]. This process is important to reduce and/or alter the lignin component, expose the internal structure of cellulose to be accessible by the cellulase [10]. Eliminating this step from the whole process could save huge operational cost. In addition, a high percentage of remaining starch in sago hampas can be easily hydrolysed by amylase to produce fermentable sugar. Therefore, the utilization of sago hampas as a raw material for fermentation operation could be cost-effective for the downstream processing of sugar production and eventually for the production of fermentation-based products, and at the same time, prevent the environmental pollution that is caused by the underutilization of sago waste.

In the production of fermentable sugars from sago hampas, this material must be gelatinized before saccharification. Gelatinization needs to be carried out to break down the hydrogen bond in the sago starch, thus, allowing the amylase to attack the α-glycosidic bond of the polysaccharides into glucose monomer [11]. Gelatinization is a simple process that applies the heat to the starch in the presence of water. As a result, the water is gradually absorbed and caused the starch granules to swell [12]. The addition of glucoamylase (EC 3.2.1.3) with a debranching enzyme such as pullulanase (EC 3.2.1.41) is practically useful as they can hydrolyse the α-1,6-glycosidic bond that links the polysaccharides chain into branches [13]. The hydrolysis process of starch by amylase takes less than 24 h [14]. Meanwhile, to fully degrade the sago hampas into fermentable sugar, the cellulase is also being used to degrade the cellulosic component. Cellulase is a mixture of enzymes composed of endoglucanase (EC 3.2.1.4), exoglucanase (EC 3.2.1.91) and β-glucosidase (EC 3.2.1.21) that act synergistically on the degradation of the β-glycosidic bond of cellulosic component into glucose monomers [15].

The low sugar concentration always obtained from enzymatic saccharification is usually not enough to initiate the fermentation process. This problem can be overcome by increasing the insoluble solid load that will enhance the fermentable sugar production, and thus, improve the efficiency of the downstream processing. However, increasing the substrate concentration can reduce the hydrolysis yield due to the high viscosity, which subsequently causes poor mixing and mass transfer [16,17]. In addition, the current process also suffers from a high cost of enzymes used in the saccharification process, especially cellulase. Therefore, the improvement of the enzymatic saccharification step is required from an economic perspective and for process feasibility. The mixture of amylase and cellulase used in the saccharification of sago hampas has been previously reported by Husin et al. [7] for the production of biobutanol. It was found that the mixture of amylase and cellulase produced higher fermentable sugar as compared to a single enzyme, either amylase or cellulase alone. However, the process has been done in batch for simultaneous saccharification and fermentation (SSF) to produce biobutanol. Although a high biobutanol production yield was obtained, the low sugar concentration produced by this operation can be improved.

Therefore, in this present study, saccharification with sequential-substrate feeding and sequential-enzymes loading has been introduced with the aim of enhancing the fermentable sugar production, and at the same time, maximize the usage of enzymes in the saccharification process. Sequential-substrate feeding is expected to provide sufficient time for enzymes to digest solid material into soluble sugar components, thus improving the capacity of the system to be loaded with a higher amount of substrate. Meanwhile, sequential-enzymes loading is expected to maximize the amount of enzymes used in the system, and technically reduce the cost of enzymes and make the process more feasible.

2. Materials and Methods

The experimental design of this study is shown in Figure 1. The sequential-substrate feeding and sequential-enzymes loading were conducted in comparison with the batch process. The process began when the gelatinized, dried and ground sago hampas mixed with acetate buffer were added with the enzymes (amylase and cellulase). Sago hampas in a total of 20 g/L was fed sequentially based on the feeding interval followed with the study on the sequential-enzymes loading by loading amylase and cellulase at a different amount. Saccharification was conducted at 60 °C, 150 rpm for 6 days or until a stationary production of fermentable sugars was obtained. Then, to optimize the mixing process, the effect of agitation speed was also conducted.

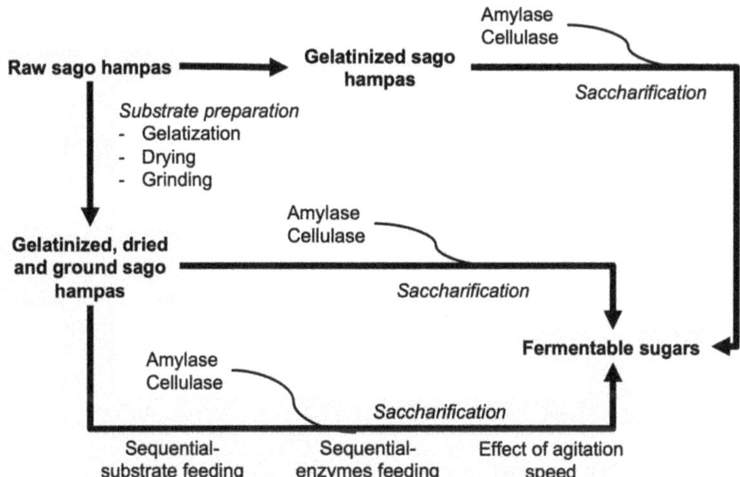

Figure 1. Schematic diagram of experimental work for enzymatic saccharification of sago hampas.

2.1. Substrate Preparation

Sago hampas supplied from River Link Sago Resources Sdn. Bhd. in Mukah, Sarawak was sun-dried for 1–2 days to drain off the excess water naturally. Then, sago hampas was gelatinised by boiling in 0.1 M of acetate buffer for 15 min. Then, the gelatinised sago hampas was oven-dried at 60 °C for 24 h and subsequently ground to pass a 1 mm screen. The moisture content of the dried samples was analysed to quantify the buffer to be added prior to the enzymatic saccharification process.

2.2. Batch Saccharification

Batch saccharification was conducted following the methods by Husin et al. [7]. An amount of 7% (w/v) of sago hampas was gelatinized in 100 mL of 0.1 M acetate buffer solution at pH 5.5 in comparison with 7% (w/v) of gelatinised, dried and ground sago hampas. The saccharification was conducted by adding Dextrozyme 71.4 U/$g_{substrate}$ of amylase (Novozymes, Bagsvaerd, Denmark) and 20 FPU/$g_{substrate}$ of Acremonium cellulase (Meiji Seika Co, Japan) into the mixture. The saccharification process was performed at 60 °C and 150 rpm up to 144 h of incubation time. All the saccharification process was performed in shaker incubator (Labwit, ZHWY-1102C) and samples were drawn out for every 24 h for the analyses.

2.3. Sacchatification with Sequential-Substrate Feeding and Sequential-Enzyme Loading

The strategies applied in this study were developed for optimising the feeding interval of the gelatinised, dried and ground sago hampas as well as the amount of enzymes loading. For this study, two sets of experiment were conducted, which are the enzymes only initially loaded and the enzymes sequentially loaded according to the amount of substrate feeding. Batch saccharification with the total substrate concentration of 20% (w/v) with 71.4 U/$g_{substrate}$ and 20 FPU/$g_{substrate}$ of amylase and cellulase, respectively, was performed as a control.

2.3.1. Feeding Interval

Table 1 illustrates the strategies behind how the feeding interval was applied. The saccharification process consists of five variables for the feeding interval, which are 0 (control), 6, 12, 24 and 36 h of interval time. The substrate was fed sequentially based on the feeding interval that makes it the total substrate loading at 20% (w/v) with the 2% (w/v) of the initial substrate for each variable except for the control. The substrate was fed only up to 72 h and prolonged the incubation for another 3 days or until the stationary production of fermentable sugar was obtained. The saccharification was performed with two sets of experiments in order to make the comparison study where the first set was the enzymes that only initially loaded while the second set was the enzymes that were loaded sequentially to per g of substrate feeding.

Table 1. Feeding strategies for the feeding interval of sago hampas on the saccharification process.

Time Interval for Substrate Feeding (h)	Initial Substrate (%)	Sequential-Substrate Feeding (%)	Total Amount of Substrate Feeding (%)
0 (control)	20	No substrate added	
6	2	1.5	
12	2	3	20
24	2	6	
36	2	9	

2.3.2. Enzymes Loading

The effect of enzymes loading were tested on the saccharification with the sequential-substrate feeding for both amylase and cellulase, with five variables as illustrated in Table 2. The optimal sequential-substrate feeding was 6% (w/v) fed sequentially at every 24 h of interval time. The study was also performed with two sets of experiments: the first set was the initially added enzymes only while the second set was the enzymes added sequentially according to the substrate feeding.

Table 2. Variables for the enzyme loading of amylase and cellulase on the sequential-substrate feeding saccharification.

Time Interval for Substrate Feeding and Enzymes Loading (h)	Sequential-Substrate Feeding (%)	Total Substrate Loading (%)	Sequential-Enzymes Loading	
			Amylase (U/g)	Cellulase (FPU/g)
			7.1	5
			14.3	10
24	6	20	71.4	15
			142.9	20
			285.7	25

2.3.3. Agitation Speed

The effect of agitation speed was performed after the optimal conditions for sequential-substrate feeding and sequential-enzymes loading were obtained. The agitation speeds were set at 60, 90, 120, 150 and 180 rpm, and saccharification with no agitation was also conducted as a control.

2.4. Analytical Procedures

The starch content was determined using iodine starch colorimetric methods by Nakamura [18]. The lignocellulosic biomass of sago hampas were determined by its three major components which are cellulose, hemicellulose and lignin using the standard procedure of acid hydrolysis method and high performance liquid chromatography (HPLC) from the National Renewable Energy Laboratory method, NREL/TP-510-42623 [19]. Total extractives content were determined following the method by the National Renewable Energy Laboratory method, NREL/TP-510-42619 [20]. The sugar monomers obtained by saccharification were analysed by high performance liquid chromatography (HPLC) (Jasco, Tokyo, Japan) equipped with a refractive index (RI) detector and a column (Shodex KS-801, Tokyo, Japan) for ligand exchange chromatography. A 100% ultrapure water was used as a mobile phase with a flow rate of 0.6 mL/min and the temperature of the column was fixed at 80 °C using oven column [7]. A statistical analysis was conducted in order to analyse the significant effect from each variable on saccharification process using an analysis of variance (ANOVA) by Statistical Analysis Software (SAS) version 9.4 and verified considering $p < 0.05$.

3. Results and Discussion

3.1. Characteristics of Sago Hampas

The composition of raw sago hampas was determined as shown in Table 3. The characterization of sago hampas in this study has been evaluated in order to ensure the quality of the substrate. All the values shown in the table are comparable to those reported previously. Starch content in sago hampas was 56.0%, while the cellulose, hemicellulose and lignin contents were of 20.7%, 11.2% and 3.1%, respectively. The value of starch content in sago hampas depends on the quality of the extraction process conducted by the sago mills [21]. Besides, both water and solvent extractives in sago hampas have a low value of 2.33% and 0.67%, respectively.

Starch, cellulose, hemicellulose and lignin are the major components of the sago hampas while extractives are the minor components. Extractives in biomass are usually the non-structural components, which can be extracted by water or other solvents. The solvents can be ethanol, acetone, benzene, hexane, dichloromethane and toluene. The compounds that are commonly extracted out from biomass are fats, waxes, phenolics, resin acids and inorganic compounds. These non-structural components of biomass could potentially interfere with the downstream analysis of the biomass sample. This may result in an error on the structural sugar values where the hydrophobic extractives could inhibit the penetration of the sample that directly caused incomplete hydrolysis [20]. Extractives could also falsely result in high values of lignin when the unhydrolyzed carbohydrates condense with the acid-insoluble lignin. Some studies reported that by removing these extractives, it showed an improvement on the enzymatic digestibility and glucose yield, respectively. Sago hampas has lower total extractives content (3.0%) as compared to other types of biomass such as corn stover (13.5%) and *Artemisia ordosica* (7.78%) [22,23]. Therefore, no pretreatment is required to remove the extractive, as this amount is not significantly affecting the saccharification process. Based on this condition of sago hampas (a high carbohydrate composition with low lignin and extractives content), this substrate has a high beneficial advantage to be used as material for the fermentation feedstock.

Table 3. Comparison of the composition of raw sago hampas with a different collection of sago hampas.

Sago Hampas Collection	Composition (%)								References
	Starch	Cellulose	Hemicellulose	Lignin	Moisture	Extractives		Others	
						Water	Solvent		
Pusa, Sarawak	49.5	26.0	14.5	7.5	n.d	n.d	n.d	2.5	[8]
Pusa, Sarawak	58.0	23.5	8.2	6.3	n.d	n.d	n.d	2.3	[24]
Mukah, Sarawak	58.0	21.0	13.4	5.4	4.7	n.d	n.d	3.13	[25]
Mukah, Sarawak	56.0	20.7	11.2	3.1	6.35	2.33	0.67	6.1	This study

n.d indicates not determined.

3.2. Saccharification of Sago Hampas

To enhance the fermentable sugar production from sago hampas, several saccharification strategies were carried out by identifying the effects of feeding interval, enzymes loading and agitation speed. The saccharification process was performed by determining the effect of preparing a substrate under wet and dried conditions followed by the sequential feeding of the dried substrate. Then, the effects of initially loaded enzymes and sequentially loaded enzymes throughout the saccharification process were also evaluated. The whole strategies were performed to determine the optimum conditions of the saccharification that can produce the highest fermentable sugar production with a low amount of enzymes loading.

3.2.1. Effect of Wet and Dried Sago Hampas

Initially, this particular experiment was carried out to determine the substrate condition used throughout the saccharification process, either in wet or dried condition. This is because in the early study of the saccharification of sago hampas, the sago hampas was gelatinised before the saccharification process, and the gelatinised sago hampas was in the wet condition. However, saccharification of sago hampas with sequential-substrate feeding must be in the dried form to ensure the consistency of the substrate feeding throughout the experiment. The gelatinization process was conducted before the saccharification process due to a high starch content in sago hampas and due to the crystalline structure of the starch. The crystallized structure of starch must be destroyed and change into the amorphous structure in order to make it susceptible to the enzyme action [21]. It works when the substrate suspension is heated in the presence of water and swelling starch granules break down the hydrogen and hydrophobic bonds [24].

In this study, the saccharification profiles of the wet and dried substrates (Figure 2) show that there is no significant difference in the sugar produced from wet and dried substrates, which produced 43.29 g/L (±2.54) and 46.23 g/L (±0.76) of sugar, respectively. However, it can be seen from the graph that the saccharification rate of the wet substrate is slightly faster as compared to the dried substrate with a slightly lower concentration of sugar being produced. It was suggested that drying the temperature also plays an important role in the characterization of sago starch in terms of drying kinetics and the equilibrium of moisture content [26]. The wet substrate might be easily degraded by amylase since the starch structure has been exposed with water, while dried substrates sometimes need the structure to be accessible by the amylase.

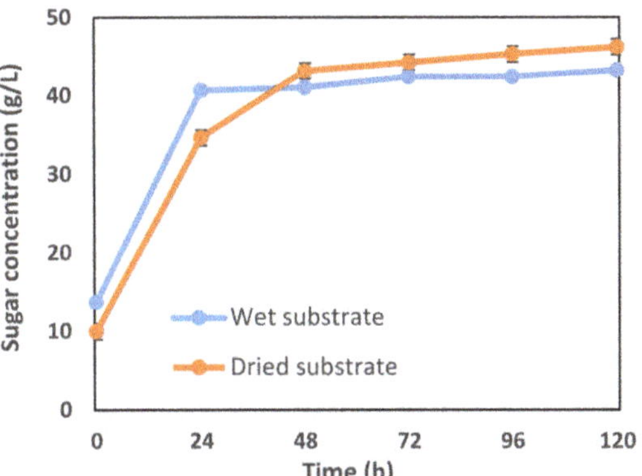

Figure 2. Effects of the wet and dried substrates used in the saccharification process of sago hampas by the mixture of amylase and cellulase.

3.2.2. Effect of Feeding Interval

In this study, 20% (w/v) of the total substrate was saccharified with the added mixture of Dextrozyme amylase (71.4 U/g) and Acremonium cellulase (20 FPU/g). However, the high substrate concentration applied in the saccharification process might lead to the high viscosity and subsequently reduced the reaction rate [27]. To saccharify a high amount of substrate, sago hampas must be loaded sequentially throughout the process to maintain the low level of viscosity and increase the accessibility of the enzymes towards the substrate [28]. To examine the effect of feeding interval and the enzymes used, several feeding intervals were conducted in two separate experiments, namely that of the only initially loaded enzymes and enzymes sequentially loaded according to the substrate feeding. All presented data are the means of triplicates $\pm$ S.D and stated using the Tukey's test with $p < 0.05$.

Figure 3a illustrates the effect of the feeding interval of the substrate with the initially loaded enzymes. This experiment was also compared with the control (without substrate feeding), where 20% (w/v) of the total substrate was added at the beginning of the saccharification. From this study, it can be observed that the control produced more sugars (80.33 g/L $\pm$ 0.02) as compared with the sequentially added substrate. The sequentially loaded substrate did not show an impressive increment in sugar production, as the enzyme activity might be alleviated throughout the process due to the lower substrate concentration at the beginning [17]. This is because sequential substrate feeding was added with only 2% (w/v) of substrate loading, whereby, a high amount of enzyme was initially added. Thus, throughout the time, most of the enzyme activity reduced and could not provide sufficient degradation capacity when the substrate was added over time. The extent of the inhibition depends on the ratio of total enzyme to the total substrate. This could be explained by the enzymes' active sites not binding with sufficient substrate at the beginning. Then, the produced sugars might occupy the empty enzyme active site and become an inhibitor to the newly added substrate. In addition, it can be seen that the pattern for the control showed that the saccharification can only be achieved until 48 h of incubation time. In comparison with the sequential-substrate feeding saccharification, the degree of hydrolysis was observed until 96 h of incubation time. Even though the sugar production from the control saccharification was higher than that of the sequential-substrate feeding saccharification, the high substrates used became significant waste, as these cannot be further hydrolysed by the enzymes. It seems that the enzyme–substrate complex has reached its maximum saturation point which is most likely due to the jamming effect phenomenon caused by the overcrowding of the substrate, with the enzyme and substrate obstructing one another [29].

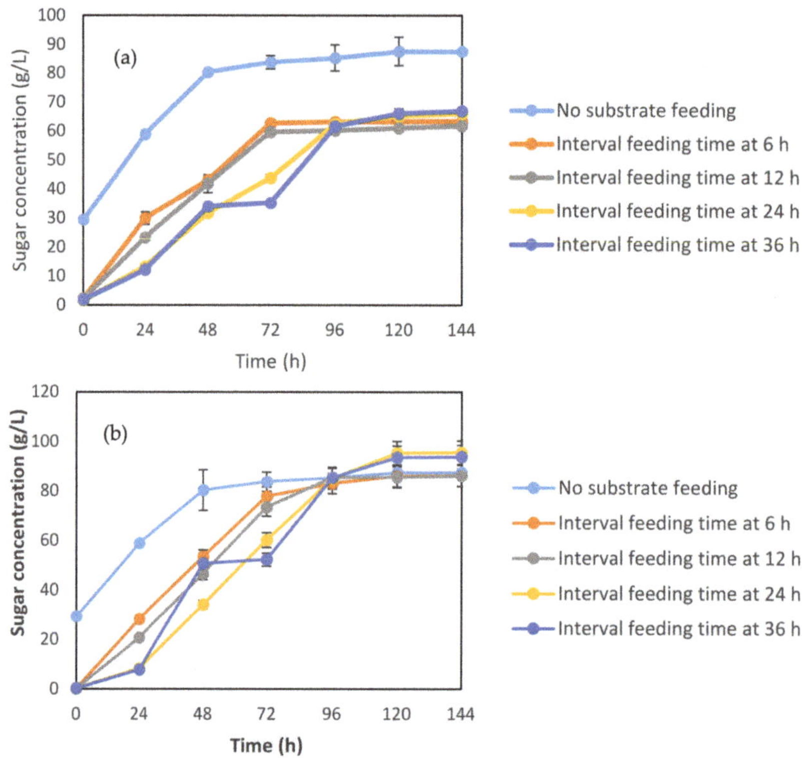

Figure 3. Effect of the feeding interval of the substrate in the saccharification process: (**a**) initially loaded enzymes only; and (**b**) sequentially loaded enzymes according to substrate feeding.

Figure 3b shows the effect of the feeding interval with sequentially loaded enzymes according to the substrate feeding. From the graph, it can be observed that the effect of feeding interval on saccharification with the sequentially loaded enzymes produced more fermentable sugars as compared with the control. From the graph, the degree of hydrolysis for sequential-substrate feeding saccharification showed that it increases gradually up to 120 h of incubation time with the addition of substrate compared with the control that reached its maximum saturation point at 48 h of incubation time. The periodical addition of substrate prolonged the process to produce more fermentable sugars [30]. In the comparison with Figure 3a, there is about 34.51% of increment for sugar production. It can also be observed that the viscosity is reduced and more runny solution can be observed. Thus, a greater fermentable sugars yield was produced. This might be due to the rate of reaction which is affected by the total number of enzymes as well as the concentration of substrate loaded accordingly [31]. The result showed that the rate of saccharification did not decrease with the increase in substrate concentration when the enzyme-to-substrate ratio was kept constant. It can be concluded that the optimal interval feeding time for sequential-substrate feeding with the sequential-enzyme loading was every 24 h, which produced the highest sugar concentration of (95.37 g/L $\pm$ 0.93) with $p < 0.05$.

There are several studies reported about the crucial parameters that affect the enzymatic saccharification, and one of the parameters is substrate-related. In this present study, the substrate concentration and feeding style have been discussed in terms of how they affect the saccharification process. The substrate features such as the substrate size, lignin structure and substrate pore surface area also play an important role in the accessibility of the substrate to enzyme [32]. This is because the lignocellulosic biomass has a complex structural arrangement, thus, it is more difficult to hydrolyse as compared to starch-based biomass. Most of the lignocellulosic biomass such as corn stover, switch grasses and forest residue need to undergo a pretreatment process prior to saccharification. This is to ensure that the lignin component was removed or reduced and/or altered to allow the interaction of substrate to enzymes.

3.2.3. Effect of Enzymes Loading

To establish an economically feasible saccharification process, an appropriate amount of enzyme used must be determined as an enzyme used in sugar production generally contributes a significantly high cost in terms of the total operational cost of converting biomass into value-added products [33]. In the previous experiment, the loaded enzyme was 71.4 U/g of amylase and 20 FPU/g of cellulase with the interval feeding time at every 24 h. In this experiment on the effect of enzymes loading, there were five variables for each enzyme range from 7.1 U/g to 285.7 U/g for amylase and 5 FPU/g to 25 FPU/g of cellulase were examined. All presented data are the means of triplicates $\pm$ S.D and stated using the Tukey's test with $p < 0.05$.

Figure 4a shows the effect of initially loading the enzyme while Figure 4b shows the effect of the sequential-enzyme loading on the saccharification with the sequential-substrate feeding. Based on Figure 4a, it can be observed that sugar production declined after 96 h of incubation time. In addition, when the enzyme was initially loaded, the inhibitors might have formed from the formation of the product, which subsequently caused the competitive inhibition [34] whereby the substrate and inhibitors compete for the same enzyme's active site [35]. Competitive inhibition occurs in one of the enzymes, in this case cellulase since it is considered the principle bottleneck for practical production from lignocellulosic materials. This situation usually occurs with high substrate concentration as the inhibitors limit the enzyme velocity in their biochemical reaction [36]. Thus, high substrate concentration might escalate the possibility of enzyme inhibition caused by product inhibitors. In addition, the availability of the enzyme's active site is limited at high substrate concentration due to the accumulation of excess substrate. Other factors that may contribute to the low degree of polysaccharide conversion at high substrate concentration, mainly because of the decrease within the reactivity of cellulosic material in the course of hydrolysis, different kinds of enzyme inactivation, and the non-specific adsorption of cellulolytic enzymes onto lignin [37]. From this study, the sugar production significantly showed the difference between the various amounts of loaded enzymes.

Meanwhile, based on Figure 4b, it showed that the performance of hydrolysis was increased with the sequentially added enzyme, as the ratio of enzyme to the substrate used is one factor that affects the saccharification [38]. When comparing these two studies, the effect of sequential-enzymes loading might reduce the inhibition of the product. In addition, the saccharification process has been prolonged up to 120 h of incubation time. However, after 120 h of incubation time, the efficiency of enzyme catalytic reaction has deprived due to the prolonged incubation time, which probably because of the enzyme has achieved its maximum enzyme thermal deactivation process after being exposed at high temperature for a long time [39]. The enzymes' reactivity is mostly associated with the enzyme-related parameters. The maximum utilisation of enzymes during saccharification is important because the enzymes represent the major contribution to the total cost of the bioconversion of biomass to value-added products. The amount of enzyme loading depends on the composition and structural arrangement of the substrate [40]. The effect of sequential-enzyme loading has resulted in approximately a 43% increment in sugar production as compared with the initially loaded enzymes. This study also showed that sequential-enzyme loading with 10 to 20 FPU/g cellulase and 14.3 to 142.9 U/g amylase did not significantly affect sugar production ($p < 0.05$). Hence, the presence of excess enzymes was a waste, since it was usually underutilized and consequently leads to the unnecessarily high cost in the saccharification process. Therefore, a low amount of enzyme loading (14.3 U/g of amylase with 10 FPU/g of cellulase) can be optimally used to produce a high concentration of (112.48 g/L $\pm$ 1.26) with $p < 0.05$.

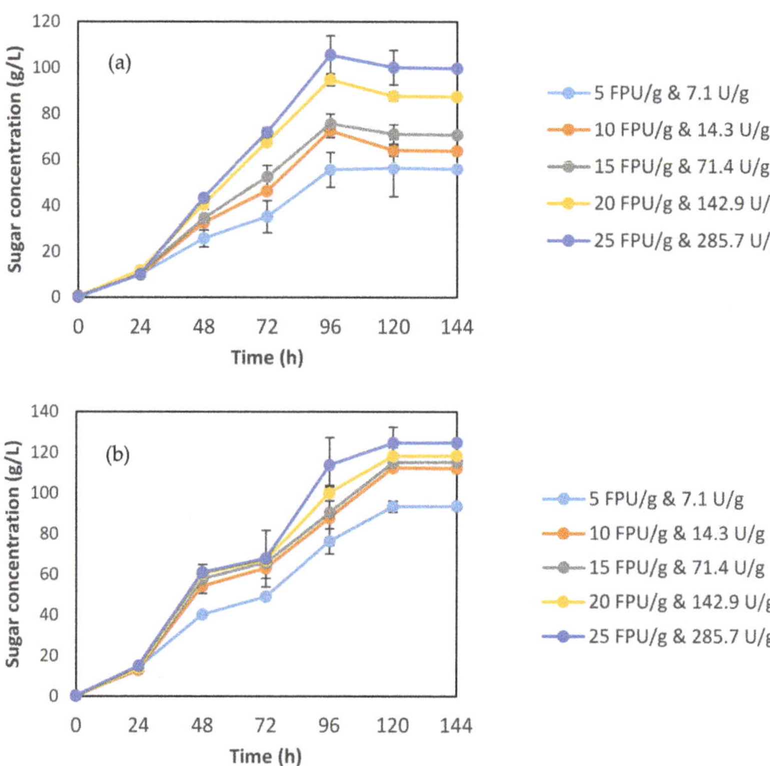

Figure 4. Effect of loading enzymes on the saccharification process: (**a**) initially loaded enzymes only; and (**b**) sequentially loaded enzymes according to substrate feeding.

3.2.4. Effect of Agitation Speed

The effect of agitation speed has been evaluated in the range of 60–180 rpm together with no agitation at the constant temperature of 60 °C for the saccharification of sequential-substrate feeding and sequential-enzyme loading. The effect of agitation speed is important to determine the relationship between the saccharification efficiency and liquid viscosity of saccharification. This is because the liquid viscosity from the saccharification process increases with the increase in the saccharification time since the substrate was added sequentially throughout the process.

The trend of fermentable sugar production with different agitation speed for 6 days of incubation is presented in Figure 5 As expected, saccharification with no agitation produced the lowest sugar concentration of (88.38 g/L $\pm$ 1.52). Increasing the agitation speed from 60 to 150 rpm had significantly increased the sugar production to (119.9 g/L $\pm$ 0.32) with p-value < 0.5. Agitation enhances the mass transfer rate during saccharification, thus improving the hydrolysis process and increasing the conversion rate of the substrate into fermentable sugars by the enzymes. It can also be observed that the agitation speed has a moderate effect on saccharification. At a 180 rpm agitation speed, the fermentable sugar production showed a slight reduction probably due to the shear stress or shear forces. It is also reported that vigorous agitation speed could have aggravated cell damage, which in turn led to the mechanical inactivation of the enzymes and contributed to the reduction in enzyme stability. Thus, sugar production was suppressed. In typical saccharification, the agitation speed of 120–150 rpm was found to be optimum. Therefore, as for this study, the agitation of 150 rpm can be considered most suitable for the saccharification of sago hampas to produce fermentable sugars. Even though the agitation speed has a moderate effect on saccharification, it still plays an important role in the cost-effective downstream processing as it can reduce energy consumption.

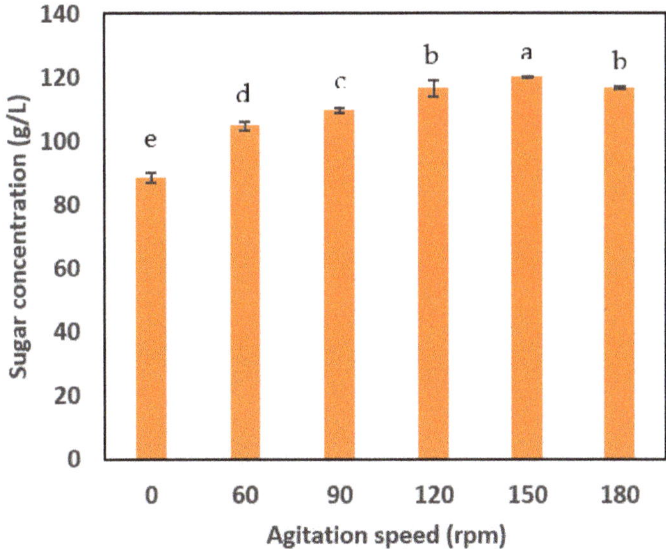

Figure 5. Effect of agitation speed on the saccharification process. All data are the means of 3 replicates ± S.D. The different alphabet indicates significant difference at $p < 0.05$. The data were stated using LSD test.

3.3. Comparison Study

This study demonstrated the enhancement of sugar production by conducting saccharification with sequential-substrate feeding and sequential-enzyme loading. The comparison of different saccharification strategies in Table 4 shows that this saccharification strategy significantly improved sugar production compared to other studies. It was 13.53% of sugar increment which could be observed when comparing the normal saccharification with sequential-substrate feeding. However, throughout the process, the normal saccharification produced more suspended solution because of the high solid–liquid ratio at the beginning of the saccharification. This highly viscous solution was caused by an ineffective heat and mass transfer due to the solution not being mixed properly, and hence, the reduction in the diffusion of the enzyme and end product [27].

In order to further enhance the fermentable sugar production, the rate of saccharification process needs to be increased. However, due to the high prices of commercial cellulase, the addition of more enzymes is not the best option. Alternatively, the rate of saccharification and fermentable sugar production can be accelerated by implementing a new strategy of feeding for both substrate and enzymes. Therefore, the sequential-substrate feeding and sequential-enzymes loading was implemented in this study. Surprisingly, the small difference in the feeding strategies had significantly improved the sugar production by 20.87% as compared to the process without sequential-enzyme loading. This approach has also reduced the amount of enzyme used for both amylase and cellulase by 20% and 50%, respectively.

Table 4. Comparison of sugar production between different strategies of saccharification using sago hampas as a substrate.

Study	Initial Substrate Concentration (g/L)	Total Substrate Loading (g/L)	Feeding Time Interval (h)	Enzyme Loading (per g of the Substrate)		* Sugar Concentration (g/L)
				Amylase (U/g)	Cellulase (U/g)	
Batch saccharification (wet substrate)	7	7	Null	71.4	20	43.29 ± 2.54
Batch saccharification (dried substrate)	7	7	Null	71.4	20	46.23 ± 0.76
Batch saccharification (control)	20	20	Null	71.4	20	80.33 ± 0.02
Saccharification with sequential-substrate feeding	2	20	Every 24 h	71.4	20	94.88 ± 2.59
Saccharification with sequential-substrate feeding and sequential-enzymes loading	2	20	Every 24 h	14.3	10	119.90 ± 0.32

* All data are the means of 3 replicates ± S.D.

The effectiveness of the enzymatic hydrolysis for starch and lignocellulosic components not only depends on the substrate concentration as a whole, but it also requires the optimum synergistic action of the amylase and cellulase components towards the substrate. Sago hampas mainly compose of starch residue (56.0%) and the starch itself is composed of linear amylose and branched amylopectin. Therefore, glucoamylase (EC 3.2.1.3) with a debranching enzyme, pullulanase (EC 3.2.1.41), was used for this study as it is an exo-acting enzyme that mainly hydrolyses an α-1,4 and α-1,6 glycosidic bond from the non-reducing ends of starch chains, which leads to the production of glucose. There are some studies which reported that glucoamylase was able to enhance the efficiency of hydrolysis and increase the substrate concentration at the active site of the enzyme catalytic centre by binding to the raw starch granules and disrupt the surface structure of starch [41,42]. Meanwhile, the lignocellulosic components in this study were hydrolysed by cellulase, which is composed of endoglucanase (EC 3.2.1.4), exoglucanase (EC 3.2.1.91) and β-glucosidase (EC 3.2.1.21) that act synergistically on the degradation of a cellulosic component into glucose monomers. During the saccharification, it is noted that the substrate level should be high enough to provide a sufficient reaction between the enzymes and substrate. Thus, in order to enhance the higher sugar production, it is must proportionally increase the rate of reaction by adding more substrate. However, by relatively adding more substrate, it contributes to the jamming effect due to enzymes needing to act on more portions of the starch and lignocellulosic components [43]. Therefore, the feeding style of substrate and enzymes must be studied in order to achieve the optimal reaction of saccharification. Hence, in this study, the efficiency of saccharification has been improved when the sequential-substrate and enzymes feeding was applied. The enzymes are able to work at their optimal level when they are supplemented according to the amount of fed substrate, instead of being added once at the beginning only.

The correlation between the analysed variables (substrate feeding, enzyme loading and agitation speed) with the sugar production are conducted. From the correlation screening, it can be seen that only enzymes loading was significantly affecting the sugar concentration with $p < 0.05$. The strength of enzymes loading and sugar concentration were associated with the R^2 value of 0.92. This is explained that by the increasing enzymes loading, which produces a higher sugar concentration. However, it should be noted that a further increase in enzymes loading is not economically practical. Both the feeding interval and agitation speed resulted in $p > 0.05$, where it does not give a significant effect on the sugar concentration. The feeding interval showed an insignificant effect in enhancing the sugar concentration, which might be due to the enzymes that engage with the substrate and were not sufficient when the interval time of the substrate feeding was too long. This will cause the alleviation of enzyme activity and result in the slowing down of the reactions [44]. Agitation speed plays an important role for an effective mixing of the substrate and enzymes. However, by increasing the agitation speed too much, there is no significant effect on the sugar concentration, where the shear forces might occur and lead to cell damage [45]. On top of that, the interaction of enzymes loading towards the feeding interval was conducted and it showed that the coefficient of determination (R^2) was 0.96. The R^2 value indicates that 96% of the variation in enzyme loading is explained by the feeding interval. Thus, the digestibility of the enzymes toward the substrate was improved as the enzymes were loaded based on the amount of substrate feeding.

In comparison with other studies, as shown in Table 5, a high substrate feeding of 20 g/L can be fed in sequential-substrate feeding and fed-batch saccharification as compared to batch saccharification, which is capable of the maximum load substrate feeding at 5–9 g/L. Previous studies have reported the decreased efficiency of hydrolysis when more than 9 g/L of solid substrate were used and this was due to product inhibition, enzyme inactivation and the decrease in substrate reactivity [46]. In addition to that, in the fed-batch saccharification [17], a similar situation was observed whereby a high increment in sugar production was obtained as compared to batch saccharification. However, no feeding strategy for enzyme loading was conducted. In comparison with the sequential-substrate-feeding and sequential-enzymes loading, fermentable sugar production from sago hampas was significantly improved even at the low amount of loaded enzymes. It shows that sugar production was affected by how the substrate and enzymes were fed. In order to increase sugar production, the rate of reaction needs to be increased by increasing the amount of substrate and enzymes in the system. However, a high solid–liquid ratio could hinder the effective heat and mass transfer and thus limit the diffusion of enzymes and the formation of end products. Therefore, the efficient saccharification process was dependent on the synergistic feeding of the substrate and enzymes.

Table 5. Comparison of sugar production by various substrates, the amount of enzyme used and different saccharification operations.

Saccharification Operations	Substrate Concentration (g/L)	Enzyme Used per g of Substrate	Sugar Concentration (g/L)	References
Batch saccharification	5 g/L oil palm empty fruit bunch	0.1 g/mL crude cellulase cocktail	12 g/L glucose	[47]
Batch saccharification	5 g/L oil palm empty fruit bunch	15 FPU/mL celluclast	31 g/L reducing sugars	[48]
Batch saccharification	9 g/L sago hampas	71.4 U/g Dextrozyme amylase + 20 FPU/g Acremonium cellulase	66.9 g/L reducing sugars	[7]
Fed-batch saccharification	20 g/L Jerusalem artichoke stalks	20 FPU/g cellulase	83.7 g/L glucose	[17]
Saccharification with sequential-substrate feeding and sequential-enzymes loading	20 g/L sago hampas	14.3 U/g Dextrozyme amylase + 10 FPU/g Acremonium cellulase	119.90 g/L glucose $\pm$ 0.32	This study

The properties of the biomass usually affected by their structure that make the hydrolysis difficult to be carried out with a higher substrate feeding. Some of the biomass with high lignin content usually needs to undergo pretreatment and this increases the total cost of bioconversion. The downstream processing usually occurs with a high cost of production. However, this cost can be reduced by minimizing the enzyme usage and maximizing the amount of substrate used. Enzymatic saccharification with a sequential-substrate feeding and sequential-enzymes loading was proven to be a promising strategy for efficient and economical saccharification. This present study is possible to implement on a large-scale processing production. It is suggested to study the proper technology that can be integrated with the present study in order to ensure the feasibility of the process.

4. Conclusions

Sago hampas has been notably known as a promising substrate for the production of fermentation-based products due to its high content of polysaccharides and the low lignin composition. In this study, the saccharification of sago hampas into fermentable sugar has been enhanced by implementing the feeding strategies of the substrate and the enzymes together with the effect of agitation speed. It can be concluded that the sequential-substrate feeding at 6 g/L for every 24 h increased the sugar production by 16% as compared to the batch process. Meanwhile, saccharification with sequential-substrate feeding and sequential-enzymes loading produced a high sugar concentration of 119.90 g/L, and at the same time reduced the amount of amylase from 71.4 $U/g_{substrate}$ to 14.4 $U/g_{substrate}$ and cellulase from 20 $FPU/g_{substrate}$ to 10 $FPU/g_{substrate}$ used in the process. Findings from this research suggest that the potential of sequential-substrate feeding and sequential-enzyme loading can be used as an alternative in improving the saccharification process of other types of substrates in order to obtain a significantly higher amount of fermentable sugar derived from biomass.

Author Contributions: Conceptualization, M.F.I. and N.H.A.; methodology, M.F.I. and N.H.A.; software, N.H.A.; validation, M.F.I., S.A.-A. and P.L.Y.; formal analysis, N.H.A.; investigation, N.H.A.; resources, N.H.A.; data curation, N.H.A. and M.F.I.; writing—original draft preparation, N.H.A.; writing—review and editing, N.H.A. and M.F.I.; visualization, N.H.A.; supervision, M.F.I., S.A.-A. and P.L.Y.; project administration, M.F.I.; funding acquisition, M.F.I. All authors have read and agreed to the published version of the manuscript.

Funding: This research was financially supported by the Geran Putra, Universiti Putra Malaysia, project number GP/2017/9559300.

Data Availability Statement: All data used to support the funding of this study are included within the article.

Acknowledgments: Highly appreciation to all members of the Environmental Biotechnology Research Group, Universiti Putra Malaysia for their kind support and help.

Conflicts of Interest: The authors declare no conflict of interest.

References

1. Bujang, K. Potential of sago for commercial production of sugars. In Proceedings of the the 10th International Sago Symposium, Bogor, Indonesia, 29–31 October 2011; pp. 1–7.
2. Pei-Lang, A.T.; Mohamed, A.M.D.; Karim, A.A. Sago starch and composition of associated components in palms of different growth stages. *Carbohydr. Polym.* **2006**, *63*, 283–286. [CrossRef]
3. Ishizaki, A. Production, purification, and health benefits of sago sugar. In Proceedings of the Concluding Remarks for the 6th International Sago Symposium; Sago Comm: Riau, Indonesia, 1997; pp. 22–24.
4. Karim, A.A.; Tie, A.P.L.; Manan, D.M.A.; Zaidul, I.S.M. Starch from the sago (*Metroxylon sagu*) palm tree-Properties, prospects, and challenges as a new industrial source for food and other uses. *Compr. Rev. Food Sci. Food Saf.* **2008**, *7*, 215–228. [CrossRef] [PubMed]
5. Awg-Adeni, D.S.; Abd-Aziz, S.; Bujang, K.B.; Hassan, M.A. Bioconversion of Sago Residue into Value Added Products. *Afr. J. Biotechnol.* **2010**, *9*, 2016–2021.
6. Ngaini, Z.; Wahi, R.; Halimatulzahara, D.; Mohd Yusoff, N.A.-N. Chemically modified sago waste for oil absorption. *Pertanika J. Sci. Technol.* **2014**, *22*, 153–161.
7. Husin, H.; Ibrahim, M.F.; Kamal Bahrin, E.; Abd-Aziz, S. Simultaneous saccharification and fermentation of sago hampas into biobutanol by Clostridium acetobutylicum ATCC 824. *Energy Sci. Eng.* **2018**, *7*, 66–75. [CrossRef]
8. Jenol, M.A.; Ibrahim, M.F.; Phang, L.Y.; Salleh, M.M. Sago biomass as a sustainable source for biohydrogen production by Clostridium butyricum A1. *BioResources* **2014**, *9*, 1007–1026. [CrossRef]
9. Chaturvedi, V.; Verma, P. An overview of key pretreatment processes employed for bioconversion of lignocellulosic biomass into biofuels and value added products. *3 Biotech* **2013**, *3*, 415–431. [CrossRef] [PubMed]
10. Canilha, L.; Chandel, A.K.; Dos Santos Milessi, S.T.; Antunes, F.A.F.; Da Costa Freitas, L.W.; Das Graças Almeida Felipe, M.; Da Silva, S.S. Bioconversion of sugarcane biomass into ethanol: An overview about composition, pretreatment methods, detoxification of hydrolysates, enzymatic saccharification, and ethanol fermentation. *J. Biomed. Biotechnol.* **2012**, *2012*. [CrossRef] [PubMed]
11. Van Der Maarel, M.J.E.C.; Van Der Veen, B.; Uitdehaag, J.C.M.; Leemhuis, H.; Dijkhuizen, L. Properties and applications of starch-converting enzymes of the α-amylase family. *J. Biotechnol.* **2002**, *94*, 137–155. [CrossRef]
12. Alcázar-Alay, S.C.; Meireles, M.A.A. Physicochemical properties, modifications and applications of starches from different botanical sources. *Food Sci. Technol.* **2015**, *35*, 215–236. [CrossRef]
13. Roy, I.; Gupta, M.N. Hydrolysis of starch by a mixture of glucoamylase and pullulanase entrapped individually in calcium alginate beads. *Enzyme Microb. Technol.* **2004**, *34*, 26–32. [CrossRef]
14. Goyal, N.; Gupta, J.K.; Soni, S.K. A novel raw starch digesting thermostable α-amylase from *Bacillus* sp. I-3 and its use in the direct hydrolysis of raw potato starch. *Enzyme Microb. Technol.* **2005**, *37*, 723–734. [CrossRef]
15. Satari, B.; Karimi, K.; Kumar, R. *Cellulose Solvent-Based Pretreatment for Enhanced Second-Generation Biofuel Production: A Review*; Royal Society of Chemistry: London, UK, 2019; Volume 3, ISBN 8415683111.
16. Gao, Y.; Xu, J.; Zhang, Y.; Liu, Y.; Liang, C. Optimization of fed-batch enzymatic hydrolysis from alkali-pretreated sugarcane bagasse for high-concentration sugar production. *Bioresour. Technol.* **2014**, *167*, 41–45. [CrossRef] [PubMed]
17. Khatun, M.M.; Li, Y.-H.; Liu, C.-G.; Zhao, X.-Q.; Bai, F.-W. Fed-batch Saccharification and ethanol fermentation of Jerusalem Artichoke Stalks by an Inulinase Producing Saccharomyces cerevisiae MK01. *R. Soc. Chem.* **2015**, *5*, 107112–107118. [CrossRef]
18. Nakamura, L.K. Lactobacillus amylovorus, a new starch-hydrolyzing species from cattle waste-corn fermentations. *Int. J. Syst. Bacteriol.* **1981**, *31*, 56–63. [CrossRef]
19. Sluiter, A.; Hames, B.; Ruiz, R.; Scarlata, C.; Sluiter, J.; Templeton, D. *Determination of Sugars, Byproducts, and Degradation Products in Liquid Fraction Process Samples Laboratory Analytical Procedure (LAP)*; Issue Date: 12/08/2006; Midwest Research Institute: Golden, CO, USA, 2008.

20. Sluiter, A.; Ruiz, R.; Scarlata, C.; Sluiter, J.; Templeton, D. *Determination of Extractives in Biomass: Laboratory Analytical Procedure (LAP)*; Issue Date 7/17/2005; Midwest Research Institute: Golden, CO, USA, 2008.

21. Awg-Adeni, D.S.; Bujang, K.B.; Hassan, M.A.; Abd-Aziz, S. Recovery of glucose from residual starch of sago hampas for bioethanol production. *Biomed Res. Int.* **2013**, *2013*, 935852. [CrossRef]

22. Li, Z.; Yu, Y.; Sun, J.; Li, D.; Huang, Y.; Feng, Y. Effect of extractives on digestibility of cellulose in corn stover with liquid hot water pretreatment. *BioResources* **2016**, *11*, 54–70. [CrossRef]

23. Wang, Y.; Wu, L.; Wang, C.; Yu, J.; Yang, Z. Investigating the influence of extractives on the oil yield and alkane production obtained from three kinds of biomass via deoxy-liquefaction. *Bioresour. Technol.* **2011**, *102*, 7190–7195. [CrossRef]

24. Vincent, M.; Jabang, E.; Nur, N.M.; Esut, E.; Unting, L.B.; Awg-Adeni, D.S. Simultaneous co-Saccharification and Fermentation of Sago Hampas for Bioethanol Production. *Agric. Eng. Int. CIGR J.* **2015**, *17*, 160–167.

25. Jenol, M.A.; Ibrahim, M.F.; Bahrin, E.K.; Kim, S.W.; Abd-Aziz, S. Direct bioelectricity generation from sago hampas by clostridium beijerinckii sr1 using microbial fuel cell. *Molecules* **2019**, *24*, 2397. [CrossRef]

26. Mustafa Kamal, M.; Baini, R.; Mohamaddan, S.; Selaman, O.S.; Ahmad Zauzi, N.; Rahman, M.R.; Abdul Rahman, N.; Chong, K.H.; Atan, M.F.; Abdul Samat, N.A.S.; et al. Effect of temperature to the properties of sago starch. *IOP Conf. Ser. Mater. Sci. Eng.* **2017**, *206*. [CrossRef]

27. Yan, S.; Yao, J.; Yao, L.; Zhi, Z.; Chen, X.; Wu, J. Fed batch enzymatic saccharification of food waste improves the sugar concentration in the hydrolysates and eventually the ethanol fermentation by saccharomyces cerevisiae H058. *Braz. Arch. Biol. Technol.* **2012**, *55*, 183–192. [CrossRef]

28. Kumar, L.; Arantes, V.; Chandra, R.; Saddler, J. The lignin present in steam pretreated softwood binds enzymes and limits cellulose accessibility. *Bioresour. Technol.* **2012**, *103*, 201–208. [CrossRef] [PubMed]

29. Bommarius, A.S.; Katona, A.; Cheben, S.E.; Patel, A.S.; Ragauskas, A.J.; Knudson, K.; Pu, Y. Cellulase kinetics as a function of cellulose pretreatment. *Metab. Eng.* **2008**, *10*, 370–381. [CrossRef]

30. Lu, C.; Dong, J.; Yang, S.-T. Butanol production from wood pulping hydrolysate in an integrated fermentation-gas stripping process. *Bioresour. Technol.* **2013**, *143*, 467–475. [CrossRef] [PubMed]

31. Van Dyk, J.S.; Pletschke, B.I. A review of lignocellulose bioconversion using enzymatic hydrolysis and synergistic cooperation between enzymes-Factors affecting enzymes, conversion and synergy. *Biotechnol. Adv.* **2012**, *30*, 1458–1480. [CrossRef] [PubMed]

32. Leu, S.Y.; Zhu, J.Y. Substrate-Related Factors Affecting Enzymatic Saccharification of Lignocelluloses: Our Recent Understanding. *Bioenergy Res.* **2013**, *6*, 405–415. [CrossRef]

33. Mathew, G.M.; Sukumaran, R.K.; Singhania, R.R.; Pandey, A. Progress in research on fungal cellulases for lignocellulose degradation. *J. Sci. Ind. Res.* **2008**, *67*, 898–907.

34. Kim, D. Physico-chemical conversion of lignocellulose: Inhibitor effects and detoxification strategies: A mini review. *Molecules* **2018**, *23*, 309. [CrossRef]

35. Yoon, J.-J.; Kim, K.-Y.; Cha, C.-J. Purification and characterization of thermostable β-glucosidase from the brown-rot basidiomycete Fomitopsis palustris grown on microcrystalline cellulose. *J. Microbiol.* **2008**, *46*, 51–55. [CrossRef] [PubMed]

36. Ojeda, K.; Kafarov, V. Exergy analysis of enzymatic hydrolysis reactors for transformation of lignocellulosic biomass to bioethanol. *Chem. Eng. J.* **2009**, *154*, 390–395. [CrossRef]

37. Mussatto, S.I.; Dragone, G.; Fernandes, M.; Milagres, A.M.F.; Roberto, I.C. The effect of agitation speed, enzyme loading and substrate concentration on enzymatic hydrolysis of cellulose from brewer's spent grain. *Cellulose* **2008**, *15*, 711. [CrossRef]

38. Hu, J.; Arantes, V.; Saddler, J.N. The enhancement of enzymatic hydrolysis of lignocellulosic substrates by the addition of accessory enzymes such as xylanase: Is it an additive or synergistic effect? *Biotechnol. Biofuels* **2011**, *4*, 1–13. [CrossRef] [PubMed]

39. Hawker, J.; Jenner, C. High temperature affects the activity of enzymes in the committed pathway of starch synthesis in developing wheat endosperm. *Aust. J. Plant Physiol.* **1993**, *20*, 197–209. [CrossRef]

40. Amit, K.; Nakachew, M.; Yilkal, B.; Mukesh, Y. A review of factors affecting enzymatic hydrolysis of pretreated lignocellulosic Biomass. *Res. J. Chem. Environ.* **2018**, *22*, 62–67.

41. Sorimachi, K.; Le Gal-Coëffet, M.F.; Williamson, G.; Archer, D.B.; Williamson, M.P. Solution structure of the granular starch binding domain of Aspergillus niger glucoamylase bound to β-cyclodextrin. *Structure* **1997**, *5*, 647–661. [CrossRef]

42. Xu, Q.S.; Yan, Y.S.; Feng, J.X. Efficient hydrolysis of raw starch and ethanol fermentation: A novel raw starch-digesting glucoamylase from Penicillium oxalicum. *Biotechnol. Biofuels* **2016**, *9*, 1–18. [CrossRef] [PubMed]

43. Eveleigh, D.; Mandels, M.; Andreotti, R.; Roche, C. Measurement of saccharifying cellulase. *Biotechnol. Biofuels* **2009**, *2*, 1–8. [CrossRef] [PubMed]

44. Robinson, P.K. Enzymes: Principles and biotechnological applications. *Essays Biochem.* **2015**, *59*, 1–41. [CrossRef]

45. Liu, Z.; Smith, S.R. Enzyme Recovery from Biological Wastewater Treatment. *Waste Biomass Valorization* **2020**. [CrossRef]

46. Chen, M.; Zhao, J.; Xia, L. Enzymatic hydrolysis of maize straw polysaccharides for the production of reducing sugars. *Carbohydr. Polym.* **2008**, *71*, 411–415. [CrossRef]

47. Ibrahim, M.F.; Linggang, S.; Jenol, M.A.; Yee, P.L.; Abd-Aziz, S. Effect of Buffering System on Acetone-Butanol-Ethanol Fermentation by Clostridium acetobutylicum ATCC 824 using Pretreated Oil Palm Empty Fruit Bunch. *BioResources* **2015**, *10*, 3890–3907. [CrossRef]

48. Salleh, M.S.M.; Ibrahim, M.F.; Roslan, A.M.; Abd-Aziz, S. Improved Biobutanol Production in 2-L Simultaneous Saccharification and Fermentation with Delayed Yeast Extract Feeding and in-situ Recovery. *Sci. Rep.* **2019**, *9*, 1–9. [CrossRef] [PubMed]

MDPI
St. Alban-Anlage 66
4052 Basel
Switzerland
Tel. +41 61 683 77 34
Fax +41 61 302 89 18
www.mdpi.com

Processes Editorial Office
E-mail: processes@mdpi.com
www.mdpi.com/journal/processes